linear algebra

SECOND EDITION

LINEAR ALGEBRA

SECOND EDITION

ROSS A. BEAUMONT
University of Washington

HARCOURT BRACE JOVANOVICH, INC.
New York Chicago San Francisco Atlanta

ISBN: 0-15-551022-3

Library of Congress Catalog Card Number: 75-183737

Printed in the United States of America

preface

Linear Algebra is designed for a course that follows introductory calculus or that is taken concurrently with the second semester of the calculus sequence. It is a concise treatment of the essential topics of linear algebra and will prepare the student for many advanced mathematics courses. For this reason, the discussion is constructed around the central theme of finite dimensional real vector spaces and their linear transformations. The field of real numbers is consistently used as the ground field of scalars; thus the fundamental results of linear algebra are presented in a form most useful in other undergraduate courses in mathematics, statistics, physics, and engineering. If an instructor desires more generality, he can, with a lecture or two on fields, adapt the book to a more general situation.

This second edition of *Linear Algebra* differs from the first in several particulars. The introductory chapter on vector methods in geometry, which serves as a source of examples for the rest of the book, has been completely rewritten with simpler notation and it contains an additional section on the cross product. Many new examples have been added throughout the text, and those taken from the first edition have been worked out in greater detail. Explanations throughout the text have been expanded wherever it seemed that this would improve the student's understanding, and several important theorems are now preceded by a description of the methods that are to be used in the proofs. Moreover, the number of exercises has been significantly increased. Exercises that give additional results of importance, as well as a few that contain definitions, are indicated by an asterisk (*).

The first edition of *Linear Algebra* has been used for a one-semester or two-quarter course, usually meeting three hours a week, and it has also served as a supplementary text in a calculus sequence in which the concepts of linear algebra are needed for a reasonable treatment of multivariate calculus and linear differential equations. The new edition is equally suited to these purposes.

This book is dedicated to my late friend and colleague Herbert S. Zuckerman, who carefully read the entire manuscript of the first edition and offered detailed and valuable criticisms. My sincere thanks are due to the many people who used the first edition of the book and made suggestions for improvement. The editorial staff of Harcourt Brace Jovanovich has been most helpful and cooperative. Finally, I owe my thanks to Mrs. Kay Kolodziej, who expertly typed the manuscript and assisted me in the chore of proofreading.

ROSS A. BEAUMONT

contents

linear algebra

SECOND EDITION

1

vector methods in geometry

1 geometric vectors

Many physical quantities such as force, displacement, velocity, and acceleration, are completely described by giving a magnitude and a direction. Moreover, it follows from the laws of physics that such quantities can be combined or ''added'' by essentially similar rules; *vectors* in two- and three-dimensional space provide a mathematical model that describes these physical systems.

If a force F acts at a point P in space, then this force can be represented geometrically by a directed line segment with initial point at P, pointing in the direction in which the force acts, and with length equal to the magnitude of the force (see Figure 1).

Consider a system of forces acting at a point P, and suppose that the given forces all act along a single line through P. We add the magnitudes of the forces acting in one direction, then add the magnitudes of the forces in the opposite direction, and subtract the smaller sum from the larger. Then a single force F, the magnitude of which is this difference, and which acts at P along the line in the direction of the larger sum, is physically equivalent to the system of forces acting at P (see Figure 2). If a positive direction is chosen on the line, then a positive or negative real number that

FIGURE 1

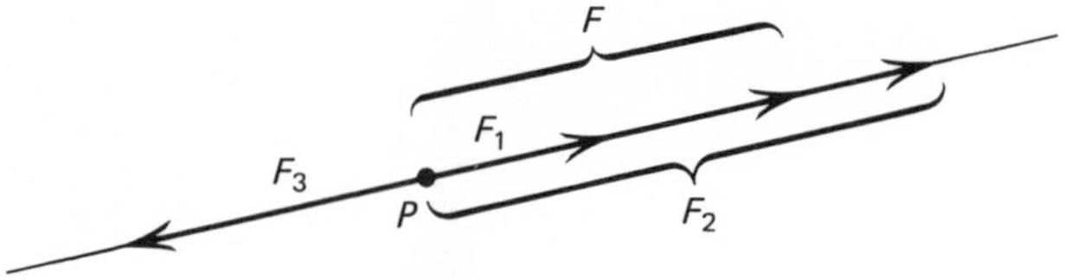

FIGURE 2

gives the magnitude of the force can be assigned to each force, and the algebra of real numbers can be used to compute the *resultant force F*.

If the several forces at the point P do not all act along the same line, then the algebra of real numbers is not adequate to compute a resultant force acting at P. By the laws of physics, two forces F_1 and F_2 acting at P are equivalent to a single force F, which has the magnitude and direction of the diagonal of the parallelogram determined by the directed line segments representing F_1 and F_2 (see Figure 3). If F_3 is a third force acting at P, then the directed line segments representing F_3 and F determine a parallelogram and F', the resultant force of F_3 and F, can be computed by the same rule used to compute F (see Figure 4). It can be shown that the directed line segment representing F' is the diagonal of the parallelepiped determined by F_1, F_2, and F_3. Hence, the same force F' would be obtained regardless of the order or grouping used to combine the three forces F_1, F_2, and F_3. Thus, the operation for combining forces enjoys some of the same algebraic properties as addition of real numbers.

The magnitude of a force F can be measured by a positive real number m. If k is any positive real number, then $k \cdot F$ is a force with magnitude km and the same direction as F, and $(-k) \cdot F$ is a force with magnitude km and direction opposite to F. If F acts at a point P, then any force acting at P along the line determined by F can be described as $r \cdot F$ for some positive or negative real number r.

Suppose that a force F_1, acting at a point P has the same magnitude and direction as a force F_2 acting at a different point Q. If P and Q are points on some rigid body, the effect of the two forces on the motion of the body will not, in general, be the same. However, in applications, it is convenient to regard F_1 and F_2 as the same force acting at two different points. That is, *two forces with the same magnitude and direction are equal.*

We have observed that forces are combined or "added" by a parallelogram law and that a force can be "multiplied" by a real number. Moreover, we have seen that two forces are regarded as equal if they have the same

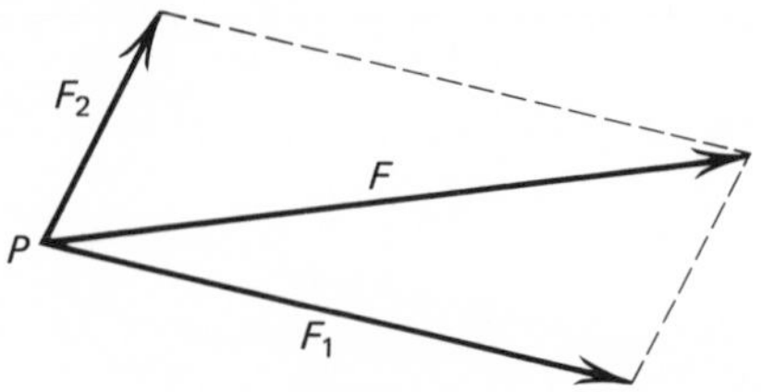

FIGURE 3

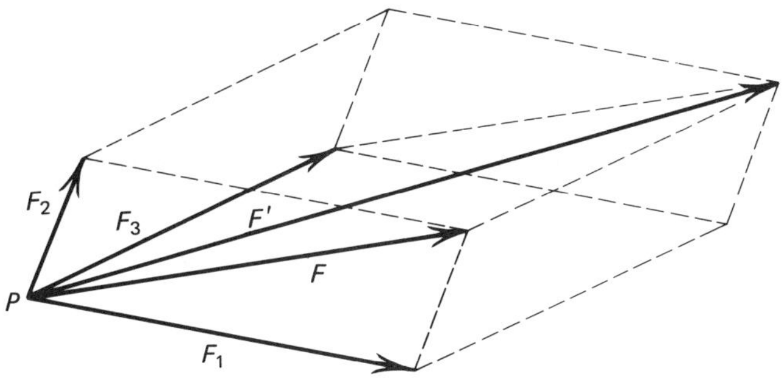

FIGURE 4

magnitude and direction. The other physical quantities mentioned above (displacement, velocity, acceleration) are added by the parallelogram law, multiplied by real numbers, and have the same definition of equality. For example, let D_1 be a displacement from a point A to point B and D_2 be a displacement from the point B to a point C. Then the single displacement D from A to C, which is equivalent to the combination of the two given displacements, is found by the same rule as that for adding two forces (see Figure 5).

Vectors form the appropriate mathematical model for these four physical systems. Our discussion of vectors in three-dimensional space assumes that the reader is familiar with elementary analytic geometry. Specifically, we will freely use the concept of a rectangular Cartesian coordinate system in the plane and in three-dimensional space. This treatment of geometric vectors will lead to interesting new ideas, mathematical tools, and important generalizations.

A directed line segment in three-dimensional space has length and direction. Each ordered pair of distinct points in space (P, Q) determines a directed line segment with initial point at P and terminal point at Q. Conversely, each directed line segment determines an ordered pair of points. Thus, it is evident that there is a one-to-one correspondence between the set of all directed line segments and the set of ordered pairs of distinct points. We denote a directed line segment from a point P to a point Q by the symbol $\overrightarrow{PQ}$. The length of $\overrightarrow{PQ}$ is written $|PQ|$. It is convenient to extend this one-to-one correspondence to the set of all possible ordered pairs of points, including even those pairs (P, P) for which the first and second member are the same point. Let (P, P) correspond to the point P, which we write as the segment $\overrightarrow{PP}$. Thus, although $\overrightarrow{PP}$ has zero length and, strictly

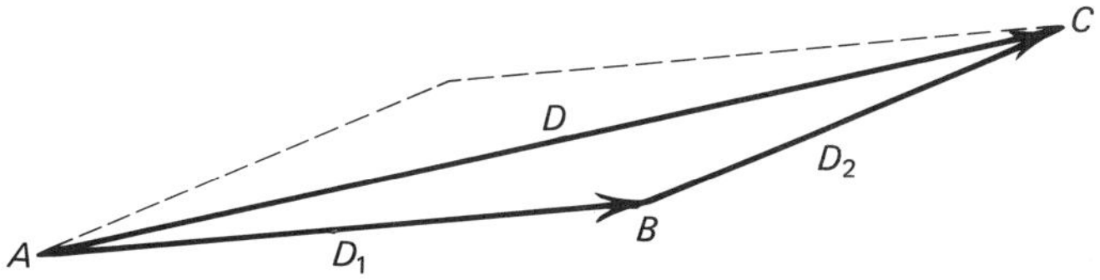

FIGURE 5

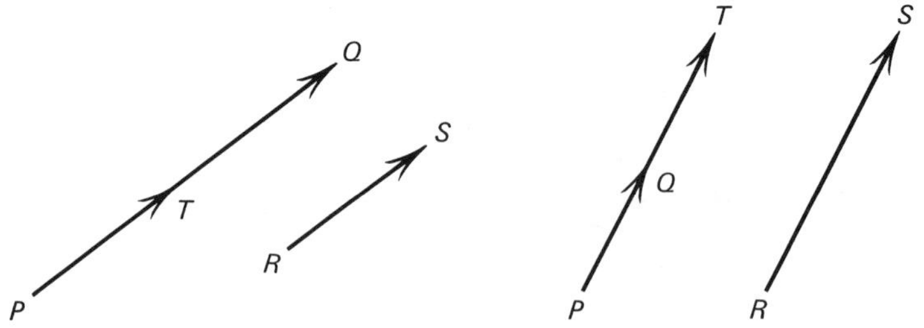

FIGURE 6

speaking, no direction, it is viewed as a directed line segment corresponding to (P, P). Such a segment $\overrightarrow{PP}$ is called a *zero directed line segment*. Consequently, a *nonzero directed line segment* is a segment $\overrightarrow{PQ}$, where P is different from Q.

Let $\overrightarrow{PQ}$ and $\overrightarrow{RS}$ be nonzero directed line segments (see Figure 6). Translate the segment $\overrightarrow{RS}$ without rotation until its initial point falls on P, and denote its new terminal point by T. Then $\overrightarrow{PQ}$ and $\overrightarrow{RS}$ have the *same direction* if and only if either T lies on $\overrightarrow{PQ}$ or Q lies on $\overrightarrow{PT}$. According to our earlier discussion, we should identify two directed line segments that have the same length and direction. Thus, if we fix a point O, each directed line segment in space can be identified with a directed line segment with its initial point at O. A directed line segment $\overrightarrow{OP}$ with initial point at O is called a *geometric vector*. Therefore, each directed line segment in space is identified with a vector, and each vector $\overrightarrow{OP}$ represents the set of all directed line segments with the same length and direction as $\overrightarrow{OP}$.

We shall study this system of geometric vectors; that is, the set of all directed line segments with the same initial point O, as motivation for the more abstract notion of a real vector space which is introduced in Chapter 2.

exercises

1.1 A jet plane has an air speed of 615 mph, and the nose of the plane is pointed due west. There is a northwest wind of 100 mph (a northwest wind is a wind from the northwest). Draw a diagram to illustrate this situation, showing the ground speed and direction of flight of the plane.

1.2 The forces shown in Figure 7 all act in the same plane at a point O. Compute the resultant force at O graphically.

1.3 Let F_1, F_2, and F_3 be forces of unit magnitude (1 lb) acting at the origin O of a rectangular Cartesian coordinate system in three-dimensional space in the directions of the positive X axis, positive Y axis, and positive Z axis, respectively (see Figure 8). Draw diagrams to illustrate the following forces:

(a) A force F acting at O with ten times the magnitude of F_1 in the same direction as F_1;

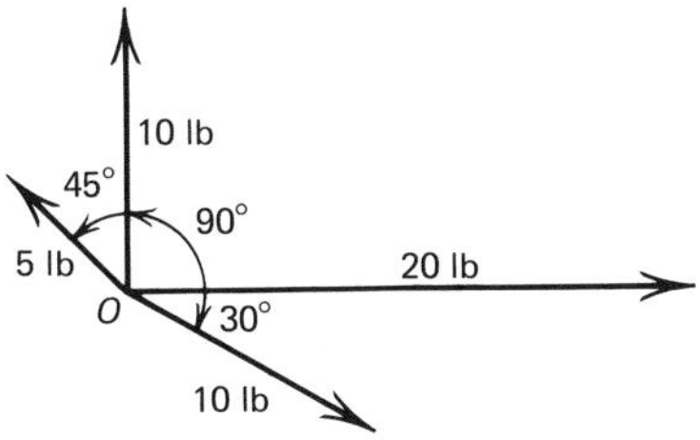

FIGURE 7

(b) a force G acting at O with three times the magnitude F_2 in the direction opposite to F_2;
(c) the resultant of the forces F and G of parts (a) and (b);
(d) a force H acting at O with $1\frac{1}{2}$ times the magnitude of F_3 in the direction of F_3;
(e) the resultant of the forces F, G, and H of parts (a), (b), and (d);
(f) the resultant of the forces F_1, F_2, and F_3.

1.4 Compute the magnitude of the resultant forces described in parts (c), (e), and (f) of Exercise 1.3.

1.5 Let F be a force acting at a point O. Sketch a pair of forces acting at O in mutually perpendicular directions which have F as their resultant. In how many ways can this be done?

1.6 Let $\overrightarrow{RS}$ be the directed line segment with initial point R having coordinates (3, 4, 2) and terminal point S having coordinates (5, −6, 7) in a rectangular Cartesian coordinate system in three-dimensional space. Sketch the vector $\overrightarrow{OP}$ which represents $\overrightarrow{RS}$. Sketch a directed line segment whose initial point has the coordinates (1, −2, −3) and which is represented by $\overrightarrow{OP}$.

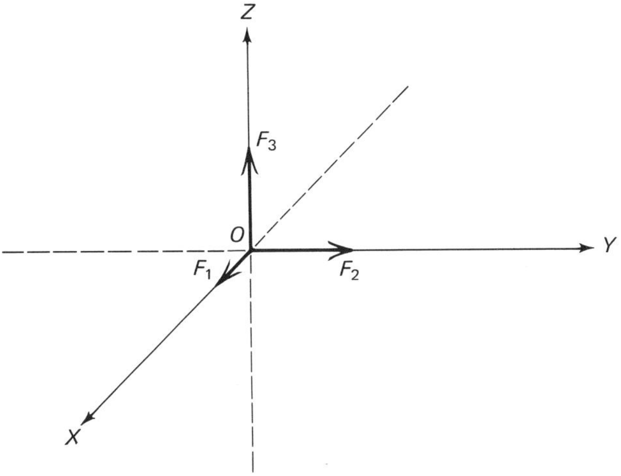

FIGURE 8

2 vector addition

If the system of geometric vectors is to be an appropriate mathematical model for the physical systems discussed in §1, the definitions of vector addition and multiplication of a vector by a real number should be motivated by the laws of physics. In particular, vectors are added according to the parallelogram rule.

Throughout the following discussion, O is a fixed point in space and each vector is a directed line segment $\overrightarrow{OP}$ with initial point at O. The *zero vector* $\overrightarrow{OO}$ will be denoted simply by $\vec{0}$. To say that a vector $\overrightarrow{OP}$ is the zero vector means, of course, that the point P coincides with O.

(2.1) definition. Let $\overrightarrow{OP}$ and $\overrightarrow{OQ}$ be vectors. Let $\overrightarrow{PR}$ be the directed line segment with initial point at P which has the same length and direction as $\overrightarrow{OQ}$. Then $\overrightarrow{OP} + \overrightarrow{OQ} = \overrightarrow{OR}$, where $\overrightarrow{OR}$ is the vector with terminal point at R.

The various possibilities that arise, namely, $\overrightarrow{OP}$ and $\overrightarrow{OQ}$ lie on the same line through O and have the same or opposite directions, or $\overrightarrow{OP}$ and $\overrightarrow{OQ}$ lie on different lines through O, are illustrated in Figure 9. In the latter case, it is clear that $\overrightarrow{OR}$ lies on the diagonal of the parallelogram determined by $\overrightarrow{OP}$ and $\overrightarrow{OQ}$. Since there is a unique directed line segment, $\overrightarrow{PR}$ with a given

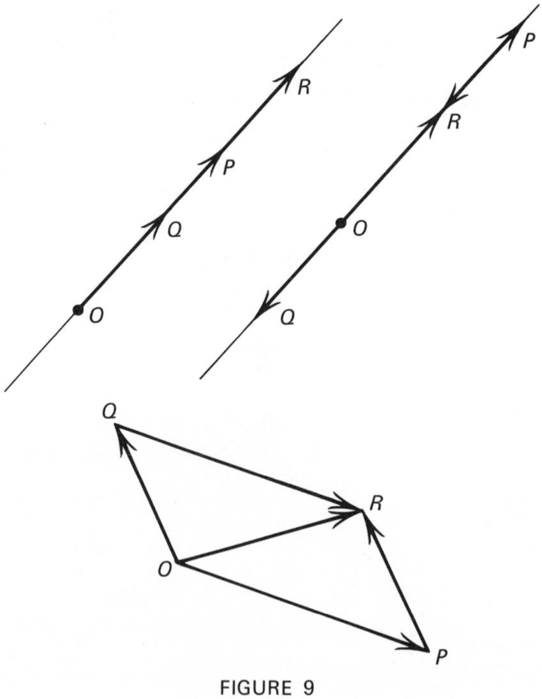

FIGURE 9

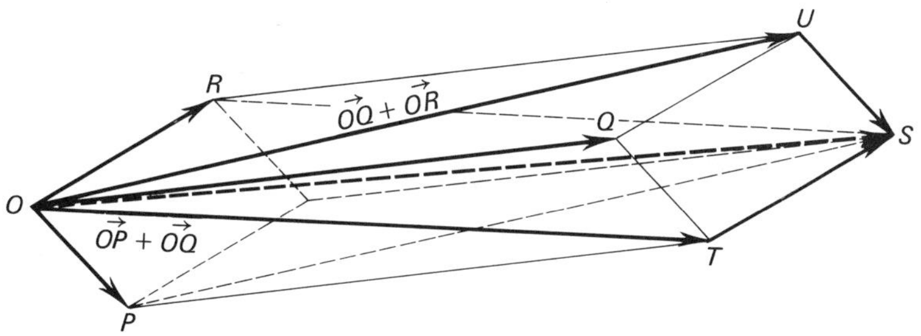

FIGURE 10

length and direction and initial point at P, vector addition is well defined.

Vector addition satisfies the same algebraic identities as addition of real numbers. It is geometrically evident that vector addition is commutative. That is, $\overrightarrow{OP} + \overrightarrow{OQ} = \overrightarrow{OQ} + \overrightarrow{OP}$. This result follows from the observation that the directed line segment $\overrightarrow{QR}$ has the same length and direction as $\overrightarrow{OP}$ (see Figure 9). Let $\overrightarrow{OP}$, $\overrightarrow{OQ}$, and $\overrightarrow{OR}$ be any three vectors. Then $(\overrightarrow{OP} + \overrightarrow{OQ}) + \overrightarrow{OR} = \overrightarrow{OP} + (\overrightarrow{OQ} + \overrightarrow{OR})$. That is, vector addition is associative. In the case where $\overrightarrow{OP}$, $\overrightarrow{OQ}$ and $\overrightarrow{OR}$ are on three different lines through O, it follows from Definition 2.1 that both $(\overrightarrow{OP} + \overrightarrow{OQ}) + \overrightarrow{OR}$ and $\overrightarrow{OP} + (\overrightarrow{OQ} + \overrightarrow{OR})$ yield a vector $\overrightarrow{OS}$ which is on the diagonal of the parallelepiped determined by $\overrightarrow{OP}$, $\overrightarrow{OQ}$, and $\overrightarrow{OR}$ (see Figure 10). The other cases where two or more of the given vectors lie on the same line through O can be easily checked.

Since vector addition is associative, we can write $\overrightarrow{OP} + \overrightarrow{OQ} + \overrightarrow{OR}$ to stand for either sum $\overrightarrow{OP} + (\overrightarrow{OQ} + \overrightarrow{OR})$ or $(\overrightarrow{OP} + \overrightarrow{OQ}) + \overrightarrow{OR}$. More generally, it follows that the grouping of terms in any indicated sum is immaterial, so that parentheses and other accumulation signs may be dropped.

It is a consequence of Definition 2.1 that the vector $\vec{0}$ plays the same role in vector addition that the number 0 plays in the addition of real numbers. That is,

$$\overrightarrow{OP} + \vec{0} = \vec{0} + \overrightarrow{OP} = \overrightarrow{OP}$$

for any vector $\overrightarrow{OP}$.

If $\overrightarrow{OP}$ is any vector, then the negative of $\overrightarrow{OP}$, which we write $-\overrightarrow{OP}$, is the vector on the same line through O as $\overrightarrow{OP}$ with the same length as $\overrightarrow{OP}$, but the opposite direction (see Figure 11). In particular, $-\vec{0} = \vec{0}$. It follows from Definition 2.1 that $\overrightarrow{OP} + (-\overrightarrow{OP}) = -\overrightarrow{OP} + \overrightarrow{OP} = \vec{0}$. Suppose that $\overrightarrow{OQ}$ is any vector such that $\overrightarrow{OP} + \overrightarrow{OQ} = \vec{0}$. Then

$$\begin{aligned}\overrightarrow{OQ} &= \vec{0} + \overrightarrow{OQ} = (-\overrightarrow{OP} + \overrightarrow{OP}) + \overrightarrow{OQ} \\ &= -\overrightarrow{OP} + (\overrightarrow{OP} + \overrightarrow{OQ}) = -\overrightarrow{OP} + 0 = -\overrightarrow{OP}.\end{aligned}$$

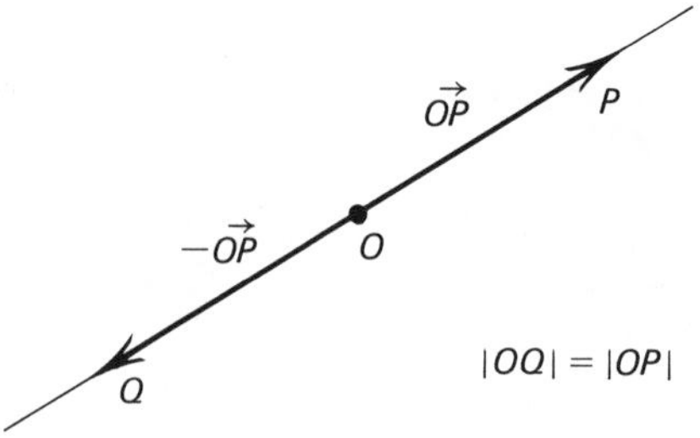

FIGURE 11

Thus, $-\overrightarrow{OP}$ is the only vector X that satisfies the vector equation $\overrightarrow{OP} + X = \vec{0}$. The usefulness of this remark is illustrated in the following examples.

EXAMPLE 1 Show that $-(-\overrightarrow{OP}) = \overrightarrow{OP}$. Using the definition of the negative of a vector, the length of $-(-\overrightarrow{OP})$ is the length of $-\overrightarrow{OP}$, which is the length of $\overrightarrow{OP}$. Moreover, the direction of $-(-\overrightarrow{OP})$ is opposite to that of $-\overrightarrow{OP}$. Thus, $-(-\overrightarrow{OP})$ has the same direction as $\overrightarrow{OP}$. Since $-(-\overrightarrow{OP})$ and $\overrightarrow{OP}$ have the same length and direction, $-(-\overrightarrow{OP}) = \overrightarrow{OP}$.

If we observe that the equation $-\overrightarrow{OP} + \overrightarrow{OP} = \vec{0}$ states that $\overrightarrow{OP}$ is a solution of the vector equation $-\overrightarrow{OP} + X = \vec{0}$, then it follows directly from the above remark that $\overrightarrow{OP} = -(-\overrightarrow{OP})$.

EXAMPLE 2 Show that $-(\overrightarrow{OP} + \overrightarrow{OQ}) = -\overrightarrow{OP} + (-\overrightarrow{OQ})$. It is sufficient to show that $-\overrightarrow{OP} + (-\overrightarrow{OQ})$ is a solution of the vector equation $(\overrightarrow{OP} + \overrightarrow{OQ}) + X = \vec{0}$. Indeed, using the associative and commutative laws of vector addition, we have

$$\begin{aligned}(\overrightarrow{OP} + \overrightarrow{OQ}) + [-\overrightarrow{OP} + (-\overrightarrow{OQ})] &= [\overrightarrow{OP} + (-\overrightarrow{OP})] + [\overrightarrow{OQ} + (-\overrightarrow{OQ})] \\ &= \vec{0} + \vec{0} = \vec{0}.\end{aligned}$$

The properties of vector addition are summarized in the following theorem.

(2.2) theorem The set of geometric vectors in three-dimensional space with initial point at O satisfy the following conditions:

(a) For any vectors $\overrightarrow{OP}$ and $\overrightarrow{OQ}$, $\overrightarrow{OP} + \overrightarrow{OQ}$ is a unique vector $\overrightarrow{OR}$;

(b) $(\overrightarrow{OP} + \overrightarrow{OQ}) + \overrightarrow{OR} = \overrightarrow{OP} + (\overrightarrow{OQ} + \overrightarrow{OR})$ for all vectors $\overrightarrow{OP}$, $\overrightarrow{OQ}$, and $\overrightarrow{OR}$;

(c) $\overrightarrow{OP} + \overrightarrow{OQ} = \overrightarrow{OQ} + \overrightarrow{OP}$ for all vectors $\overrightarrow{OP}$ and $\overrightarrow{OQ}$;

(d) there is a vector $\vec{0}$ such that $\overrightarrow{OP} + \vec{0} = \overrightarrow{OP}$ for every vector $\overrightarrow{OP}$;

(e) if $\overrightarrow{OP}$ is any vector, then there is a vector $-\overrightarrow{OP}$ such that $\overrightarrow{OP} + (-\overrightarrow{OP}) = \vec{0}$.

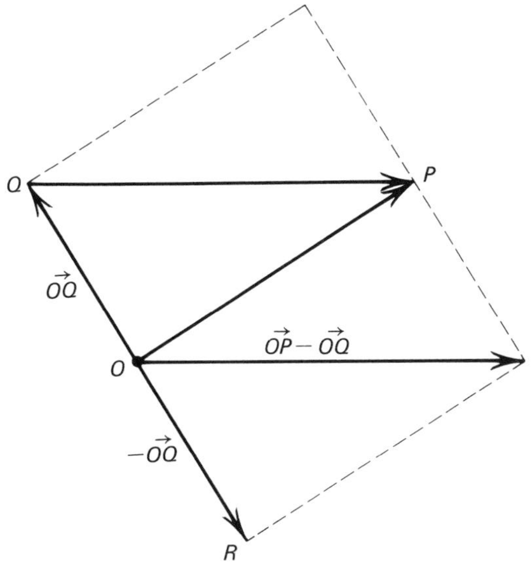

FIGURE 12

Subtraction of vectors is defined by

$$\overrightarrow{OP} - \overrightarrow{OQ} = \overrightarrow{OP} + (-\overrightarrow{OQ}).$$

We observe that $\overrightarrow{OP} - \overrightarrow{OQ}$ has the same length and direction as the directed line segment $\overrightarrow{QP}$ from the terminal point of $\overrightarrow{OQ}$ to the terminal point of $\overrightarrow{OP}$ (see Figure 12).

exercises

2.1 Let $\overrightarrow{OP}$, $\overrightarrow{OQ}$, and $\overrightarrow{OR}$ be the vectors illustrated in Figure 13, where $|OQ| = 3|OP|$ and $|OR| = \frac{1}{2}|OQ|$. Draw figures which show

(a) $\overrightarrow{OP}$, $\overrightarrow{OQ}$, and $\overrightarrow{OP} + \overrightarrow{OQ}$

(b) $\overrightarrow{OP}$, $\overrightarrow{OQ}$, and $\overrightarrow{OQ} + \overrightarrow{OP}$

(c) $\overrightarrow{OQ}$, $\overrightarrow{OR}$ and $\overrightarrow{OQ} + \overrightarrow{OR}$

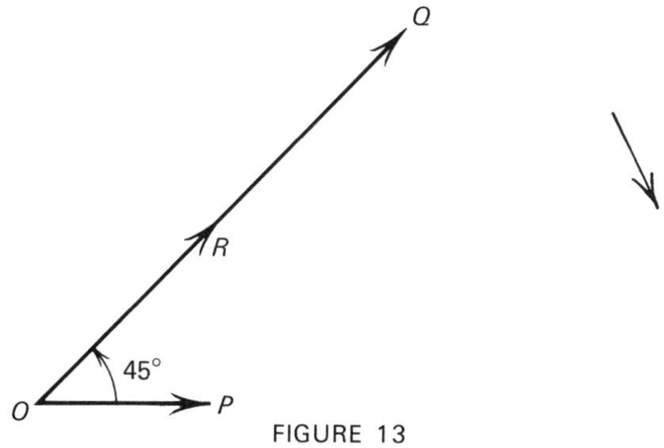

FIGURE 13

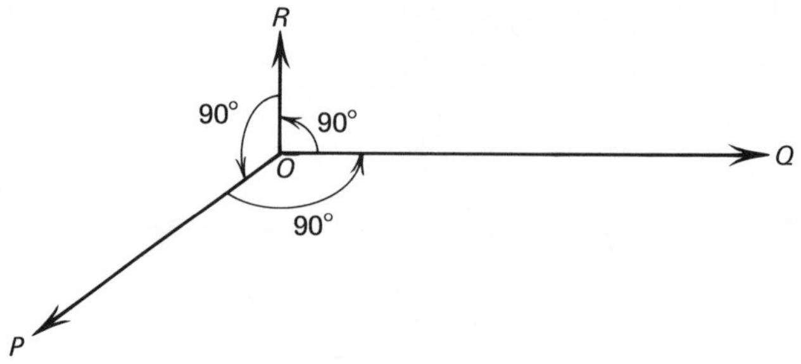

FIGURE 14

(d) $\overrightarrow{OP}$, $\overrightarrow{OQ}$, $-\overrightarrow{OP}$, and $-\overrightarrow{OQ}$

(e) $\overrightarrow{OP}$, $\overrightarrow{OQ}$, $\overrightarrow{OP} - \overrightarrow{OQ}$, and $\overrightarrow{OQ} - \overrightarrow{OP}$

(f) $\overrightarrow{OP}$, $\overrightarrow{OQ}$, $\overrightarrow{OR}$, and $(\overrightarrow{OR} + \overrightarrow{OQ}) + \overrightarrow{OP}$

(g) $\overrightarrow{OP}$, $\overrightarrow{OQ}$, $\overrightarrow{OR}$, and $(\overrightarrow{OP} + \overrightarrow{OQ}) + \overrightarrow{OR}$

(h) $\overrightarrow{OP}$, $\overrightarrow{OQ}$, $\overrightarrow{OR}$, and $\overrightarrow{OR} + (\overrightarrow{OP} + \overrightarrow{OQ})$

(i) $\overrightarrow{OP}$, $\overrightarrow{OQ}$, $\overrightarrow{OR}$, and $\overrightarrow{OR} - (\overrightarrow{OQ} - \overrightarrow{OP})$

(j) $\overrightarrow{OP}$, $\overrightarrow{OQ}$, $\overrightarrow{OR}$, and $(\overrightarrow{OR} - \overrightarrow{OQ}) - \overrightarrow{OP}$

2.2 Let $\overrightarrow{OP}$, $\overrightarrow{OQ}$, and $\overrightarrow{OR}$ be the vectors illustrated in Figure 14, where $|OQ| = 4|OR|$ and $|OP| = 5|OR|$. Draw figures which show

(a) $\overrightarrow{OP}$, $\overrightarrow{OQ}$, $\overrightarrow{OR}$, and $\overrightarrow{OP} + \overrightarrow{OQ} + \overrightarrow{OR}$

(b) $\overrightarrow{OP}$, $\overrightarrow{OQ}$, $\overrightarrow{OR}$, $-\overrightarrow{OQ}$ and $(\overrightarrow{OP} - \overrightarrow{OQ}) + \overrightarrow{OR}$

(c) $\overrightarrow{OP}$, $\overrightarrow{OQ}$, $\overrightarrow{OR}$, $-\overrightarrow{OP}$, $-\overrightarrow{OQ}$, and $\overrightarrow{OR} - (\overrightarrow{OP} + \overrightarrow{OQ})$

2.3 Prove that vector subtraction satisfies the following identities:

(a) $\overrightarrow{OP} - (\overrightarrow{OQ} + \overrightarrow{OR}) = \overrightarrow{OP} - \overrightarrow{OQ}) - \overrightarrow{OR}$

(b) $\overrightarrow{OP} - (\overrightarrow{OQ} - \overrightarrow{OR}) = (\overrightarrow{OP} - \overrightarrow{OQ}) + \overrightarrow{OR} = (\overrightarrow{OP} + \overrightarrow{OR}) - \overrightarrow{OQ}$
$= \overrightarrow{OP} + (\overrightarrow{OR} - \overrightarrow{OQ})$

2.4 Prove that $\overrightarrow{OQ} - \overrightarrow{OP}$ is the unique solution of the vector equation $\overrightarrow{OP} + X = \overrightarrow{OQ}$.

3 scalar multiplication

The second vector operation suggested by the examples of §1 is *scalar multiplication,* a process by which a vector is multiplied by a real number to yield another vector.

(3.1) definition Let r be a real number and let $\overrightarrow{OP}$ be any vector. The vector $r \cdot \overrightarrow{OP}$ is a vector with length $|r|\,|OP|$. If $r > 0$, $r \cdot \overrightarrow{OP}$ has the same direction as $\overrightarrow{OP}$; if $r < 0$, $r \cdot \overrightarrow{OP}$ has the direction opposite to $\overrightarrow{OP}$; if $r = 0$, $r \cdot \overrightarrow{OP} = \vec{0}$.

In the above definition, $|r|$ is the absolute value of the real number r. That is, $|r| = r$ if $r \geq 0$, and $|r| = -r$, if $r < 0$. Note that if $\overrightarrow{OP}$ is the zero vector, then $|OP| = 0$, so that the length of $r \cdot \overrightarrow{OP}$ is 0. Thus, if $\overrightarrow{OP} = \vec{0}$, then $r \cdot \overrightarrow{OP} = \vec{0}$.

The vector $r \cdot \overrightarrow{OP}$ can be described as the vector which is "r times as long as $\overrightarrow{OP}$" with the same or opposite direction as $\overrightarrow{OP}$ according as r is positive or negative. When r is negative, $r \cdot \overrightarrow{OP}$ is the negative of the vector $|r| \cdot \overrightarrow{OP}$; that is, $r \cdot \overrightarrow{OP} = -(|r| \cdot \overrightarrow{OP})$. Figure 15 shows some examples of scalar multiplication.

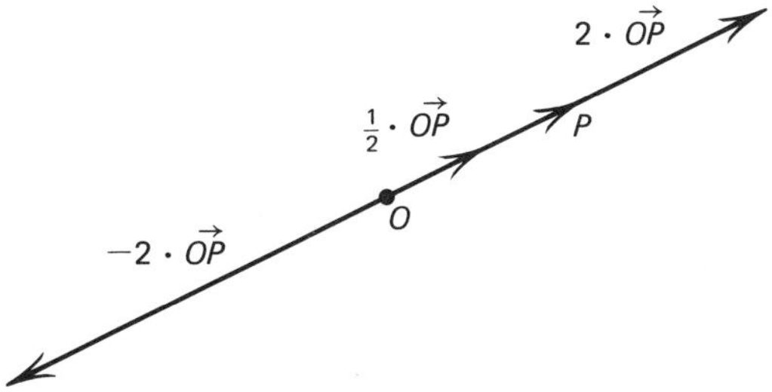

FIGURE 15

Scalar multiplication satisfies several useful algebraic identities, which are listed in the following theorem.

(3.2) theorem Let $\overrightarrow{OP}$ and $\overrightarrow{OQ}$ be vectors and let r and s be real numbers Then

(a) $r \cdot (s \cdot \overrightarrow{OP}) = s \cdot (r \cdot \overrightarrow{OP}) = (rs) \cdot \overrightarrow{OP}$;

(b) $(r + s) \cdot \overrightarrow{OP} = r \cdot \overrightarrow{OP} + s \cdot \overrightarrow{OP}$;

(c) $r \cdot (\overrightarrow{OP} + \overrightarrow{OQ}) = r \cdot \overrightarrow{OP} + r \cdot \overrightarrow{OQ}$;

(d) $(-r) \cdot \overrightarrow{OP} = -(r \cdot \overrightarrow{OP})$;

(e) $1 \cdot \overrightarrow{OP} = \overrightarrow{OP}$.

The proof of Theorem 3.2 is left as an exercise (Exercise 3.3).

EXAMPLE 1 Let $\overrightarrow{OP}$ be any vector. Then $2 \cdot (3 \cdot \overrightarrow{OP}) = 3 \cdot (2 \cdot \overrightarrow{OP}) = 6 \cdot \overrightarrow{OP}$; $(-\frac{1}{2}) \cdot (2 \cdot \overrightarrow{OP}) = (-1) \cdot \overrightarrow{OP} = -\overrightarrow{OP}$; $(-8) \cdot [(-\frac{1}{4}) \cdot \overrightarrow{OP}] = 2 \cdot \overrightarrow{OP}$; $6 \cdot \overrightarrow{OP} + (-2) \cdot \overrightarrow{OP} = 4 \cdot \overrightarrow{OP}$; $\frac{1}{2} \cdot \overrightarrow{OP} + \frac{5}{2} \cdot \overrightarrow{OP} = 3 \cdot \overrightarrow{OP}$.

A collection of vectors is said to be *collinear* if they lie on the same line through O. In particular, the zero vector $\vec{0}$ is collinear with any other vector. Definition 3.1 leads to the following theorem concerning collinear vectors.

(3.3) theorem If $\overrightarrow{OP}$ is any vector, the set of all vectors of the form $r \cdot \overrightarrow{OP}$, where r is a real number, is a collection of collinear vectors. Moreover, if

$\overrightarrow{OP} \neq \vec{0}$, then every vector collinear with $\overrightarrow{OP}$ has the form $r \cdot \overrightarrow{OP}$ for some real number r.

PROOF It follows at once from Definition 3.1 that for any real number r, $\overrightarrow{OP}$ and $r \cdot \overrightarrow{OP}$, are collinear vectors. That is, every vector of the form $r \cdot \overrightarrow{OP}$ is on the same line through O as $\overrightarrow{OP}$, so that this collection of vectors is collinear. (Of course if $\overrightarrow{OP} = \vec{0}$, then $r \cdot \overrightarrow{OP} = \vec{0}$ for every real number r.) This proves the first statement of the theorem.

Suppose that $\overrightarrow{OP} \neq 0$ and that $\overrightarrow{OQ}$ is a vector collinear with $\overrightarrow{OP}$. By Definition 3.1, the vector $(|OQ|/|OP|) \cdot \overrightarrow{OP}$ has length $(|OQ|/|OP|)\ |OP| = |OQ|$. Since $\overrightarrow{OP}$ and $\overrightarrow{OQ}$ lie on the same line through O, it follows that $\overrightarrow{OQ} = (|OQ|/|OP|) \cdot \overrightarrow{OP}$ if $\overrightarrow{OP}$ and $\overrightarrow{OQ}$ have the same direction. By Theorem 3.2(d) $(-r) \cdot \overrightarrow{OP} = -(r \cdot \overrightarrow{OP})$. Thus, if $\overrightarrow{OP}$ and $\overrightarrow{OQ}$ have opposite directions, then $\overrightarrow{OQ} = (-|OQ|/|OP|) \cdot \overrightarrow{OP}$. In either case, $\overrightarrow{OQ} = r \cdot \overrightarrow{OP}$ for a suitable real number r. ∎

Theorem 3.3 can be used to obtain a useful test for the collinearity of two vectors.

(3.4) corollary The vectors $\overrightarrow{OP}$ and $\overrightarrow{OQ}$ are collinear if and only if there are real numbers r and s, not both zero, such that $r \cdot \overrightarrow{OP} + s \cdot \overrightarrow{OQ} = \vec{0}$.

PROOF Suppose first that $\overrightarrow{OP}$ and $\overrightarrow{OQ}$ are collinear. If $\overrightarrow{OP} = \vec{0}$, then we can choose $r = 1$ and $s = 0$, in which case, $1 \cdot \overrightarrow{OP} + \vec{0} \cdot \overrightarrow{OQ} = \overrightarrow{OP} = \vec{0}$. If $\overrightarrow{OP} \neq \vec{0}$, then by Theorem 3.3, $\overrightarrow{OQ} = r \cdot \overrightarrow{OP}$ for some real number r. Therefore, $r \cdot \overrightarrow{OP} - \overrightarrow{OQ} = \vec{0}$. Since $(-1) \cdot \overrightarrow{OQ} = -\overrightarrow{OQ}$ by Theorem 3.2 (d) and (e), we can choose $s = -1$, and obtain $r \cdot \overrightarrow{OP} + (-1) \cdot \overrightarrow{OQ} = \vec{0}$.

Conversely, suppose that there are real numbers r and s such that $r \cdot \overrightarrow{OP} + s \cdot \overrightarrow{OQ} = \vec{0}$. If $r \neq 0$, then by Theorem 3.2,

$$\frac{1}{r} \cdot (r \cdot \overrightarrow{OP} + s \cdot \overrightarrow{OQ}) = \frac{1}{r} \cdot (r \cdot \overrightarrow{OP}) + \frac{1}{r} \cdot (s \cdot \overrightarrow{OQ})$$

$$= 1 \cdot \overrightarrow{OP} + \frac{s}{r} \cdot \overrightarrow{OQ} = \frac{1}{r} \cdot \vec{0} = \vec{0}.$$

Therefore, $\overrightarrow{OP} = (-s/r) \cdot \overrightarrow{OQ}$. By Theorem 3.3, $\overrightarrow{OP}$ and $\overrightarrow{OQ}$ are collinear. If $r = 0$, then $s \neq 0$, and a similar argument shows that $\overrightarrow{OQ} = (-r/s) \cdot \overrightarrow{OP}$, so that as before, $\overrightarrow{OP}$ and $\overrightarrow{OQ}$ are collinear. ∎

EXAMPLE 2 Let $\overrightarrow{OP}$ and $\overrightarrow{OQ}$ be nonzero collinear vectors with the same direction such that $\overrightarrow{OQ}$ is five times as long as $\overrightarrow{OP}$. Then $\overrightarrow{OQ} = 5 \cdot \overrightarrow{OP}$, and

$5 \cdot \overrightarrow{OP} + (-1) \cdot \overrightarrow{OQ} = \vec{0}$. If $\overrightarrow{OP}$ and $\overrightarrow{OQ}$ have opposite directions, then $\overrightarrow{OQ} = (-5) \cdot \overrightarrow{OP}$, and $5 \cdot \overrightarrow{OP} + 1 \cdot \overrightarrow{OQ} = \vec{0}$.

A collection of vectors is *coplanar* if they lie in the same plane through O. Results similar to those obtained for collinear vectors can be derived for coplanar vectors.

(3.5) theorem If $\overrightarrow{OP}$ and $\overrightarrow{OQ}$ are any vectors, the set of all vectors of the form $r \cdot \overrightarrow{OP} + s \cdot \overrightarrow{OQ}$, where r and s are real numbers, is a collection of coplanar vectors. Moreover, if $\overrightarrow{OP}$ and $\overrightarrow{OQ}$ are noncollinear vectors, then every vector coplanar with $\overrightarrow{OP}$ and $\overrightarrow{OQ}$ has the form $r \cdot \overrightarrow{OP} + s \cdot \overrightarrow{OQ}$ for some real numbers r and s.

PROOF It is clear that any pair of vectors is coplanar. Since $r \cdot \overrightarrow{OP}$ is collinear with $\overrightarrow{OP}$, and $s \cdot \overrightarrow{OQ}$ is collinear with $\overrightarrow{OQ}$, for real numbers r and s, it follows that $r \cdot \overrightarrow{OP}$ and $s \cdot \overrightarrow{OQ}$ lie in the same plane through O as $\overrightarrow{OP}$ and $\overrightarrow{OQ}$. By Definition 2.1, the vector $r \cdot \overrightarrow{OP} + s \cdot \overrightarrow{OQ}$ lies in the same plane through O as $r \cdot \overrightarrow{OP}$ and $s \cdot \overrightarrow{OQ}$. Thus, every vector of the form $r \cdot \overrightarrow{OP} + s \cdot \overrightarrow{OQ}$ is coplanar with $\overrightarrow{OP}$ and $\overrightarrow{OQ}$. This proves the first statement of the theorem.

Now suppose that $\overrightarrow{OP}$ and $\overrightarrow{OQ}$ are noncollinear. In particular, this implies that $\overrightarrow{OP} \neq \vec{0}$ and $\overrightarrow{OQ} \neq \vec{0}$, since $\vec{0}$ is collinear with any vector. Let $\overrightarrow{OR}$ be a vector coplanar with $\overrightarrow{OP}$ and $\overrightarrow{OQ}$. If $\overrightarrow{OR}$ and $\overrightarrow{OP}$ are collinear, then by Theorem 3.3 $\overrightarrow{OR} = r \cdot \overrightarrow{OP} = r \cdot \overrightarrow{OP} + 0 \cdot \overrightarrow{OQ}$ for some r. Similarly, if $\overrightarrow{OR}$ and $\overrightarrow{OQ}$ are collinear, than $\overrightarrow{OR} = s \cdot \overrightarrow{OQ} = 0 \cdot \overrightarrow{OP} + s \cdot \overrightarrow{OQ}$ for some s. Hence, we have shown that if $\overrightarrow{OR}$ is collinear with either $\overrightarrow{OP}$ or $\overrightarrow{OQ}$, then $\overrightarrow{OR} = r \cdot \overrightarrow{OP} + s \cdot \overrightarrow{OQ}$. Next, we consider the case where $\overrightarrow{OR}$ is not collinear with either $\overrightarrow{OP}$ or $\overrightarrow{OQ}$. In this case $\overrightarrow{OR}$, $\overrightarrow{OP}$, and $\overrightarrow{OQ}$ lie on three distinct lines ℓ, ℓ_1, and ℓ_2 respectively, in the plane through O determined by ℓ_1 and ℓ_2 (see Figure 16). Let ℓ_1' and ℓ_2' be lines through R parallel to ℓ_1 and ℓ_2 respectively. Then the lines ℓ_1, ℓ_1', ℓ_2, and ℓ_2' determine a parallelogram $OBRA$ which has $\overrightarrow{OR}$ as its diagonal. According to Definition 2.1, $\overrightarrow{OR} = \overrightarrow{OB} + \overrightarrow{OA}$. Since $\overrightarrow{OB}$ is collinear with $\overrightarrow{OP}$, $\overrightarrow{OB} = r \cdot \overrightarrow{OP}$ for some real number r. Similarly, since $\overrightarrow{OA}$ is collinear with $\overrightarrow{OQ}$, $\overrightarrow{OA} = s \cdot \overrightarrow{OQ}$ for some real number s. Therefore, $\overrightarrow{OR} = \overrightarrow{OB} + \overrightarrow{OA} = r \cdot \overrightarrow{OP} + s \cdot \overrightarrow{OQ}$. ∎

(3.6) corollary The vectors $\overrightarrow{OP}$, $\overrightarrow{OQ}$, and $\overrightarrow{OR}$ are coplanar if and only if there are real numbers r, s, and t, not all zero, such that $r \cdot \overrightarrow{OP} + s \cdot \overrightarrow{OQ} + t \cdot \overrightarrow{OR} = \vec{0}$.

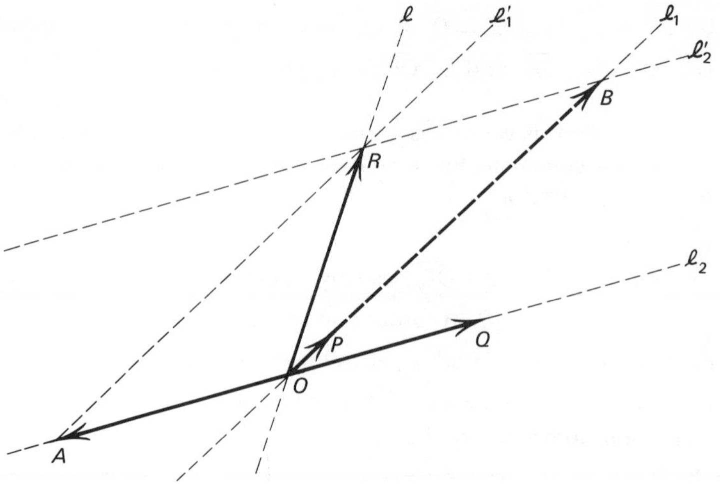

FIGURE 16

PROOF Suppose first that $\overrightarrow{OP}$, $\overrightarrow{OQ}$, and $\overrightarrow{OR}$ are coplanar. If $\overrightarrow{OP}$ and $\overrightarrow{OQ}$ are collinear, then by Corollary 3.4 there are real numbers r and s, not both zero, such that $r \cdot \overrightarrow{OP} + s \cdot \overrightarrow{OQ} = \vec{0}$. In this case, we can choose $t = 0$ and obtain $r \cdot \overrightarrow{OP} + s \cdot \overrightarrow{OQ} + 0 \cdot \overrightarrow{OR} = \vec{0}$. We have obtained the desired result when $\overrightarrow{OP}$ and $\overrightarrow{OQ}$ are collinear. Now assume that $\overrightarrow{OP}$ and $\overrightarrow{OQ}$ are noncollinear. By Theorem 3.5, $\overrightarrow{OR} = r \cdot \overrightarrow{OP} + s \cdot \overrightarrow{OQ}$ for suitable real numbers r and s. Selecting $t = -1$, we have $r \cdot \overrightarrow{OP} + s \cdot \overrightarrow{OQ} + (-1) \cdot \overrightarrow{OR} = \vec{0}$.

Conversely, suppose that $r \cdot \overrightarrow{OP} + s \cdot \overrightarrow{OQ} + t \cdot \overrightarrow{OR} = \vec{0}$, where r, s, and t are not all zero. If $r \neq 0$, then by Theorem 3.2

$$\frac{1}{r} \cdot (r \cdot \overrightarrow{OP} + s \cdot \overrightarrow{OQ} + t \cdot \overrightarrow{OR}) = \frac{1}{r} \cdot (r \cdot \overrightarrow{OP}) + \frac{1}{r} \cdot (s \cdot \overrightarrow{OQ})$$
$$+ \frac{1}{r} \cdot (t \cdot \overrightarrow{OR}) = 1 \cdot \overrightarrow{OP} + \frac{s}{r} \cdot \overrightarrow{OQ} + \frac{t}{r} \cdot \overrightarrow{OR} = \vec{0}.$$

Therefore, $\overrightarrow{OP} = (-s/r) \cdot \overrightarrow{OQ} + (-t/r) \cdot \overrightarrow{OR}$. By the first part of Theorem 3.5, $\overrightarrow{OP}$, $\overrightarrow{OQ}$, and $\overrightarrow{OR}$ are coplanar. If $r = 0$, then $s \neq 0$ or $t \neq 0$, and a similar argument shows that $\overrightarrow{OQ} = (-r/s) \cdot \overrightarrow{OP} + (-t/s) \cdot \overrightarrow{OR}$ or $\overrightarrow{OR} = (-r/t) \cdot \overrightarrow{OP} + (-s/t) \cdot \overrightarrow{OQ}$. In any case, it follows from Theorem 3.5 that $\overrightarrow{OP}$, $\overrightarrow{OQ}$, and $\overrightarrow{OR}$ are coplanar. ∎

EXAMPLE 3 Let $\overrightarrow{OP}$, $\overrightarrow{OQ}$, and $\overrightarrow{OR}$ be coplanar vectors as shown in Figure 17, where $|OP| = \frac{1}{2}|OQ|$ and $|OR| = 3|OP|$. We have $\overrightarrow{OR} = \overrightarrow{OS} + \overrightarrow{OT}$. Moreover, $|OS| = |OT| = \sqrt{2}/2|OR| = 3\sqrt{2}/2|OP| = 3\sqrt{2}/4|OQ|$. There-

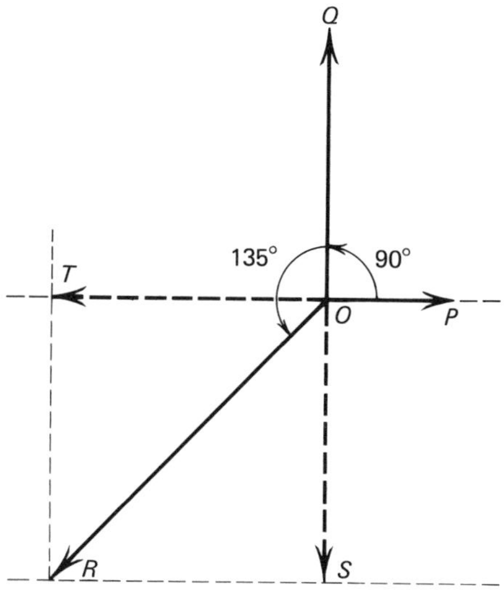

FIGURE 17

fore, $\overrightarrow{OS} = (-3\sqrt{2}/4) \cdot \overrightarrow{OQ}$ and $\overrightarrow{OT} = (-3\sqrt{2}/2) \cdot \overrightarrow{OP}$. Thus, $\overrightarrow{OR} = (-3\sqrt{2}/4) \cdot \overrightarrow{OQ} + (-3\sqrt{2}/2) \cdot \overrightarrow{OP}$, so that

$$(3\sqrt{2}/2) \cdot \overrightarrow{OP} + (3\sqrt{2}/4) \cdot \overrightarrow{OQ} + 1 \cdot \overrightarrow{OR} = \vec{0}.$$

exercises

3.1 Let $\overrightarrow{OP}$ be any nonzero vector. Draw figures showing $\overrightarrow{OP}$ and $\frac{1}{2} \cdot \overrightarrow{OP}$, $\overrightarrow{OP}$ and $5 \cdot \overrightarrow{OP}$, $\overrightarrow{OP}$ and $(-0.3) \cdot \overrightarrow{OP}$, $\overrightarrow{OP}$ and $(\sqrt{2}) \cdot \overrightarrow{OP}$, $\overrightarrow{OP}$ and $(-\sqrt{3}) \cdot \overrightarrow{OP}$, $\overrightarrow{OP}$ and $0 \cdot \overrightarrow{OP}$.

3.2 Let $\overrightarrow{OP}$ and $\overrightarrow{OQ}$ be the vectors shown in Figure 18, where $|OP| = 3|OQ|$. Draw figures which show

(a) $\overrightarrow{OP}$, $\overrightarrow{OQ}$, and $2 \cdot \overrightarrow{OP} + 3 \cdot \overrightarrow{OQ}$

(b) $\overrightarrow{OP}$, $\overrightarrow{OQ}$, and $\frac{1}{2} \cdot \overrightarrow{OP} + 5 \cdot \overrightarrow{OQ}$

(c) $\overrightarrow{OP}$, $\overrightarrow{OQ}$, and $1 \cdot \overrightarrow{OP} + (-1.5) \cdot \overrightarrow{OQ}$

(d) $\overrightarrow{OP}$, $\overrightarrow{OQ}$, and $[\frac{1}{3} \cdot \overrightarrow{OP} + 2 \cdot \overrightarrow{OQ}] + [(-\frac{4}{3}) \cdot \overrightarrow{OP} + (-1) \cdot \overrightarrow{OQ}]$

(e) $\overrightarrow{OP}$, $\overrightarrow{OQ}$, and $3 \cdot \overrightarrow{OP} - 5 \cdot \overrightarrow{OQ}$

(f) $\overrightarrow{OP}$, $\overrightarrow{OQ}$, and $[(2.5) \cdot \overrightarrow{OP} + (0.5) \cdot \overrightarrow{OQ}] - [3 \cdot \overrightarrow{OP} + (2.5) \cdot \overrightarrow{OQ}]$

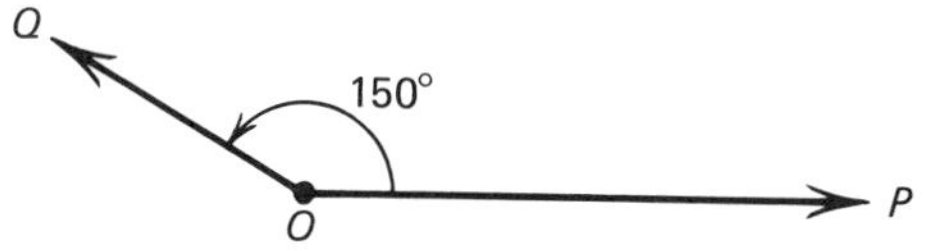

FIGURE 18

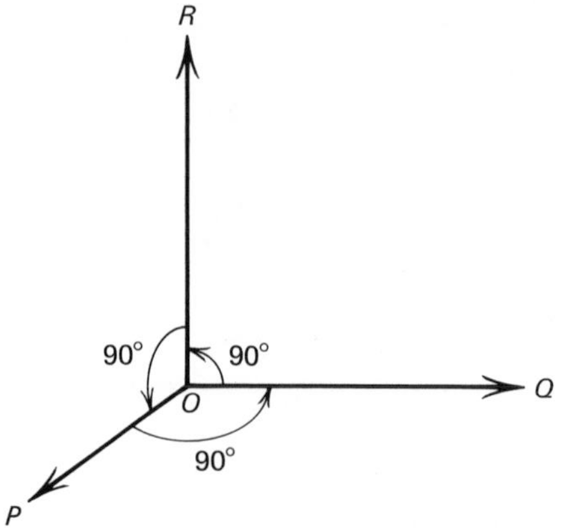

FIGURE 19

3.3 Prove Theorem 3.2

3.4 Show that if n is a positive integer, then

$$n \cdot \overrightarrow{OP} = \underbrace{\overrightarrow{OP} + \cdots + \overrightarrow{OP}.}_{n \text{ terms}}$$

3.5 Prove that $r \cdot (-\overrightarrow{OP}) = (-r) \cdot \overrightarrow{OP}$.

3.6 Prove that $r \cdot \overrightarrow{OP} = \overrightarrow{0}$ implies that either $r = 0$ or $\overrightarrow{OP} = \overrightarrow{0}$.

3.7 Let $\overrightarrow{OP}$, $\overrightarrow{OQ}$, and $\overrightarrow{OR}$ be the vectors shown in Figure 19, where $|OP| = |OQ| = |OR|$. Draw figures which show

(a) $\overrightarrow{OP}$, $\overrightarrow{OQ}$, $\overrightarrow{OR}$, and $2 \cdot \overrightarrow{OP} + 3 \cdot \overrightarrow{OQ} + 4 \cdot \overrightarrow{OR}$

(b) $\overrightarrow{OP}$, $\overrightarrow{OQ}$, $\overrightarrow{OR}$, and $\frac{1}{2} \cdot \overrightarrow{OP} + (-2) \cdot \overrightarrow{OQ} + \frac{5}{2} \cdot \overrightarrow{OR}$

3.8 Let $\overrightarrow{OP}$ and $\overrightarrow{OQ}$ be nonzero collinear vectors such that $|OQ| = 5|OP|$ and $\overrightarrow{OQ}$ has direction opposite to that of $\overrightarrow{OP}$. Find r such that $\overrightarrow{OP} = r \cdot \overrightarrow{OQ}$. Find r and s such that $r \cdot \overrightarrow{OP} + s \cdot \overrightarrow{OQ} = \overrightarrow{0}$.

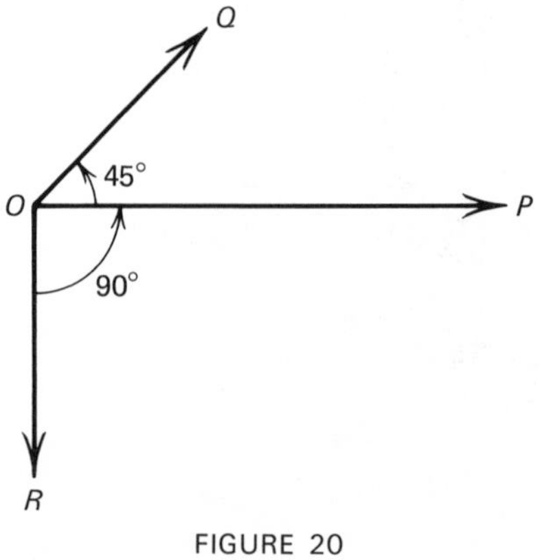

FIGURE 20

3.9 Let $\overrightarrow{OP}$, $\overrightarrow{OQ}$, and $\overrightarrow{OR}$ be the coplanar vectors shown in Figure 20, where $|OP| = 2|OQ| = 3|OR|$. Find r, s, and t such that $r \cdot \overrightarrow{OP} + s \cdot \overrightarrow{OQ} + t \cdot \overrightarrow{OR} = \overrightarrow{0}$.

4 inner multiplication

There is a third vector operation in which two vectors are combined to yield a real number. Let $\overrightarrow{OP}$ and $\overrightarrow{OQ}$ be nonzero vectors. The angle θ, where $0° \leq \theta \leq 180°$, measured from the direction of $\overrightarrow{OP}$ to the direction of $\overrightarrow{OQ}$ is called the *angle between the vectors* $\overrightarrow{OP}$ and $\overrightarrow{OQ}$ (See Figure 21). If either $\overrightarrow{OP} = \overrightarrow{0}$ or $\overrightarrow{OQ} = \overrightarrow{0}$, the angle θ between $\overrightarrow{OP}$ and $\overrightarrow{OQ}$ is defined to be $90°$.

(4.1) definition Let $\overrightarrow{OP}$ and $\overrightarrow{OQ}$ be vectors and let θ be the angle between them. The *inner product* of $\overrightarrow{OP}$ and $\overrightarrow{OQ}$ is the real number $|OP|\,|OQ| \cos \theta$, which is denoted by

$$\overrightarrow{OP} \circ \overrightarrow{OQ}.$$

The inner product of vectors is sometimes called the *dot product* or *scalar product* (Since the result is a real number).

EXAMPLE 1 Suppose that $\overrightarrow{OP}$ and $\overrightarrow{OQ}$ are vectors such that $|OQ| = 4|OP|$ and the angle between $\overrightarrow{OP}$ and $\overrightarrow{OQ}$ is $30°$. Then $\overrightarrow{OP} \circ \overrightarrow{OQ} = |OP|\,|OQ| \cos 30° = 4|OP|^2 (\sqrt{3}/2) = 2\sqrt{3}|OP|^2$.

EXAMPLE 2 Suppose that $\overrightarrow{OP}$ and $\overrightarrow{OQ}$ are vectors of length 2 and 3, respectively, and that $\overrightarrow{OP} \circ \overrightarrow{OQ} = -3\sqrt{2}$. Then $\overrightarrow{OP} \circ \overrightarrow{OQ} = (2)(3) \cos \theta = -3\sqrt{2}$. Hence, $\cos \theta = -\sqrt{2}/2$. Since $0° \leq \theta \leq 180°$, $\theta = 135°$

EXAMPLE 3 IF $\overrightarrow{OP}$ and $\overrightarrow{OQ}$ are any vectors and θ is the angle between $\overrightarrow{OP}$ and $\overrightarrow{OQ}$, the number $q = |OQ| \cos \theta$ is called the *component* of $\overrightarrow{OQ}$ *on* $\overrightarrow{OP}$

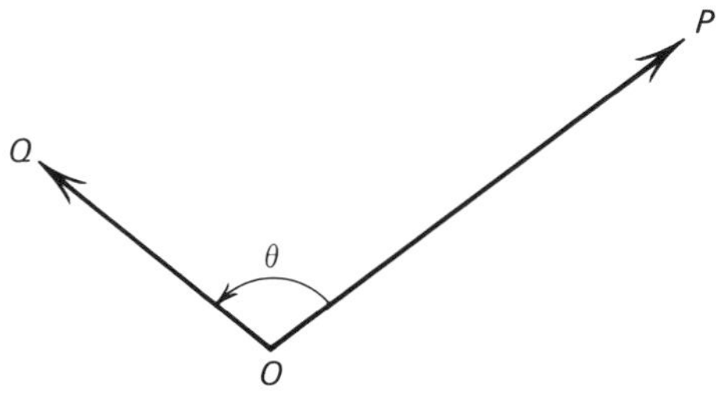

FIGURE 21

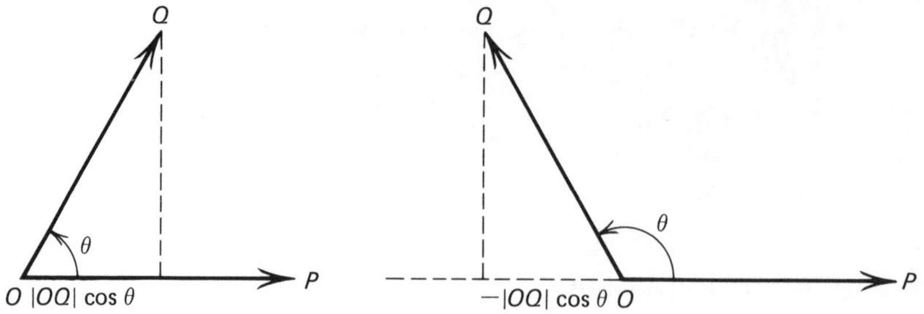

FIGURE 22

(see Figure 22). The component of $\overrightarrow{OQ}$ on $\overrightarrow{OP}$ is numerically equal to the length of the projection of $\overrightarrow{OQ}$ onto the line of $\overrightarrow{OP}$. Thus, $\overrightarrow{OP} \circ \overrightarrow{OQ} = |OP|\,|OQ| \cos \theta = |OP|q$ is the product of the length of $\overrightarrow{OP}$ and the component of $\overrightarrow{OQ}$ on $\overrightarrow{OP}$.

It follows directly from Definition 4.1 that inner multiplication is commutative. That is,

$$\overrightarrow{OP} \circ \overrightarrow{OQ} = \overrightarrow{OQ} \circ \overrightarrow{OP}. \tag{4–1}$$

The length of a vector can be expressed in terms of the inner product. By Definition 4.1, $\overrightarrow{OP} \circ \overrightarrow{OP} = |OP|\,|OP| \cos 0° = |OP|^2$. Therefore,

$$\text{length of } \overrightarrow{OP} = (\overrightarrow{OP} \circ \overrightarrow{OP})^{1/2}. \tag{4–2}$$

Two vectors are *orthogonal* if the angle between them is 90°. In other words, two vectors are orthogonal if they lie on perpendicular lines through O. Notice that the zero vector is orthogonal to every vector, since we define the angle between the zero vector and any other vector to be 90°. Orthogonality of vectors can be characterized in terms of the inner product.

(4.2) theorem The vectors $\overrightarrow{OP}$ and $\overrightarrow{OQ}$ are orthogonal if and only if $\overrightarrow{OP} \circ \overrightarrow{OQ} = 0$.

PROOF If $\overrightarrow{OP}$ and $\overrightarrow{OQ}$ are orthogonal, then the angle between them is 90°. Thus, $\overrightarrow{OP} \circ \overrightarrow{OQ} = |OP|\,|OQ| \cos 90° = 0$. Conversely, suppose that $\overrightarrow{OP} \circ \overrightarrow{OQ} = 0$. If either $\overrightarrow{OP}$ or $\overrightarrow{OQ}$ is the zero vector, then $\overrightarrow{OP}$ and $\overrightarrow{OQ}$ are orthogonal, as stated above. If both $\overrightarrow{OP}$ and $\overrightarrow{OQ}$ are nonzero vectors, then $|OP| \neq 0$ and $|OQ| \neq 0$. Therefore, $\overrightarrow{OP} \circ \overrightarrow{OQ} = |OP|\,|OQ| \cos \theta = 0$ implies $\cos \theta = 0$. Since $0° \leq \theta \leq 180°$, $\cos \theta = 0$ yields $\theta = 90°$. Therefore, $\overrightarrow{OP}$ and $\overrightarrow{OQ}$ lie on perpendicular lines through O. That is, $\overrightarrow{OP}$ and $\overrightarrow{OQ}$ are orthogonal. ∎

Further properties of the inner product will be derived after we have discussed the coordinates of vectors.

exercises

4.1 Let $\overrightarrow{OP}$ and $\overrightarrow{OQ}$ be vectors with the given lengths and given angle θ between them. Compute $\overrightarrow{OP} \circ \overrightarrow{OQ}$ in each case.

(a) $|OP| = \frac{1}{2}$, $|OQ| = 4$, $\theta = 45°$

(b) $|OP| = 6$, $|OQ| = 3$, $\theta = 60°$

(c) $|OP| = 1.5$, $|OQ| = 2.5$, $\theta = 150°$

(d) $|OP| = 4.612$, $|OQ| = 7.375$, $\theta = 90°$

(e) $|OP| = 2$, $|OQ| = \frac{1}{3}$, $\theta = 180°$

(f) $|OP| = 1$, $|OQ| = 1$, $\theta = 120°$

4.2 Find the angle between the vectors $\overrightarrow{OP}$ and $\overrightarrow{OQ}$ if $|OP| = 5$, $|OQ| = \frac{1}{2}$, and $\overrightarrow{OP} \circ \overrightarrow{OQ} = -\frac{5}{4}$.

4.3 What is the possible range of values for $\overrightarrow{OP} \circ \overrightarrow{OQ}$ if $|OP| = 2$ and $|OQ| = 3$?

4.4 Prove that two nonzero vectors $\overrightarrow{OP}$ and $\overrightarrow{OQ}$ are collinear if and only if $\overrightarrow{OP} \circ \overrightarrow{OQ}$ is equal to plus or minus the product of the lengths of $\overrightarrow{OP}$ and $\overrightarrow{OQ}$.

***4.5** Let $\overrightarrow{OP}$, $\overrightarrow{OQ}$, and $\overrightarrow{OR}$ be any vectors. Let q be the component of $\overrightarrow{OQ}$ on $\overrightarrow{OP}$ and r be the component of $\overrightarrow{OR}$ on $\overrightarrow{OP}$ (see Example 3). Show that the component of $\overrightarrow{OQ} + \overrightarrow{OR}$ on $\overrightarrow{OP}$ is $q + r$. Use this result to derive the vector identity $\overrightarrow{OP} \circ (\overrightarrow{OQ} + \overrightarrow{OR}) = \overrightarrow{OP} \circ \overrightarrow{OQ} + \overrightarrow{OP} \circ \overrightarrow{OR}$.

5 the cross product

In addition to the inner product, there is another product defined for geometric vectors in three-dimensional space which is useful in applications. The cross product of two vectors is a vector operation in which two vectors are combined to yield another vector.

5.1 definition Let $\overrightarrow{OP}$ and $\overrightarrow{OQ}$ be vectors and let θ be the angle between them. The *cross product* of $\overrightarrow{OP}$ and $\overrightarrow{OQ}$, denoted by

$$\overrightarrow{OP} \times \overrightarrow{OQ}$$

is a vector of length $|OP|\,|OQ| \sin\theta$. If $\overrightarrow{OP} \neq \overrightarrow{0}$, $\overrightarrow{OQ} \neq \overrightarrow{0}$, and $0° < \theta < 180°$, then $\overrightarrow{OP} \times \overrightarrow{OQ}$ is perpendicular to the plane determined by $\overrightarrow{OP}$ and $\overrightarrow{OQ}$ and is so directed that a counterclockwise rotation about $\overrightarrow{OP} \times \overrightarrow{OQ}$ through the angle θ carries $\overrightarrow{OP}$ into $\overrightarrow{OQ}$ (see Figure 23).

Just as the inner product is sometimes called the scalar product, the cross product is sometimes called the *vector product*.

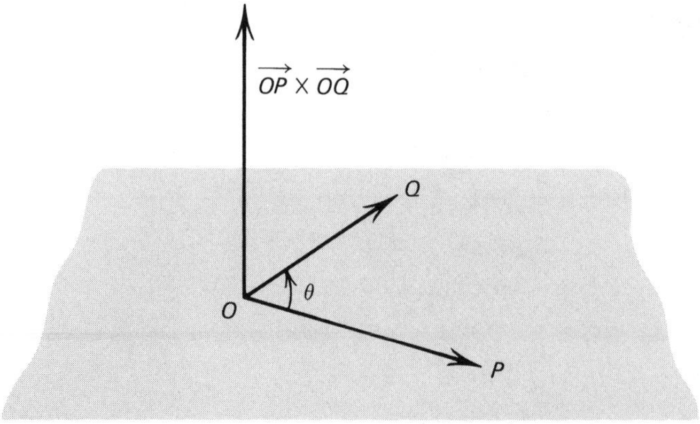

FIGURE 23

Since the length of $\overrightarrow{OP} \times \overrightarrow{OQ}$ is $|OP|\,|OQ| \sin \theta$, it follows that $\overrightarrow{OP} \times \overrightarrow{OQ} = \overrightarrow{0}$ if and only if $\overrightarrow{OP}$ and $\overrightarrow{OQ}$ are collinear. When $\overrightarrow{OP} \times \overrightarrow{OQ}$ is not the zero vector, its direction perpendicular to the plane containing $\overrightarrow{OP}$ and $\overrightarrow{OQ}$ can be described by saying that the vectors $\overrightarrow{OP}$, $\overrightarrow{OQ}$, and $\overrightarrow{OP} \times \overrightarrow{OQ}$ form a right-handed system. That is, if the fingers of the right hand indicate the direction of the angle θ from $\overrightarrow{OP}$ to $\overrightarrow{OQ}$, then the thumb points in the direction of $\overrightarrow{OP} \times \overrightarrow{OQ}$.

EXAMPLE 1 Let $\overrightarrow{OP}$ and $\overrightarrow{OQ}$ be vectors such that $|OP| = 2$, $|OQ| = 5$, and the angle between them is $\theta = 30°$. Then the length of $\overrightarrow{OP} \times \overrightarrow{OQ}$ is $|OP|\,|OQ| \sin \theta = (2)(5)(\tfrac{1}{2}) = 5$. The vector $\overrightarrow{OP} \times \overrightarrow{OQ}$ is shown in Figure 24.

It is interesting that the cross product is an example of a useful operation which does not satisfy either the commutative or associative laws. In fact, it follows at once from Definition 5.1 that

$$\overrightarrow{OQ} \times \overrightarrow{OP} = -(\overrightarrow{OP} \times \overrightarrow{OQ}).$$

To see that the associative law does not hold in general, we note that $\overrightarrow{OP} \times (\overrightarrow{OQ} \times \overrightarrow{OR})$ is perpendicular to $\overrightarrow{OQ} \times \overrightarrow{OR}$, and $\overrightarrow{OQ} \times \overrightarrow{OR}$ is perpendicular to the plane containing $\overrightarrow{OQ}$ and $\overrightarrow{OR}$. Thus $\overrightarrow{OP} \times (\overrightarrow{OQ} \times \overrightarrow{OR})$ lies in the plane containing $\overrightarrow{OQ}$ and $\overrightarrow{OR}$. On the other hand, $(\overrightarrow{OP} \times \overrightarrow{OQ}) \times \overrightarrow{OR}$ is perpendicular to $\overrightarrow{OP} \times \overrightarrow{OQ}$, and $\overrightarrow{OP} \times \overrightarrow{OQ}$ is perpendicular to the plane containing $\overrightarrow{OP}$ and $\overrightarrow{OQ}$. Thus, $(\overrightarrow{OP} \times \overrightarrow{OQ}) \times \overrightarrow{OR}$ lies in the plane containing $\overrightarrow{OP}$ and $\overrightarrow{OQ}$. Thus, except in certain special cases,

$$\overrightarrow{OP} \times (\overrightarrow{OQ} \times \overrightarrow{OR}) \neq (\overrightarrow{OP} \times \overrightarrow{OQ}) \times \overrightarrow{OR}.$$

There are useful algebraic identities connecting the cross product with scalar multiplication and vector addition.

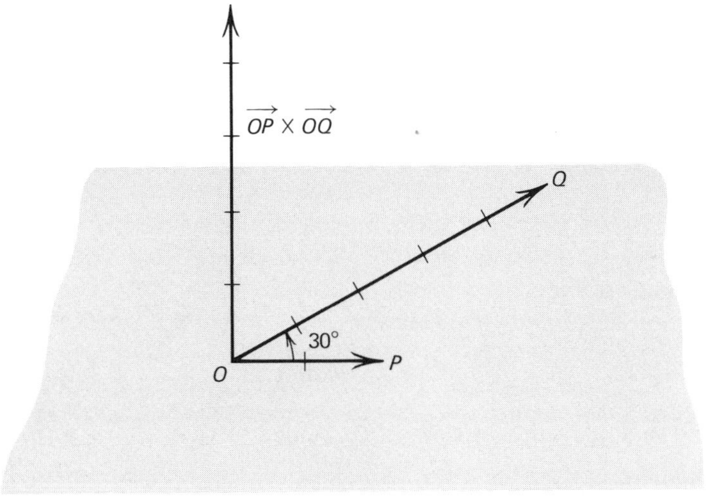

FIGURE 24

(5.2) theorem Let $\overrightarrow{OP}$ and $\overrightarrow{OQ}$ be vectors and let r be a real number. Then

$$(r \cdot \overrightarrow{OP}) \times \overrightarrow{OQ} = \overrightarrow{OP} \times (r \cdot \overrightarrow{OQ}) = r \cdot (\overrightarrow{OP} \times \overrightarrow{OQ}).$$

PROOF We will first show that $\overrightarrow{OP} \times (r \cdot \overrightarrow{OQ}) = r \cdot (\overrightarrow{OP} \times \overrightarrow{OQ})$. The other half of the identity follows easily.

If $\overrightarrow{OP}$ and $\overrightarrow{OQ}$ are collinear, then so are $\overrightarrow{OP}$ and $r \cdot \overrightarrow{OQ}$. In this case, as well as when $r = 0$, both vectors $\overrightarrow{OP} \times (r \cdot \overrightarrow{OQ})$ and $r \cdot (\overrightarrow{OP} \times \overrightarrow{OQ})$ are $\vec{0}$, and therefore, equal. Next, we consider the case where $\overrightarrow{OP}$ and $\overrightarrow{OQ}$ are noncollinear and $r \neq 0$. In this case $\overrightarrow{OP} \neq \vec{0}$, $\overrightarrow{OQ} \neq \vec{0}$, $r \neq 0$, and $0° < \theta < 180°$, where θ is the angle measured from $\overrightarrow{OP}$ to $\overrightarrow{OQ}$.

We note that since $\overrightarrow{OQ}$ and $r \cdot \overrightarrow{OQ}$ are collinear, the plane determined by $\overrightarrow{OP}$ and $\overrightarrow{OQ}$ is the same as the plane determined by $\overrightarrow{OP}$ and $r \cdot \overrightarrow{OQ}$. Therefore, both $\overrightarrow{OP} \times (r \cdot \overrightarrow{OQ})$ and $r \cdot (\overrightarrow{OP} \times \overrightarrow{OQ})$ lie on a line through O perpendicular to this plane.

If $r > 0$, then $\overrightarrow{OQ}$ and $r \cdot \overrightarrow{OQ}$ have the same direction, so that the angle measured from $\overrightarrow{OP}$ to $r \cdot \overrightarrow{OQ}$ is θ (see Figure 25). Thus, $\overrightarrow{OP} \times (r \cdot \overrightarrow{OQ})$ has

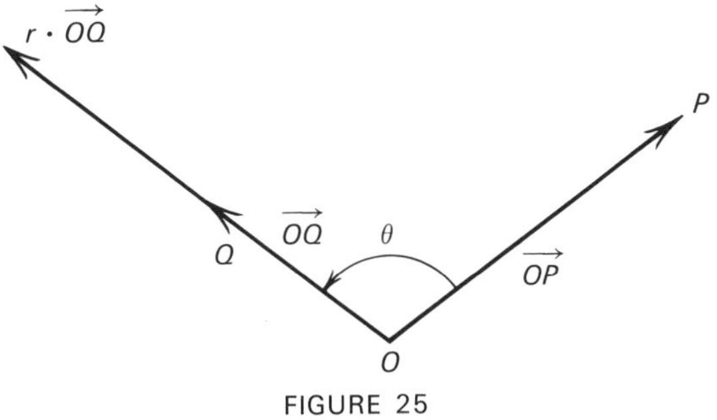

FIGURE 25

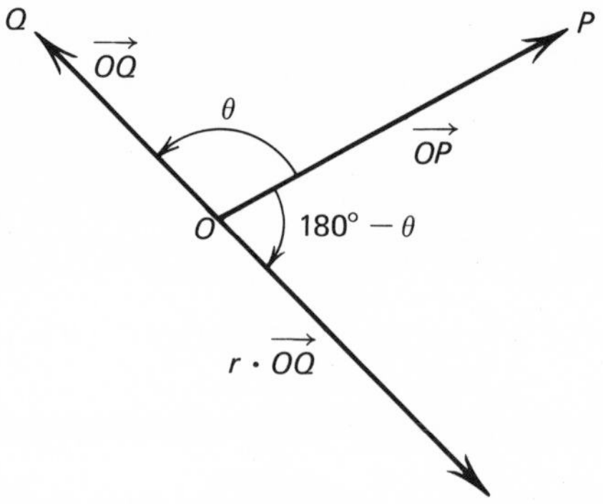

FIGURE 26

the same direction as $\overrightarrow{OP} \times \overrightarrow{OQ}$. Since $r > 0$, $r \cdot (\overrightarrow{OP} \times \overrightarrow{OQ})$ has the same direction as $\overrightarrow{OP} \times \overrightarrow{OQ}$. Therefore, in this case, $\overrightarrow{OP} \times (r \cdot \overrightarrow{OQ})$ and $r \cdot (\overrightarrow{OP} \times \overrightarrow{OQ})$ have the same direction. Finally, the length of $\overrightarrow{OP} \times (r \cdot \overrightarrow{OQ})$ is $|OP|\,(r|OQ|) \sin\theta = r(|OP|\,|OQ| \sin\theta)$, which is the length of $r \cdot (\overrightarrow{OP} \times \overrightarrow{OQ})$. Hence, $\overrightarrow{OP} \times (r \cdot \overrightarrow{OQ}) = r \cdot (\overrightarrow{OP} \times \overrightarrow{OQ})$.

If $r < 0$, then $\overrightarrow{OQ}$ and $r \cdot \overrightarrow{OQ}$ have opposite directions, and the angle from $\overrightarrow{OP}$ to $r \cdot \overrightarrow{OQ}$ is $180° - \theta$, measured in the opposite sense from θ (see Figure 26). Thus, the direction of $\overrightarrow{OP} \times (r \cdot \overrightarrow{OQ})$ is opposite to the direction of $\overrightarrow{OP} \times \overrightarrow{OQ}$. However, since $r < 0$, the direction of $r \cdot (\overrightarrow{OP} \times \overrightarrow{OQ})$ is opposite to the direction of $\overrightarrow{OP} \times \overrightarrow{OQ}$. Thus, $\overrightarrow{OP} \times (r \cdot \overrightarrow{OQ})$ and $r \cdot (\overrightarrow{OP} \times \overrightarrow{OQ})$ have the same direction. The length of $\overrightarrow{OP} \times (r \cdot \overrightarrow{OQ})$ is $|OP|\;(|r|\;|OQ|)$ $\sin(180° - \theta) = |r|\;(|OP|\;|OQ| \sin\theta)$, which is the length of $r \cdot (\overrightarrow{OP} \times \overrightarrow{OQ})$. Hence, in this case also, $\overrightarrow{OP} \times (r \cdot \overrightarrow{OQ}) = r \cdot (\overrightarrow{OP} \times \overrightarrow{OQ})$.

To derive the other identity $(r \cdot \overrightarrow{OP}) \times \overrightarrow{OQ} = r \cdot (\overrightarrow{OP} \times \overrightarrow{OQ})$, we use the fact that $\overrightarrow{OQ} \times \overrightarrow{OP} = -(\overrightarrow{OP} \times \overrightarrow{OQ})$ for any vectors $\overrightarrow{OP}$ and $\overrightarrow{OQ}$, as well as the identity just derived. Hence, we have

$$\begin{aligned}(r \cdot \overrightarrow{OP}) \times \overrightarrow{OQ} &= -[\overrightarrow{OQ} \times (r \cdot \overrightarrow{OP})] = -[r \cdot (\overrightarrow{OQ} \times \overrightarrow{OP})] \\ &= -\{r \cdot [-(\overrightarrow{OP} \times \overrightarrow{OQ})]\} = -\{-[r \cdot (\overrightarrow{OP} \times \overrightarrow{OQ})]\} \\ &= r \cdot (\overrightarrow{OP} \times \overrightarrow{OQ}).\end{aligned}$$

∎

To prove that cross multiplication distributes over vector addition, that is, $\overrightarrow{OP} \times (\overrightarrow{OQ} + \overrightarrow{OR}) = \overrightarrow{OP} \times \overrightarrow{OQ} + \overrightarrow{OP} \times \overrightarrow{OR}$ for any vectors $\overrightarrow{OP}$, $\overrightarrow{OQ}$, and $\overrightarrow{OR}$, we observe that the cross product $\overrightarrow{OP} \times \overrightarrow{OQ}$ can be regarded as a rule (or function) which sends the vector $\overrightarrow{OQ}$ into the vector $\overrightarrow{OP} \times \overrightarrow{OQ}$. This rule can be described geometrically as a combination of three simple processes.

Let π be a plane through O perpendicular to $\overrightarrow{OP}$. First project $\overrightarrow{OQ}$ onto the plane π to obtain a vector $\overrightarrow{OR}$ (see Figure 27). The length of $\overrightarrow{OR}$ is $|OQ| \cos(90° - \theta) = |OQ| \sin\theta$. Next, rotate $\overrightarrow{OR}$ by 90° in the plane π counterclockwise about $\overrightarrow{OP}$ to obtain a vector $\overrightarrow{OS}$ which has length $|OQ| \sin\theta$ and direction perpendicular to the plane determined by $\overrightarrow{OP}$ and $\overrightarrow{OQ}$, and such that $\overrightarrow{OP}$, $\overrightarrow{OQ}$, and $\overrightarrow{OS}$ form a right-handed system. Finally, the vector $|OP| \cdot \overrightarrow{OS}$ has length $|OP|\,|OQ| \sin\theta$ and the proper direction so that $|OP| \cdot \overrightarrow{OS} = \overrightarrow{OP} \times \overrightarrow{OQ}$.

Thus, if we let P_π denote the operation which projects a vector onto a plane π through O perpendicular to $\overrightarrow{OP}$, and let T denote a rotation of 90° in this plane in a counterclockwise sense about $\overrightarrow{OP}$, we have

$$\overrightarrow{OP} \times \overrightarrow{OQ} = |OP| \cdot T[P_\pi(\overrightarrow{OQ})].$$

That is, first $\overrightarrow{OQ}$ is projected onto π to obtain $P_\pi(\overrightarrow{OQ})$, then $P_\pi(\overrightarrow{OQ})$ is rotated 90° about $\overrightarrow{OP}$ to obtain $T[P_\pi(\overrightarrow{OQ})]$, and finally, the vector $T[P_\pi(\overrightarrow{OQ})]$ is multiplied by the real number $|OP|$ to obtain $|OP| \cdot T[P_\pi(\overrightarrow{OQ})] = \overrightarrow{OP} \times \overrightarrow{OQ}$.

Now suppose that for any vectors $\overrightarrow{OQ}$ and $\overrightarrow{OR}$, the operations P_π and T have the property $P_\pi(\overrightarrow{OQ} + \overrightarrow{OR}) = P_\pi(\overrightarrow{OQ}) + P_\pi(\overrightarrow{OR})$, $T(\overrightarrow{OQ} + \overrightarrow{OR}) = T(\overrightarrow{OQ}) + T(\overrightarrow{OR})$. Then using Theorem 3.2(c), we have

$$\begin{aligned}\overrightarrow{OP} \times (\overrightarrow{OQ} + \overrightarrow{OR}) &= |OP| \cdot T[P_\pi(\overrightarrow{OQ} + \overrightarrow{OR})] = |OP| \cdot T[P_\pi(\overrightarrow{OQ}) + P_\pi(\overrightarrow{OR})] \\ &= |OP| \cdot \{T[P_\pi(\overrightarrow{OQ})] + T[P_\pi(\overrightarrow{OR})]\} \\ &= |OP| \cdot T[P_\pi(\overrightarrow{OQ})] + |OP| \cdot T[P_\pi(\overrightarrow{OR})] \\ &= \overrightarrow{OP} \times \overrightarrow{OQ} + \overrightarrow{OP} \times \overrightarrow{OR}.\end{aligned}$$

The fact that $P_\pi(\overrightarrow{OQ} + \overrightarrow{OR}) = P_\pi(\overrightarrow{OQ}) + P_\pi(\overrightarrow{OR})$ follows from the obser-

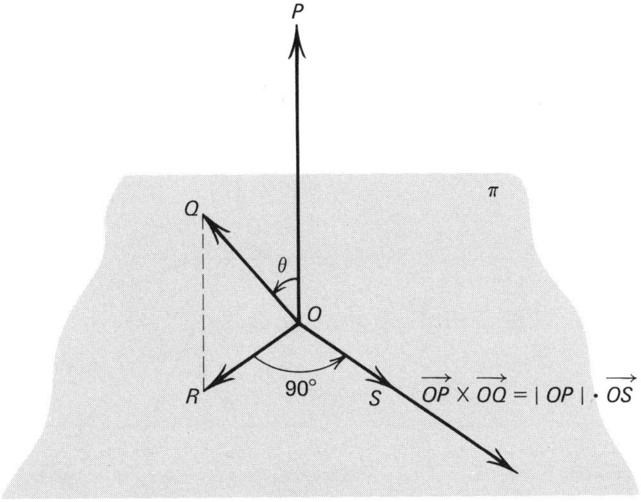

FIGURE 27

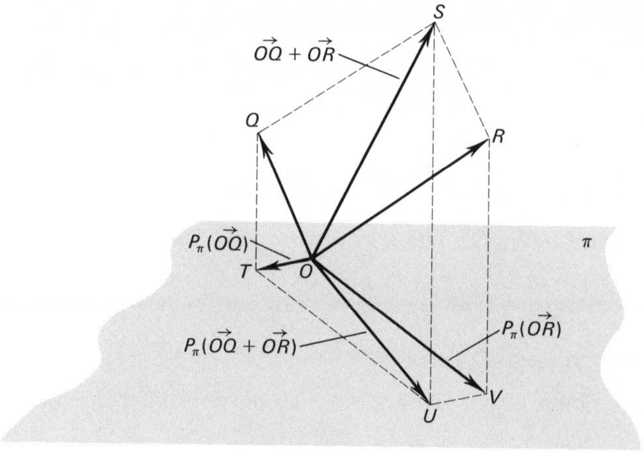

FIGURE 28

vation that the process of projecting vectors onto a plane π containing O, projects a parallelogram into a parallelogram. Thus, in Figure 28, the parallelogram $ORSQ$ is projected into the parallelogram $OVUT$.

Finally, rotating vectors in a plane π containing O through 90° about an axis perpendicular to π at O, sends a parallelogram determined by two vectors into a parallelogram determined by the vectors after rotation (see Figure 29). That is, $T(\overrightarrow{OQ} + \overrightarrow{OR}) = T(\overrightarrow{OQ}) + T(\overrightarrow{OR})$.

We restate for reference the important property of the cross product that we have proved above.

(5.3) theorem Let $\overrightarrow{OP}$, $\overrightarrow{OQ}$, and $\overrightarrow{OR}$ be any vectors. Then

$$\overrightarrow{OP} \times (\overrightarrow{OQ} + \overrightarrow{OR}) = \overrightarrow{OP} \times \overrightarrow{OQ} + \overrightarrow{OP} \times \overrightarrow{OR}.$$

Combining the results of Theorems 5.2 and 5.3 we obtain the following fact.

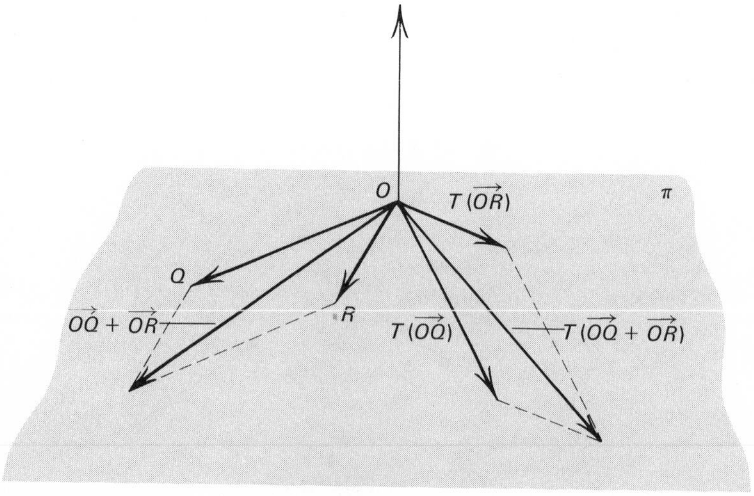

FIGURE 29

(5.4) corollary Let $\overrightarrow{OP}$, $\overrightarrow{OQ}$, and $\overrightarrow{OR}$ be any vectors, and let r and s be real numbers. Then

$$\overrightarrow{OP} \times (r \cdot \overrightarrow{OQ} + s \cdot \overrightarrow{OR}) = r \cdot (\overrightarrow{OP} \times \overrightarrow{OQ}) + s \cdot (\overrightarrow{OP} \times \overrightarrow{OR}),$$

and

$$(r \cdot \overrightarrow{OP} + s \cdot \overrightarrow{OQ}) \times \overrightarrow{OR} = r \cdot (\overrightarrow{OP} \times \overrightarrow{OR}) + s \cdot (\overrightarrow{OQ} \times \overrightarrow{OR}).$$

PROOF By Theorems 5.3 and 5.2, we have

$$\begin{aligned}\overrightarrow{OP} \times (r \cdot \overrightarrow{OQ} + s \cdot \overrightarrow{OR}) &= \overrightarrow{OP} \times (r \cdot \overrightarrow{OQ}) + \overrightarrow{OP} \times (s \cdot \overrightarrow{OR}) \\ &= r \cdot (\overrightarrow{OP} \times \overrightarrow{OQ}) + s \cdot (\overrightarrow{OP} \times \overrightarrow{OQ}).\end{aligned}$$

The second identity follows from the one just derived:

$$\begin{aligned}(r \cdot \overrightarrow{OP} + s \cdot \overrightarrow{OQ}) \times \overrightarrow{OR} &= -[\overrightarrow{OR} \times (r \cdot \overrightarrow{OP} + s \cdot \overrightarrow{OQ})] \\ &= -[r \cdot (\overrightarrow{OR} \times \overrightarrow{OP}) + s \cdot (\overrightarrow{OR} \times \overrightarrow{OQ})] \\ &= -\{r \cdot [-(\overrightarrow{OP} \times \overrightarrow{OR})] + s \cdot [-(\overrightarrow{OQ} \times \overrightarrow{OR})]\} \\ &= -[(-r) \cdot (\overrightarrow{OP} \times \overrightarrow{OR}) + (-s) \cdot (\overrightarrow{OQ} \times \overrightarrow{OR})] \\ &= r \cdot (\overrightarrow{OP} \times \overrightarrow{OR}) + s \cdot (\overrightarrow{OQ} \times \overrightarrow{OR}).\end{aligned}$$ ∎

The property of the cross product given in Corollary 5.4 states that this operation is *linear*. Thus, the cross product provides an example of what is called a linear transformation. Linear transformations are studied in detail in Chapter 4.

exercises

5.1 Draw figures showing $\overrightarrow{OP}$, $\overrightarrow{OQ}$, and $\overrightarrow{OP} \times \overrightarrow{OQ}$ in the following cases

(a) $|OP| = 3$, $|OQ| = 6$, $\theta = 60°$

(b) $|OP| = \frac{1}{2}$, $|OQ| = \frac{1}{3}$, $\theta = 90°$

(c) $|OP| = 3$, $|OQ| = 5$, $\theta = 135°$

(d) $|OP| = 1$, $|OQ| = 1$, $\theta = 180°$

(e) $|OP| = 5$, $|OQ| = 10$, $\theta = 150°$

5.2 Let $\overrightarrow{OP}$, $\overrightarrow{OQ}$, and $\overrightarrow{OR}$ be the coplanar vectors illustrated in Figure 30. Draw

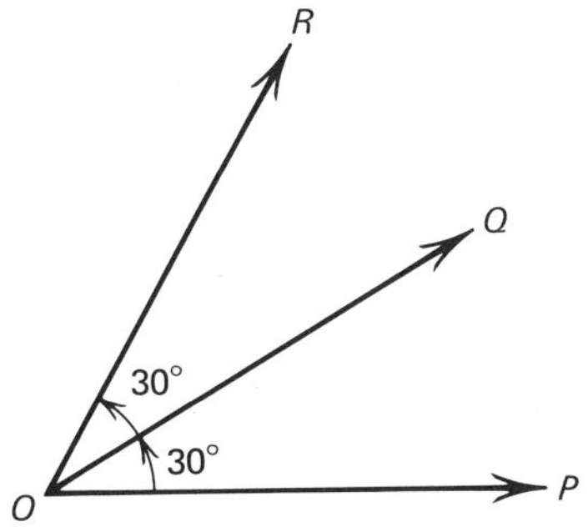

FIGURE 30

a figure showing $\overrightarrow{OP}$, $\overrightarrow{OQ}$, $\overrightarrow{OR}$, $\overrightarrow{OP} \times (\overrightarrow{OQ} \times \overrightarrow{OR})$, and $(\overrightarrow{OP} \times \overrightarrow{OQ}) \times \overrightarrow{OR}$. Assume that $|OP| = |OQ| = |OR| = 1$.

5.3 Suppose that $\overrightarrow{OP} \neq \overrightarrow{0}$ and $\overrightarrow{OQ} \neq \overrightarrow{0}$ are orthogonal and that $\overrightarrow{OR} \neq \overrightarrow{0}$ is perpendicular to the plane determined by $\overrightarrow{OP}$ and $\overrightarrow{OQ}$. Show that $\overrightarrow{OP} \times (\overrightarrow{OQ} \times \overrightarrow{OR}) = \overrightarrow{0} = (\overrightarrow{OP} \times \overrightarrow{OQ}) \times \overrightarrow{OR}$.

5.4 Show that if $\overrightarrow{OP}$, $\overrightarrow{OQ}$, and $\overrightarrow{OR}$ are coplanar, then $\overrightarrow{OP} \times (\overrightarrow{OQ} \times \overrightarrow{OR})$ and $(\overrightarrow{OP} \times \overrightarrow{OQ}) \times \overrightarrow{OR}$ lie in the plane containing $\overrightarrow{OP}$, $\overrightarrow{OQ}$, and $\overrightarrow{OR}$.

5.5 Prove that $\overrightarrow{OP}$ and $\overrightarrow{OQ}$ are collinear if and only if $\overrightarrow{OP} \times \overrightarrow{OQ} = \overrightarrow{0}$.

5.6 Which of the following products are defined? $(\overrightarrow{OP} \cdot \overrightarrow{OQ}) \cdot \overrightarrow{OR}$, $(\overrightarrow{OP} \circ \overrightarrow{OQ}) \cdot \overrightarrow{OR}$, $\overrightarrow{OP} \circ (\overrightarrow{OQ} \cdot \overrightarrow{OR})$, $(\overrightarrow{OP} \times \overrightarrow{OQ}) \cdot \overrightarrow{OR}$, $(\overrightarrow{OP} \circ \overrightarrow{OQ}) \circ \overrightarrow{OR}$, $\overrightarrow{OP} \circ (\overrightarrow{OQ} \times \overrightarrow{OR})$, $(\overrightarrow{OP} \times \overrightarrow{OQ}) \circ \overrightarrow{OR}$, $(\overrightarrow{OP} \times \overrightarrow{OQ}) \times \overrightarrow{OR}$. Of those products which are defined, which are vectors and which are real numbers?

6 coordinates

We have defined the system of geometric vectors in three-dimensional space as the set of all directed line segments with initial point at some fixed point O in space. In the previous sections, the vector operations of addition, scalar multiplication, inner product, and cross product were defined geometrically. The introduction of a rectangular Cartesian coordinate system with origin at the point O, enables us to describe a vector $\overrightarrow{OP}$ by the coordinates of its terminal point P. Moreover, all the vector operations can be expressed algebraically in terms of coordinates. In particular, this has the advantage of permitting algebraic derivations of the properties of the operations, which in many cases are much simpler than the corresponding geometric arguments.

A rectangular Cartesian coordinate system establishes a one-to-one correspondence between the points in space and ordered triples (x, y, z) of real numbers, where the coordinates x, y, and z of a point P are the directed distances of P from the YZ, XZ, and XY planes, respectively (Figure 31). It follows that there is a one-to-one correspondence between the system of geometric vectors and ordered triples of real numbers, where a vector $\overrightarrow{OP}$ corresponds to the coordinate triple (x, y, z) of its endpoint P.

The introduction of a coordinate system establishes a unit of length for geometric vectors. A *unit vector* is a vector of length one. We denote by **I**, **J**, and **K** the unit vectors in the positive directions of the X, Y, and Z axes, respectively. Suppose that (x, y, z) is the coordinate triple of the terminal point P of a vector $\overrightarrow{OP}$. The projection of the point P on the X axis is the point of intersection P_x of the X axis and a plane through P perpendicular to the X axis (see Figure 32). The directed distance along

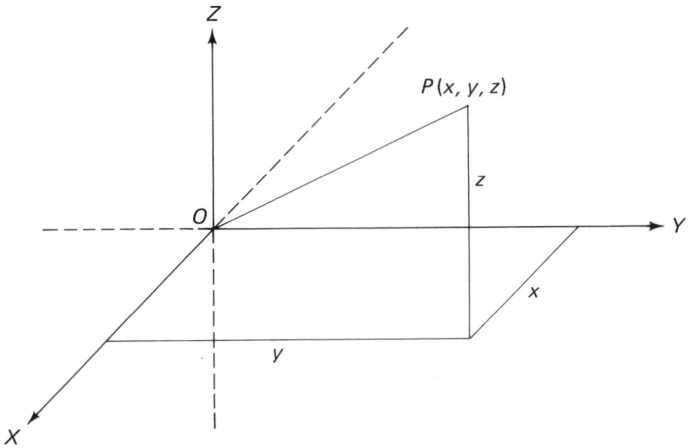

FIGURE 31

the X axis of P_x from O is clearly the same as the directed distance of P from the YZ plane, since P_x lies in a plane parallel to the YZ plane. Therefore, this directed distance of P_x from O is x, the x coordinate of P. Since $\mathbf{I}$ is a unit vector in the positive direction of the X axis, it follows from Definition 3.1 that $x \cdot \mathbf{I}$ is a vector along the X axis with terminal point P_x. In fact, the length of $x \cdot \mathbf{I}$ is $|x|\,|\mathbf{I}| = |x|$, and $x \cdot \mathbf{I}$ points in the positive direction of the X axis if x is positive and in the negative direction of the X axis if x is negative. Similarly, $y \cdot \mathbf{J}$ and $z \cdot \mathbf{K}$ are vectors along the Y and Z axes, with terminal points P_y and P_z, the projections of P on the Y and Z axes, respectively.

According to the definition of vector addition, $x \cdot \mathbf{I} + y \cdot \mathbf{J} + z \cdot \mathbf{K}$ is the vector with terminal point at the opposite vertex from O of the parallelepiped determined by $x \cdot \mathbf{I}$, $y \cdot \mathbf{J}$, and $z \cdot \mathbf{K}$. Since this opposite vertex is the point P, we have (see Figure 33)

$$\overrightarrow{OP} = x \cdot \mathbf{I} + y \cdot \mathbf{J} + z \cdot \mathbf{K}. \tag{6–1}$$

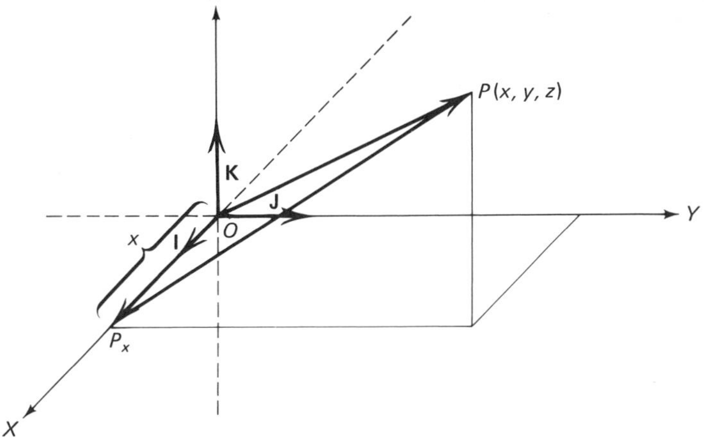

FIGURE 32

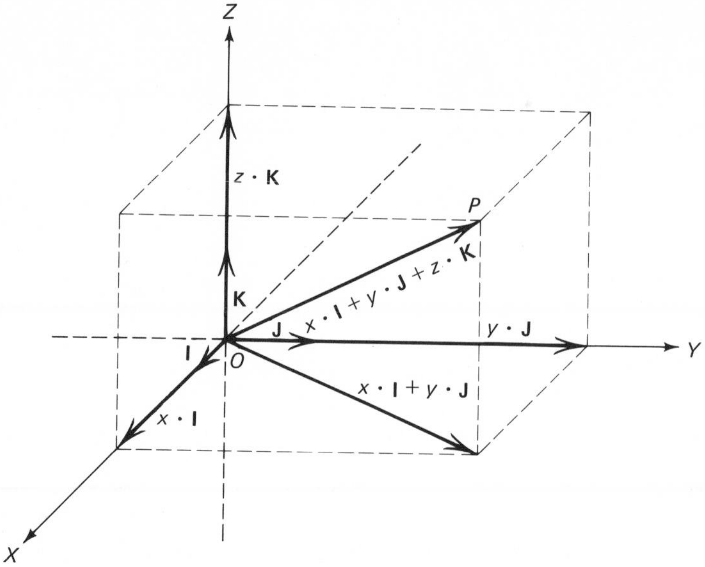

FIGURE 33

We say that equation (6–1) expresses $\overrightarrow{OP}$ as a *linear combination* of the vectors **I**, **J**, and **K**. In general, if $\overrightarrow{OP_1}, \overrightarrow{OP_2}, \ldots, \overrightarrow{OP_n}$ is any set of vectors and $r_1, r_2, \ldots, r_n$ are real numbers, then

$$r_1 \cdot \overrightarrow{OP_1} + r_2 \cdot \overrightarrow{OP_2} + \cdots + r_n \cdot \overrightarrow{OP_n}$$

is called a linear combination of the given vectors. Theorem 3.5 concerning coplanar vectors can be stated using the language of linear combinations of vectors: The set of all linear combinations of two vectors $\overrightarrow{OP}$ and $\overrightarrow{OQ}$ is a coplanar set. Moreover, if a coplanar set of vectors contains two noncollinear vectors $\overrightarrow{OP}$ and $\overrightarrow{OQ}$, then every vector in the set can be expressed as a linear combination of $\overrightarrow{OP}$ and $\overrightarrow{OQ}$.

The results of our discussion concerning the introduction of a coordinate system are summarized in the following theorem.

(6.2) theorem Let **I**, **J**, and **K** be unit vectors in the positive directions of the X, Y, and Z axes of a rectangular Cartesian coordinate system in three-dimensional space. Then each vector $\overrightarrow{OP}$ can be expressed as a linear combination

$$\overrightarrow{OP} = x \cdot \mathbf{I} + y \cdot \mathbf{J} + z \cdot \mathbf{K}$$

of **I**, **J**, and **K**. Moreover, this expression for $\overrightarrow{OP}$ is unique. That is, if $\overrightarrow{OP} = x' \cdot \mathbf{I} + y' \cdot \mathbf{J} + z' \cdot \mathbf{K}$ is any other expression for $\overrightarrow{OP}$ as a linear combination of **I**, **J**, and **K**, then $x' = x$, $y' = y$, and $z' = z$.

PROOF The fact that $\overrightarrow{OP} = x \cdot \mathbf{I} + y \cdot \mathbf{J} + z \cdot \mathbf{K}$ for real numbers x, y, and z was proved in the above discussion. Hence, suppose that $\overrightarrow{OP} =$

$x' \cdot \mathbf{I} + y' \cdot \mathbf{J} + z' \cdot \mathbf{K}$ is another expression for $\overrightarrow{OP}$. Then, using the properties of scalar multiplication given in Theorem 3.2, we obtain

$$\begin{aligned}\vec{0} = \overrightarrow{OP} - \overrightarrow{OP} &= (x \cdot \mathbf{I} + y \cdot \mathbf{J} + z \cdot \mathbf{K}) - (x' \cdot \mathbf{I} + y' \cdot \mathbf{J} + z' \cdot \mathbf{K}) \\ &= (x \cdot \mathbf{I} - x' \cdot \mathbf{I}) + (y \cdot \mathbf{J} - y' \cdot \mathbf{J}) + (z \cdot \mathbf{K} - z' \cdot \mathbf{K}) \\ &= (x - x') \cdot \mathbf{I} + (y - y') \cdot \mathbf{J} + (z - z') \cdot \mathbf{K}.\end{aligned}$$

If the numbers $x - x'$, $y - y'$, and $z - z'$ are not all zero, then by Corollary 3.6, the above equation implies that **I, J,** and **K** are coplanar. But this is a contradiction since **I, J,** and **K** are not coplanar. Therefore, $x - x' = y - y' = z - z' = 0$. That is, $x = x'$, $y = y'$, and $z = z'$. ▮

Since each vector $\overrightarrow{OP}$ has a unique expression $\overrightarrow{OP} = x \cdot \mathbf{I} + y \cdot \mathbf{J} + z \cdot \mathbf{K}$, it is convenient to regard the ordered triple of vectors $\{\mathbf{I}, \mathbf{J}, \mathbf{K}\}$ as a coordinate system for the set of all geometric vectors. Then the ordered triple of real numbers (x, y, z) is called the *coordinates of* $\overrightarrow{OP}$ *with respect to* $\{\mathbf{I}, \mathbf{J}, \mathbf{K}\}$. Of course, the coordinates of $\overrightarrow{OP}$ with respect to $\{\mathbf{I}, \mathbf{J}, \mathbf{K}\}$ are the same as the coordinates of the point P in the given rectangular Cartesian coordinate system.

Using arguments similar to those employed in proving Theorem 6.2, it is possible to show (see Exercise 6.5) that if $\overrightarrow{OU}$, $\overrightarrow{OV}$, and $\overrightarrow{OW}$ are any three noncoplanar vectors, and if $\overrightarrow{OP}$ is any vector, then $\overrightarrow{OP}$ can be uniquely expressed as a linear combination of $\overrightarrow{OU}$, $\overrightarrow{OV}$, and $\overrightarrow{OW}$; that is, $\overrightarrow{OP} = u \cdot \overrightarrow{OU} + v \cdot \overrightarrow{OV} + w \cdot \overrightarrow{OW}$. Thus, in the sense of the previous paragraph, any ordered triple $\{\overrightarrow{OU}, \overrightarrow{OV}, \overrightarrow{OW}\}$ of noncoplanar vectors would serve as a coordinate system for the set of geometric vectors. The triple (u, v, w) from the expression $\overrightarrow{OP} = u \cdot \overrightarrow{OU} + v \cdot \overrightarrow{OV} + w \cdot \overrightarrow{OW}$ would be the coordinates of $\overrightarrow{OP}$ with respect to $\{\overrightarrow{OU}, \overrightarrow{OV}, \overrightarrow{OW}\}$ (see Figure 34).

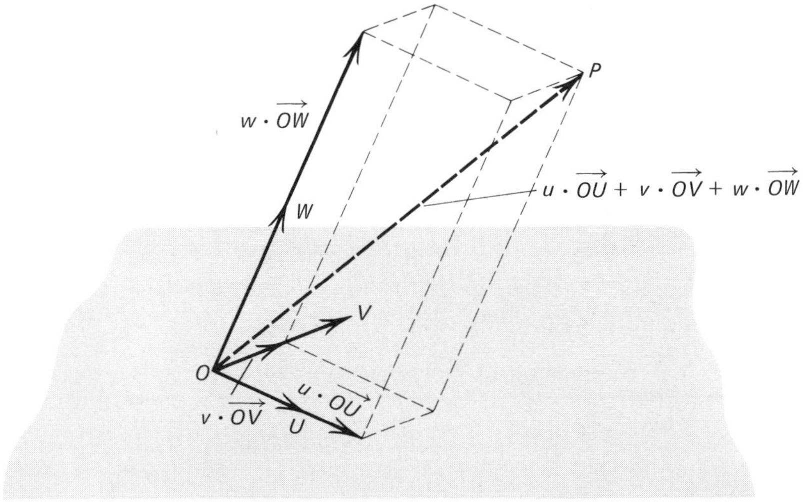

FIGURE 34

exercises

6.1 Sketch the vectors $\overrightarrow{OP}$, where the point P has the following coordinates in a rectangular Cartesian coordinate system: $(2, -1, 3)$; $(-5, 0, 1)$; $(0, 0, -3)$; $(1, 1, 1)$; $(7, 6, 4)$; $(-1, -2, 0)$.

6.2 Express each vector in Exercise 6.1 as a linear combination of the vectors **I**, **J**, and **K**.

6.3 Express the vector $\overrightarrow{OP} = 2 \cdot \mathbf{I} + 3 \cdot \mathbf{J}$ as a linear combination of the vectors $\mathbf{I} + \mathbf{J}$ and $\mathbf{I} - \mathbf{J}$.

6.4 Express the vector $\overrightarrow{OP} = 3 \cdot \mathbf{I} + (-1) \cdot \mathbf{J} + 5 \cdot \mathbf{K}$ as a linear combination of the vectors $\mathbf{I}$, $\mathbf{I} + \mathbf{J}$, and $\mathbf{I} + \mathbf{J} + \mathbf{K}$.

***6.5** Prove that if $\overrightarrow{OU}$, $\overrightarrow{OV}$ and $\overrightarrow{OW}$ are any three noncoplanar vectors, then any vector $\overrightarrow{OP}$ has a unique expression $\overrightarrow{OP} = u \cdot \overrightarrow{OU} + v \cdot \overrightarrow{OV} + w \cdot \overrightarrow{OW}$.

6.6 Let $\overrightarrow{OP} = 2 \cdot \mathbf{I} + 3 \cdot \mathbf{J} + 4 \cdot \mathbf{K}$, $\overrightarrow{OQ} = 5 \cdot \mathbf{I} + (-3) \cdot \mathbf{J} + (-2) \cdot \mathbf{K}$, $\overrightarrow{OR} = 1 \cdot \mathbf{I} + 0 \cdot \mathbf{J} + 1 \cdot \mathbf{K}$, and $\overrightarrow{OS} = (-2) \cdot \mathbf{I} + 2 \cdot \mathbf{J} + 3 \cdot \mathbf{K}$. Find real numbers r, s, t, u such that $r \cdot \overrightarrow{OP} + s \cdot \overrightarrow{OQ} + t \cdot \overrightarrow{OR} + u \cdot \overrightarrow{OS} = \overrightarrow{0}$.

6.7 Prove that if $\overrightarrow{OP}$, $\overrightarrow{OQ}$, $\overrightarrow{OR}$, and $\overrightarrow{OS}$ are any four vectors, then there exist real numbers r, s, t, u, not all zero, such that $r \cdot \overrightarrow{OP} + s \cdot \overrightarrow{OQ} + t \cdot \overrightarrow{OR} + u \cdot \overrightarrow{OS} = \overrightarrow{0}$.

7 vector operations in terms of coordinates

We now turn to a discussion of vector operations in terms of coordinates with respect to $\{\mathbf{I}, \mathbf{J}, \mathbf{K}\}$.

(7.1) theorem Let $\overrightarrow{OP}$ have coordinates (x_1, y_1, z_1) and $\overrightarrow{OQ}$ have coordinates (x_2, y_2, z_2). Then,

(a) $\overrightarrow{OP} + \overrightarrow{OQ}$ has coordinates $(x_1 + x_2, y_1 + y_2, z_1 + z_2)$;

(b) $r \cdot \overrightarrow{OP}$ has coordinates (rx_1, ry_1, rz_1).

PROOF We have $\overrightarrow{OP} = x_1 \cdot \mathbf{I} + y_1 \cdot \mathbf{J} + z_1 \cdot \mathbf{K}$ and $\overrightarrow{OQ} = x_2 \cdot \mathbf{I} + y_2 \cdot \mathbf{J} + z_2 \cdot \mathbf{K}$. Using the properties of addition and scalar multiplication, we obtain

$$\begin{aligned}\overrightarrow{OP} + \overrightarrow{OQ} &= (x_1 \cdot \mathbf{I} + y_1 \cdot \mathbf{J} + z_1 \cdot \mathbf{K}) + (x_2 \cdot \mathbf{I} + y_2 \cdot \mathbf{J} + z_2 \cdot \mathbf{K})\\ &= (x_1 \cdot \mathbf{I} + x_2 \cdot \mathbf{I}) + (y_1 \cdot \mathbf{J} + y_2 \cdot \mathbf{J}) + (z_1 \cdot \mathbf{K} + z_2 \cdot \mathbf{K})\\ &= (x_1 + x_2) \cdot \mathbf{I} + (y_1 + y_2) \cdot \mathbf{J} + (z_1 + z_2) \cdot \mathbf{K}.\end{aligned}$$

Thus, the coordinates of $\overrightarrow{OP} + \overrightarrow{OQ}$ are $(x_1 + x_2, y_1 + y_2, z_1 + z_2)$. Furthermore,

$$\begin{aligned}r \cdot \overrightarrow{OP} &= r \cdot (x_1 \cdot \mathbf{I} + y_1 \cdot \mathbf{J} + z_1 \cdot \mathbf{K})\\ &= r \cdot (x_1 \cdot \mathbf{I}) + r \cdot (y_1 \cdot \mathbf{J}) + r \cdot (z_1 \cdot \mathbf{K})\\ &= (rx_1) \cdot \mathbf{I} + (ry_1) \cdot \mathbf{J} + (rz_1) \cdot \mathbf{K},\end{aligned}$$

using Theorem 3.2. Therefore, the coordinates of $r \cdot \overrightarrow{OP}$ are (rx_1, ry_1, rz_1). ∎

EXAMPLE 1 Let $\overrightarrow{OP}$ have coordinates $(-4, 2, -1)$ and $\overrightarrow{OQ}$ have coordinates $(2, 0, -3)$. Then the coordinates of $\overrightarrow{OP} + \overrightarrow{OQ}$ are $(-2, 2, -4)$. The coordinates of $\frac{1}{2} \cdot \overrightarrow{OP}$ are $(-2, 1, -\frac{1}{2})$. The coordinates of $(-3) \cdot \overrightarrow{OQ}$ are $(-6, 0, 9)$.

It follows from Theorem 7.1 that the vector operations of addition and scalar multiplication can be performed algebraically in terms of the coordinates of the vectors. We note that the coordinates of $\vec{0}$ are $(0, 0, 0)$, and that if the coordinates of $\overrightarrow{OP}$ are (x, y, z), then the coordinates of $-\overrightarrow{OP}$ are $(-x, -y, -z)$.

EXAMPLE 2 Let $\overrightarrow{OP}$ have coordinates $(-1, 2, -3)$, $\overrightarrow{OQ}$ have coordinates $(2, -1, 4)$, and $\overrightarrow{OR}$ have coordinates $(5, 6, 2)$. Then the coordinates of $5 \cdot \overrightarrow{OP} + 2 \cdot \overrightarrow{OQ} + (-1) \cdot \overrightarrow{OR}$ are $(-6, 2, -9)$.

EXAMPLE 3 Let $\overrightarrow{QP}$ be a directed line segment in space, where the coordinates of Q are (x_Q, y_Q, z_Q) and the coordinates of P are (x_P, y_P, z_P). What are the coordinates of a vector which has the same length and direction as $\overrightarrow{QP}$? We recall that $\overrightarrow{OP} - \overrightarrow{OQ}$ is the vector with the same length and direction as $\overrightarrow{QP}$ (see Figure 12). The coordinates of $\overrightarrow{OP} - \overrightarrow{OQ} = \overrightarrow{OP} + (-\overrightarrow{OQ})$ are $(x_P - x_Q, y_P - y_Q, z_P - z_Q)$.

The inner product of two vectors has a simple algebraic expression in terms of the coordinates of the vectors.

(7.2) theorem Let $\overrightarrow{OP}$ have coordinates (x_1, y_1, z_1) and let $\overrightarrow{OQ}$ have coordinates (x_2, y_2, z_2). Then

$$\overrightarrow{OP} \circ \overrightarrow{OQ} = x_1x_2 + y_1y_2 + z_1z_2.$$

PROOF By the law of cosines, we have

$$|PQ|^2 = |OP|^2 + |OQ|^2 - 2|OP|\,|OQ| \cos \theta$$

where θ is the angle between $\overrightarrow{OP}$ and $\overrightarrow{OQ}$ (see Figure 35). Therefore,

$$\overrightarrow{OP} \circ \overrightarrow{OQ} = |OP|\,|OQ| \cos \theta = |OP|\,|OQ| \left(\frac{|PQ|^2 - |OP|^2 - |OQ|^2}{-2|OP|\,|OQ|} \right)$$

$$= \frac{|OP|^2 + |OQ|^2 - |PQ|^2}{2}.$$

By the distance formula,

$$|OP|^2 = x_1^2 + y_1^2 + z_1^2, \quad |OQ|^2 = x_2^2 + y_2^2 + z_2^2,$$

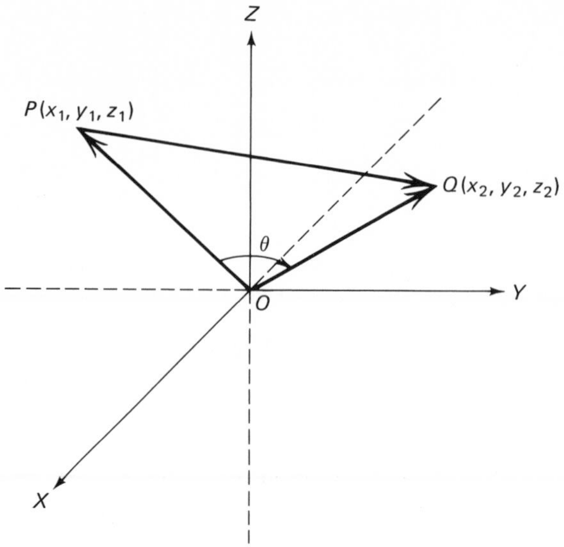

FIGURE 35

and

$$|PQ|^2 = (x_2 - x_1)^2 + (y_2 - y_1)^2 + (z_2 - z_1)^2.$$

Therefore,

$$\overrightarrow{OP} \circ \overrightarrow{OQ}$$
$$= \frac{x_2^1 + y_1^2 + z_1^2 + x_2^2 + y_2^2 + z_2^2 - (x_2 - x_1)^2 - (y_2 - y_1)^2 - (x_2 - x_1)^2}{2}$$
$$= x_1x_2 + y_1y_2 + z_1z_2.$$

■

EXAMPLE 4 Let $\overrightarrow{OP}$ have coordinates (2, 0, 1) and $\overrightarrow{OQ}$ have coordinates (6, −7, 2). Then

$$\overrightarrow{OP} \circ \overrightarrow{OQ} = (-2)(6) + (0)(-7) + (1)(2) = -12 + 0 + 2 = -10.$$

Also

$$|OP| = \sqrt{(2)^2 + (0)^2 + (1)^2} = \sqrt{5}$$

and

$$|OQ| = \sqrt{(6)^2 + (-7)^2 + (2)^2} = \sqrt{89}.$$

The angle θ between the vectors $\overrightarrow{OP}$ and $\overrightarrow{OQ}$ is determined by $\cos\theta = \overrightarrow{OP} \circ \overrightarrow{OQ}/|OP|\,|OQ| = -10/\sqrt{5}\sqrt{89} = -10/\sqrt{445} = -0.47404$. Thus, $\theta = 118°18'$.

EXAMPLE 5 Let $\overrightarrow{OP}$ have coordinates (7, −2, −9) and $\overrightarrow{OQ}$ have coordinates $(1, \frac{1}{2}, \frac{2}{3})$. Then $\overrightarrow{OP} \circ \overrightarrow{OQ} = 7(1) + (-2)\frac{1}{2} + (-9)\frac{2}{3} = 0$. Hence, the vectors $\overrightarrow{OP}$ and $\overrightarrow{OQ}$ are orthogonal.

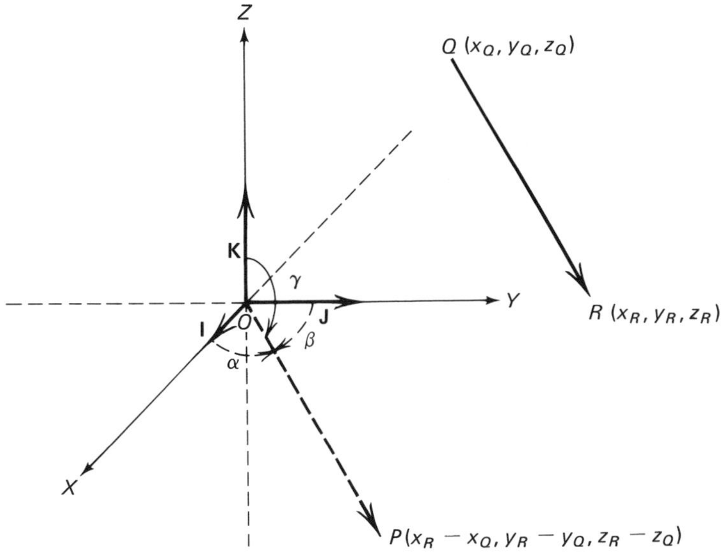

FIGURE 36

EXAMPLE 6 Let $\overrightarrow{QR}$ be a directed line segment and let $\overrightarrow{OP}$ be the vector with the same length and direction as $\overrightarrow{QR}$. The *direction angles* of $\overrightarrow{QR}$ are the angles α, β, γ, where α is the angle between **I** and $\overrightarrow{OP}$, β is the angle between **J** and $\overrightarrow{OP}$, and γ is the angle between **K** and $\overrightarrow{OP}$ (see Figure 36). We have $\mathbf{I} \circ \overrightarrow{OP} = |OP| \cos \alpha = |QR| \cos \alpha$, $\mathbf{J} \circ \overrightarrow{OP} = |OP| \cos \beta = |QR| \cos \beta$, and $\mathbf{K} \circ \overrightarrow{OP} = |OP| \cos \gamma = |QR| \cos \gamma$. Suppose the point R has coordinates (x_R, y_R, z_R) and Q has coordinates (x_Q, y_Q, z_Q). Then the vector $\overrightarrow{OP}$ has coordinates $(x_R - x_Q, y_R - y_Q, z_R - z_Q)$ [see Example 3]. The length of $\overrightarrow{QR}$ (and $\overrightarrow{OP}$) is

$$\sqrt{(x_R - x_Q)^2 + (y_R - y_Q)^2 + (z_R - z_Q)^2}.$$

Moreover,

$$\mathbf{I} \circ \overrightarrow{OP} = 1(x_R - x_Q) + 0(y_R - y_Q) + 0(z_R - z_Q) = x_R - x_Q,$$
$$\mathbf{J} \circ \overrightarrow{OP} = 0(x_R - x_Q) + 1(y_R - y_Q) + 0(z_R - z_Q) = y_R - y_Q,$$

and

$$\mathbf{K} \circ \overrightarrow{OP} = 0(x_R - x_Q) + 0(y_R - y_Q) + 1(z_R - z_Q) = z_R - z_Q.$$

Therefore,

$$\cos \alpha = \frac{\mathbf{I} \circ \overrightarrow{OP}}{|QR|} = \frac{x_R - x_Q}{\sqrt{(x_R - x_Q)^2 + (y_R - y_Q)^2 + (z_R - z_Q)^2}},$$

$$\cos \beta = \frac{\mathbf{J} \circ \overrightarrow{OP}}{|QR|} = \frac{y_R - y_Q}{\sqrt{(x_R - x_Q)^2 + (y_R - y_Q)^2 + (z_R - z_Q)^2}},$$

and

$$\cos \gamma = \frac{\mathbf{K} \circ \overrightarrow{OP}}{|QR|} = \frac{z_R - z_Q}{\sqrt{(x_R - x_Q)^2 + (y_R - y_Q)^2 + (z_R - z_Q)^2}}.$$

Theorem 7.2 is useful in deriving identities connecting inner multiplication with vector addition and scalar multiplication.

(7.3) corollary Let $\overrightarrow{OP}$, $\overrightarrow{OQ}$, and $\overrightarrow{OR}$ be any vectors. Then

$$\overrightarrow{OP} \circ (\overrightarrow{OQ} + \overrightarrow{OR}) = \overrightarrow{OP} \circ \overrightarrow{OQ} + \overrightarrow{OP} \circ \overrightarrow{OR}.$$

PROOF Let $\overrightarrow{OP}$ have coordinates (x_1, y_1, z_1), $\overrightarrow{OQ}$ have coordinates (x_2, y_2, z_2), and $\overrightarrow{OR}$ have coordinates (x_3, y_3, z_3). Then by Theorem 7.1(a), $\overrightarrow{OQ} + \overrightarrow{OR}$ has coordinates $(x_2 + x_3, y_2 + y_3, z_2 + z_3)$. By Theorem 7.2,

$$\begin{aligned}\overrightarrow{OP} \circ (\overrightarrow{OQ} + \overrightarrow{OR}) &= x_1(x_2 + x_3) + y_1(y_2 + y_3) + z_1(z_2 + z_3)\\ &= (x_1x_2 + y_1y_2 + z_1z_2) + (x_1x_3 + y_1y_3 + z_1z_3)\\ &= \overrightarrow{OP} \circ \overrightarrow{OQ} + \overrightarrow{OP} \circ \overrightarrow{OR}.\end{aligned}$$ ∎

Corollary 7.3 expresses the fact that inner multiplication is distributive with respect to vector addition.

(7.4) corollary Let $\overrightarrow{OP}$ and $\overrightarrow{OQ}$ be any vectors and let r be a real number. Then

$$(r \cdot \overrightarrow{OP}) \circ \overrightarrow{OQ} = \overrightarrow{OP} \circ (r \cdot \overrightarrow{OQ}) = r(\overrightarrow{OP} \circ \overrightarrow{OQ}).$$

PROOF Let $\overrightarrow{OP}$ have coordinates (x_1, y_1, z_1) and $\overrightarrow{OQ}$ have coordinates (x_2, y_2, z_2). Then the coordinates of $r \cdot \overrightarrow{OP}$ are (rx_1, ry_1, rz_1) by Theorem 7.1(b). Using Theorem 7.2, we have

$$\begin{aligned}(r \cdot \overrightarrow{OP}) \circ \overrightarrow{OQ} &= (rx_1)x_2 + (ry_1)y_2 + (rz_1)z_2\\ &= r(x_1x_2 + y_1y_2 + z_1z_2) = r(\overrightarrow{OP} \circ \overrightarrow{OQ}).\end{aligned}$$

The other part of the identity is proved in a similar manner. ∎

Using Corollary 5.4, we can obtain an expression for the cross product of two vectors in terms of the coordinates of the vectors. We note that it follows from Definition 5.1 that $\mathbf{I} \times \mathbf{I} = \mathbf{J} \times \mathbf{J} = \mathbf{K} \times \mathbf{K} = \vec{0}$, $\mathbf{I} \times \mathbf{J} = \mathbf{K}$, $\mathbf{J} \times \mathbf{K} = \mathbf{I}$, and $\mathbf{K} \times \mathbf{I} = \mathbf{J}$. Moreover, $\mathbf{J} \times \mathbf{I} = -(\mathbf{I} \times \mathbf{J}) = -\mathbf{K}$, $\mathbf{K} \times \mathbf{J} = -(\mathbf{J} \times \mathbf{K}) = -\mathbf{I}$, and $\mathbf{I} \times \mathbf{K} = -(\mathbf{K} \times \mathbf{I}) = -\mathbf{J}$.

(7.5) theorem Let $\overrightarrow{OP}$ have coordinates (x_1, y_1, z_1) and $\overrightarrow{OQ}$ have coordinates (x_2, y_2, z_2). Then $\overrightarrow{OP} \times \overrightarrow{OQ}$ has coordinates $(y_1z_2 - y_2z_1, x_2z_1 - x_1z_2, x_1y_2 - x_2y_1)$.

PROOF We have $\overrightarrow{OP} = x_1 \cdot \mathbf{I} + y_1 \cdot \mathbf{J} + z_1 \cdot \mathbf{K}$ and $\overrightarrow{OQ} = x_2 \cdot \mathbf{I} + y_2 \cdot \mathbf{J} + z_2 \cdot \mathbf{K}$. Then, by Corollary 5.4 and Theorem 3.2,

$$\begin{aligned}
\overrightarrow{OP} \times \overrightarrow{OQ} &= \overrightarrow{OP} \times (x_2 \cdot \mathbf{I} + y_2 \cdot \mathbf{J} + z_2 \cdot \mathbf{K}) \\
&= x_2 \cdot (\overrightarrow{OP} \times \mathbf{I}) + y_2 \cdot (\overrightarrow{OP} \times \mathbf{J}) + z_2 \cdot (\overrightarrow{OP} \times \mathbf{K}) \\
&= x_2 \cdot [(x_1 \cdot \mathbf{I} + y_1 \cdot \mathbf{J} + z_1 \cdot \mathbf{K}) \times \mathbf{I}] \\
&\quad + y_2 \cdot [(x_1 \cdot \mathbf{I} + y_1 \cdot \mathbf{J} + z_1 \cdot \mathbf{K}) \times \mathbf{J}] \\
&\quad + z_2 \cdot [(x_1 \cdot \mathbf{I} + y_1 \cdot \mathbf{J} + z_1 \cdot \mathbf{K}) \times \mathbf{K}] \\
&= x_2 \cdot [x_1 \cdot (\mathbf{I} \times \mathbf{I}) + y_1 \cdot (\mathbf{J} \times \mathbf{I}) + z_1 \cdot (\mathbf{K} \times \mathbf{I})] \\
&\quad + y_2 \cdot [x_1 \cdot (\mathbf{I} \times \mathbf{J}) + y_1 \cdot (\mathbf{J} \times \mathbf{J}) + z_1 \cdot (\mathbf{K} \times \mathbf{J})] \\
&\quad + z_2 \cdot [x_1 \cdot (\mathbf{I} \times \mathbf{K}) + y_1 \cdot (\mathbf{J} \times \mathbf{K}) + z_1 \cdot (\mathbf{K} \times \mathbf{K})] \\
&= x_2 \cdot [y_1 \cdot (-\mathbf{K}) + z_1 \cdot \mathbf{J}] + y_2 \cdot [x_1 \cdot \mathbf{K} + z_1 \cdot (-\mathbf{I})] \\
&\quad + z_2 \cdot [x_1 \cdot (-\mathbf{J}) + y_1 \cdot \mathbf{I}] \\
&= (-x_2 y_1) \cdot \mathbf{K} + (x_2 z_1) \cdot \mathbf{J} + (y_2 x_1) \cdot \mathbf{K} + (-y_2 z_1) \cdot \mathbf{I} \\
&\quad + (-z_2 x_1) \cdot \mathbf{J} + (z_2 y_1) \cdot \mathbf{I} \\
&= (y_1 z_2 - y_2 z_1) \cdot \mathbf{I} + (x_2 z_1 - x_1 z_2) \cdot \mathbf{J} + (x_1 y_2 - x_2 y_1) \cdot \mathbf{K}.
\end{aligned}$$

Thus, the coordinates of $\overrightarrow{OP} \times \overrightarrow{OQ}$ are

$$(y_1 z_2 - y_2 z_1, x_2 z_1 - x_1 z_2, x_1 y_2 - x_2 y_1). \qquad ∎$$

EXAMPLE 7 Let $\overrightarrow{OP}$ have coordinates $(-2, 1, \frac{1}{3})$ and $\overrightarrow{OQ}$ have coordinates $(5, \frac{1}{2}, 0)$. Then $\overrightarrow{OP} \times \overrightarrow{OQ}$ has coordinates $((1)(0) - (\frac{1}{2})(\frac{1}{3}), (5)(\frac{1}{3}) - (-2)(0), (-2)(\frac{1}{2}) - (5)(1)) = (-\frac{1}{6}, \frac{5}{3}, -6)$. The vectors are shown in Figure 37.

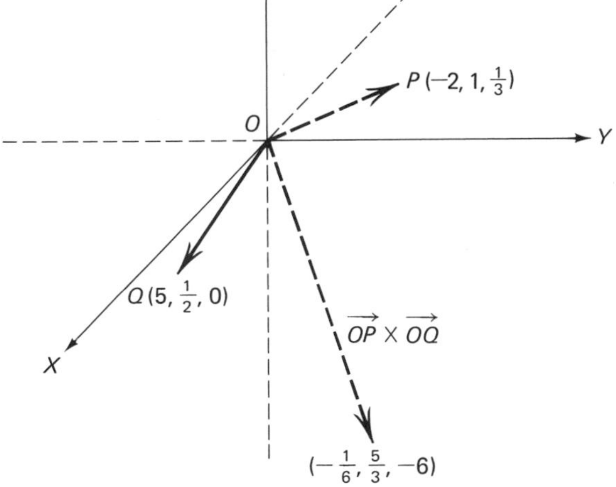

FIGURE 37

EXAMPLE 8 Determinants are discussed in detail in Chapter 6. Undoubtedly most readers are familiar with the notion of a determinant from elementary algebra courses. The expression for the coordinates of the cross product, derived in Theorem 7.5, can be conveniently given in determinant form. If $\overrightarrow{OP}$ has coordinates (x_1, y_1, z_1) and $\overrightarrow{OQ}$ has coordinates (x_2, y_2, z_2), then

$$\overrightarrow{OP} \times \overrightarrow{OQ} = \begin{vmatrix} x_1 & x_2 & \mathbf{I} \\ y_1 & y_2 & \mathbf{J} \\ z_1 & z_2 & \mathbf{K} \end{vmatrix}$$

$$= x_1y_2 \cdot \mathbf{K} + y_1z_2 \cdot \mathbf{I} + x_2z_1 \cdot \mathbf{J} - y_2z_1 \cdot \mathbf{I} - x_2y_1 \cdot \mathbf{K} - x_1z_2 \cdot \mathbf{J}$$
$$= (y_1z_2 - y_2z_1) \cdot \mathbf{I} + (x_2z_1 - x_1z_2) \cdot \mathbf{J} + (x_1y_2 - x_2y_1) \cdot \mathbf{K}.$$

Let $\overrightarrow{OP}$ have coordinates $(-1, 0, 7)$ and $\overrightarrow{OQ}$ have coordinates $(3, -6, 4)$. Then

$$\overrightarrow{OP} \times \overrightarrow{OQ} = \begin{vmatrix} -1 & 3 & \mathbf{I} \\ 0 & -6 & \mathbf{J} \\ 7 & 4 & \mathbf{K} \end{vmatrix}$$

$$= 6 \cdot \mathbf{K} + 0 \cdot \mathbf{I} + 21 \cdot \mathbf{J} + 42 \cdot \mathbf{I} + 0 \cdot \mathbf{K} + 4 \cdot \mathbf{J}$$
$$= 42 \cdot \mathbf{I} + 25 \cdot \mathbf{J} + 6 \cdot \mathbf{K}.$$

That is, the coordinates of $\overrightarrow{OP} \times \overrightarrow{OQ}$ are $(42, 25, 6)$.

exercises

7.1 Let $\overrightarrow{OP}$ have coordinates $(2, -1, 7)$ and $\overrightarrow{OQ}$ have coordinates $(-3, 6, 1)$. Find the coordinates of the following vectors: $\overrightarrow{OP} + \overrightarrow{OQ}$; $\overrightarrow{OP} - \overrightarrow{OQ}$; $(-5) \cdot \overrightarrow{OP}$; $6 \cdot \overrightarrow{OQ}$; $2 \cdot \overrightarrow{OP} + (-3) \cdot \overrightarrow{OQ}$; $(-3) \cdot \overrightarrow{OP} + 4 \cdot \overrightarrow{OQ}$; $\overrightarrow{OQ} - \overrightarrow{OP}$, $5 \cdot \overrightarrow{OP} + 10 \cdot \overrightarrow{OQ}$.

7.2 Let $\overrightarrow{OP}$ have coordinates $(-1, 0, 3)$, $\overrightarrow{OQ}$ have coordinates $(\sqrt{2}, \frac{1}{2}, -5)$, and $\overrightarrow{OR}$ have coordinates $(0, -7, \sqrt{3})$. Find the coordinates of the following vectors: $5 \cdot \overrightarrow{OP} - \sqrt{2} \cdot \overrightarrow{OQ}$; $\overrightarrow{OP} + \overrightarrow{OQ} + \overrightarrow{OR}$; $3 \cdot (2 \cdot \overrightarrow{OP} + 6 \cdot \overrightarrow{OR})$; $\frac{1}{2} \cdot (-\overrightarrow{OP} + \overrightarrow{OR}) + 7 \cdot \overrightarrow{OQ}$; $2 \cdot \overrightarrow{OP} + \frac{1}{3} \cdot \overrightarrow{OQ} + \frac{2}{5} \cdot \overrightarrow{OR}$.

7.3 Let $\overrightarrow{OP} = \mathbf{I} + 2 \cdot \mathbf{J} + 3 \cdot \mathbf{K}$, $\overrightarrow{OQ} = 3 \cdot \mathbf{I} + 2 \cdot \mathbf{J} + \mathbf{K}$, and $\overrightarrow{OR} = \mathbf{I} + \mathbf{J} + \mathbf{K}$.

(a) Find the coordinates of the following vectors: $\overrightarrow{OP} - \overrightarrow{OQ} + \overrightarrow{OR}$; $5 \cdot \overrightarrow{OP} + 2 \cdot \overrightarrow{OQ} + (-3) \cdot \overrightarrow{OR}$; $\frac{1}{2} \cdot (4 \cdot \overrightarrow{OP} + 6 \cdot \overrightarrow{OR})$; $3 \cdot (2 \cdot \overrightarrow{OP} + (-1) \cdot \overrightarrow{OQ} + 5 \cdot \overrightarrow{OR}) - 4 \cdot (\overrightarrow{OP} + 2 \cdot \overrightarrow{OQ} + 3 \cdot \overrightarrow{OR})$.

(b) For each vector in part (a), draw a figure showing $\overrightarrow{OP}$, $\overrightarrow{OQ}$, $\overrightarrow{OR}$ and that vector.

7.4 Let $\overrightarrow{OP} = \mathbf{I} + \mathbf{J} + \mathbf{K}$, $\overrightarrow{OQ} = 2 \cdot \mathbf{I} + 3 \cdot \mathbf{J}$, $\overrightarrow{OR} = 3 \cdot \mathbf{I} + 5 \cdot \mathbf{J} - 2 \cdot \mathbf{K}$, and $\overrightarrow{OS} = -\mathbf{J} + \mathbf{K}$. Show that the vectors $\overrightarrow{OP} - \overrightarrow{OQ}$ and $\overrightarrow{OR} - \overrightarrow{OS}$ are collinear.

7.5 Show that the vectors **I** − **J**, **J** − **K**, and **K** − **I** are coplanar.

7.6 Let the point P have coordinates (1, 0, 3), the point Q have coordinates $(-5, \frac{1}{2}, -2)$, the point R have coordinates $(1 - \sqrt{5}, 3\sqrt{2}, -3)$, and the point S have coordinates $(\sqrt{5}, 2\sqrt{2}, 4)$. Find the coordinates of the vectors which have the same length and direction as the directed line segments $\overrightarrow{PQ}$, $\overrightarrow{RS}$, $\overrightarrow{SP}$, $\overrightarrow{QS}$, $\overrightarrow{RP}$, and $\overrightarrow{PS}$.

7.7 Compute $\overrightarrow{OP} \circ \overrightarrow{OQ}$ for each of the following pairs of vectors:

(a) $\overrightarrow{OP} = 2 \cdot \mathbf{I} + 3 \cdot \mathbf{J} + (-5) \cdot \mathbf{K}$, $\overrightarrow{OQ} = \mathbf{I} + (-2) \cdot \mathbf{J} + 6 \cdot \mathbf{K}$

(b) $\overrightarrow{OP} = \mathbf{I} + \mathbf{J} - \mathbf{K}$, $\overrightarrow{OQ} = -\mathbf{I} - \mathbf{J} + 2 \cdot \mathbf{K}$

(c) $\overrightarrow{OP} = \frac{1}{2} \cdot \mathbf{I} + \frac{1}{3} \cdot \mathbf{J} + \frac{1}{5} \cdot \mathbf{K}$, $\overrightarrow{OQ} = 4 \cdot \mathbf{I} + 9 \cdot \mathbf{J} + 25 \cdot \mathbf{K}$

(d) $\overrightarrow{OP} = \sqrt{2} \cdot \mathbf{I} + \sqrt{3} \cdot \mathbf{J} + \sqrt{5} \cdot \mathbf{K}$, $\overrightarrow{OQ} = \sqrt{30} \cdot \overrightarrow{OP}$

(e) $\overrightarrow{OP} = 5 \cdot \mathbf{I} + 7 \cdot \mathbf{J} + (-11) \cdot \mathbf{K}$, $\overrightarrow{OQ} = \mathbf{I} + \frac{7}{5} \cdot \mathbf{J} + \frac{11}{5} \cdot \mathbf{K}$

7.8 Find the cosine of the angle between each pair of vectors given in Exercise 7.7.

7.9 Find the direction cosines of the directed line segments $\overrightarrow{PQ}$, $\overrightarrow{RS}$, $\overrightarrow{SP}$, $\overrightarrow{QS}$, $\overrightarrow{RP}$, and $\overrightarrow{PS}$ given in Exercise 7.6.

7.10 Let $\overrightarrow{OP}$ and $\overrightarrow{OQ}$ be vectors and let r and s be real numbers. Prove that

$$(r \cdot \overrightarrow{OP}) \circ (s \cdot \overrightarrow{OQ}) = rs(\overrightarrow{OP} \circ \overrightarrow{OQ}).$$

7.11 Show that the coordinates of a vector $\overrightarrow{OP}$ are $(\mathbf{I} \circ \overrightarrow{OP}, \mathbf{J} \circ \overrightarrow{OP}, \mathbf{K} \circ \overrightarrow{OP})$.

7.12 Verify the identity in Corollary 4.5 for the vectors $\overrightarrow{OP}$, $\overrightarrow{OQ}$, and $\overrightarrow{OR}$ given in Exercise 7.2.

7.13 Prove the following vector identity:

$$(r \cdot \overrightarrow{OP} + s \cdot \overrightarrow{OQ}) \circ (t \cdot \overrightarrow{OR} + u \cdot \overrightarrow{OS}) = rt(\overrightarrow{OP} \circ \overrightarrow{OR}) + ru(\overrightarrow{OP} \circ \overrightarrow{OS}) + st(\overrightarrow{OQ} \circ \overrightarrow{OR}) + su(\overrightarrow{OQ} \circ \overrightarrow{OS}).$$

7.14 Let $\overrightarrow{OP}$ and $\overrightarrow{OQ}$ be vectors such that $\overrightarrow{OP} \circ \overrightarrow{OQ} = t$ and $|OQ| = r \neq 0$. Prove that $\overrightarrow{OP} - (t/r^2) \cdot \overrightarrow{OQ}$ and $\overrightarrow{OQ}$ are orthogonal. Express $\overrightarrow{OP}$ as the sum of two orthogonal vectors.

7.15 Let $\overrightarrow{OU}$, $\overrightarrow{OV}$, and $\overrightarrow{OW}$ be noncoplanar vectors. Suppose that

$$\overrightarrow{OP} = u_1 \cdot \overrightarrow{OU} + v_1 \cdot \overrightarrow{OV} + w_1 \cdot \overrightarrow{OW}$$

and

$$\overrightarrow{OQ} = u_2 \cdot \overrightarrow{OU} + v_2 \cdot \overrightarrow{OV} + w_2 \cdot \overrightarrow{OW}$$

(see Exercise 6.5). Give an example to show that the formula

$$\overrightarrow{OP} \circ \overrightarrow{OQ} = u_1u_2 + v_1v_2 + w_1w_2$$

does not hold in general.

7.16 Show that the square of the length of the vector $\overrightarrow{OP} + \overrightarrow{OQ}$ is $|OP|^2 + |OQ|^2 + 2(\overrightarrow{OP} \circ \overrightarrow{OQ})$. Use this result to show that the inner product of two vectors can be expressed in terms of lengths, sums, and scalar multiples of the vectors.

7.17 Show that if $\overrightarrow{OP}$ and $\overrightarrow{OQ}$ are orthogonal and $\overrightarrow{OP}$ and $\overrightarrow{OR}$ are orthogonal, then $\overrightarrow{OP}$ is orthogonal to every vector which is a linear combination of $\overrightarrow{OQ}$ and $\overrightarrow{OR}$.

7.18 Express $\overrightarrow{OP} \times \overrightarrow{OQ}$ as a linear combination of **I**, **J**, and **K** for the pairs of vectors $\overrightarrow{OP}$ and $\overrightarrow{OQ}$ given in Exercise 7.7.

7.19 Compute $(\overrightarrow{OP} \times \overrightarrow{OQ}) \circ \overrightarrow{OR}$ and $\overrightarrow{OP} \circ (\overrightarrow{OQ} \times \overrightarrow{OR})$ for the vectors $\overrightarrow{OP}$, $\overrightarrow{OQ}$, and $\overrightarrow{OR}$ given in Exercise 7.2.

***7.20** Prove that $(\overrightarrow{OP} \times \overrightarrow{OQ}) \circ \overrightarrow{OR} = \overrightarrow{OP} \circ (\overrightarrow{OQ} \times \overrightarrow{OR})$ for any three vectors $\overrightarrow{OP}$, $\overrightarrow{OQ}$, and $\overrightarrow{OR}$.

7.21 Prove that $\overrightarrow{OP} \times (\overrightarrow{OQ} \times \overrightarrow{OR}) = (\overrightarrow{OP} \circ \overrightarrow{OR}) \cdot \overrightarrow{OQ} - (\overrightarrow{OP} \circ \overrightarrow{OQ}) \cdot \overrightarrow{OR}$ for any three vectors $\overrightarrow{OP}$, $\overrightarrow{OQ}$, and $\overrightarrow{OR}$.

7.22 Use the result in Exercise 7.21 to prove that $(\overrightarrow{OP} \times \overrightarrow{OQ}) \times \overrightarrow{OR} = (\overrightarrow{OR} \circ \overrightarrow{OP}) \cdot \overrightarrow{OQ} - (\overrightarrow{OR} \circ \overrightarrow{OQ}) \cdot \overrightarrow{OP}$.

7.23 Use the identities given in Exercises 7.21 and 7.22 to find an example of vectors $\overrightarrow{OP}$, $\overrightarrow{OQ}$, and $\overrightarrow{OR}$ such that $\overrightarrow{OP} \times (\overrightarrow{OQ} \times \overrightarrow{OR}) \neq (\overrightarrow{OP} \times \overrightarrow{OQ}) \times \overrightarrow{OR}$.

8 applications of geometric vectors

In this section we present several examples of geometric and physical problems which can be stated and solved using the language of geometric vectors.

The fact that the length of $r \cdot \overrightarrow{OP}$ is $|r|\,|OP|$ (see Definition 3.1) can be used in geometric problems that involve the division of a line segment in a given ratio.

EXAMPLE 1 Let P and Q be points in space, where P has coordinates (x_P, y_P, z_P) and Q has coordinates (x_Q, y_Q, z_Q). Let R be a point on the line segment joining Q and P such that $|QR|/|QP| = r$, where r is a real number, and $0 \leq r \leq 1$. We wish to find the coordinates of R in terms of the coordinates of P and Q. Denote the coordinates of R by (x, y, z) [see Figure 38]. Recall that a vector $\overrightarrow{OS}$ with the same length and direction as $\overrightarrow{QP}$ is the vector $\overrightarrow{OP} - \overrightarrow{OQ}$. Let T be a point on $\overrightarrow{OS}$ such that TR is parallel to $\overrightarrow{OQ}$. Then $|OT|/|OS| = |QR|/|QP| = r$. By the definitions of vector addition and scalar multiplication, $\overrightarrow{OR} = \overrightarrow{OQ} + \overrightarrow{OT} = \overrightarrow{OQ} + r \cdot \overrightarrow{OS} = \overrightarrow{OQ} + r \cdot (\overrightarrow{OP} - \overrightarrow{OQ})$. The coordinates (x, y, z) of R are the coordinates of the vector $\overrightarrow{OR} = \overrightarrow{OQ} + r \cdot (\overrightarrow{OP} - \overrightarrow{OQ})$. By Theorem 7.1, this latter vector has coordinates $(x_Q + r(x_P - x_Q),\ y_Q + r(y_P - y_Q),\ z_Q + r(z_P - z_Q))$, or $((1 - r)x_Q + rx_P,\ (1 - r)y_Q + ry_P,\ (1 - r)z_Q + rz_P)$. Therefore, $x =$

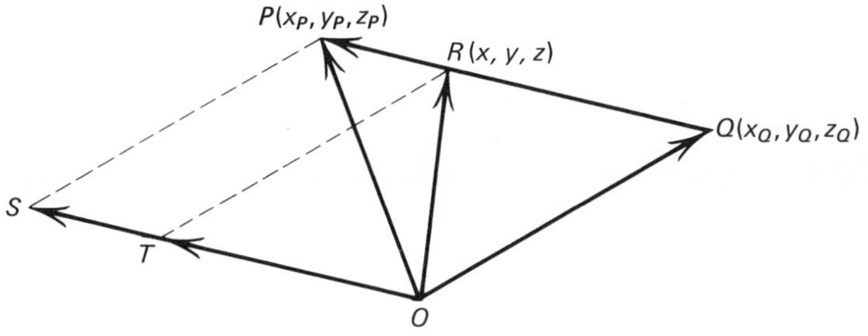

FIGURE 38

$(1 - r)x_Q + rx_P$, $y = (1 - r)y_Q + ry_P$, and $z = (1 - r)z_Q + rz_P$. In particular, if R bisects the line segment QP, then $r = \frac{1}{2}$, and $x = \frac{1}{2}(x_Q + x_P)$, $y = \frac{1}{2}(y_Q + y_P)$, and $z = \frac{1}{2}(z_Q + z_P)$.

EXAMPLE 2 Prove that the diagonals of a parallelogram bisect each other Let $OQSP$ be a parallelogram with diagonals OS and QP intersecting at R (see Figure 39). Let $\overrightarrow{OU}$ be the vector with the same length and direction as the directed line segment $\overrightarrow{QP}$. Then $\overrightarrow{OU} = \overrightarrow{OP} - \overrightarrow{OQ}$. Let T be a point on $\overrightarrow{OU}$ such that TR is parallel to $\overrightarrow{OQ}$. Now $\overrightarrow{OR} = r \cdot \overrightarrow{OS}$ and $\overrightarrow{OT} = s \cdot \overrightarrow{OU}$, and it is sufficient to show that $r = s = \frac{1}{2}$. We have

$$\overrightarrow{OR} = r \cdot \overrightarrow{OS} = r \cdot (\overrightarrow{OQ} + \overrightarrow{OP}) = r \cdot \overrightarrow{OQ} + r \cdot \overrightarrow{OP},$$

and

$$\begin{aligned}\overrightarrow{OR} = \overrightarrow{OQ} + \overrightarrow{OT} = \overrightarrow{OQ} + s \cdot \overrightarrow{OU} &= \overrightarrow{OQ} + s \cdot (\overrightarrow{OP} - \overrightarrow{OQ}) \\ &= (1 - s) \cdot \overrightarrow{OQ} + s \cdot \overrightarrow{OP}.\end{aligned}$$

Therefore,

$$r \cdot \overrightarrow{OQ} + r \cdot \overrightarrow{OP} = (1 - s) \cdot \overrightarrow{OQ} + s \cdot \overrightarrow{OP}$$

or

$$(r + s - 1) \cdot \overrightarrow{OQ} + (r - s) \cdot \overrightarrow{OP} = \vec{0}.$$

If either $r + s - 1 \neq 0$ or $r - s \neq 0$, then $\overrightarrow{OQ}$ and $\overrightarrow{OP}$ are collinear by Corollary 3.4. But $\overrightarrow{OQ}$ and $\overrightarrow{OP}$ are not collinear. Hence, $r + s - 1 = 0$ and $r - s = 0$. Solving these equations, we find $r = s = \frac{1}{2}$.

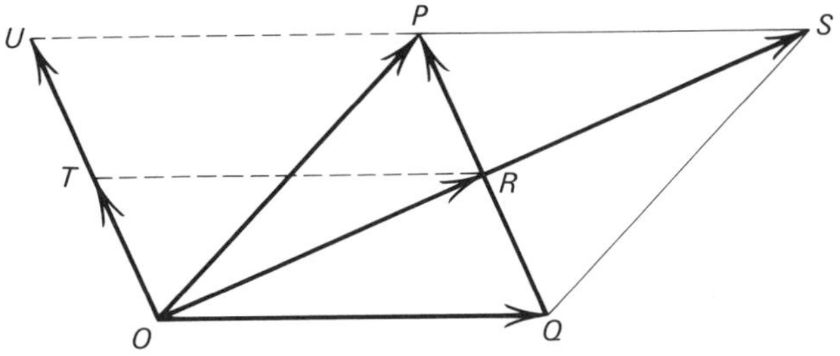

FIGURE 39

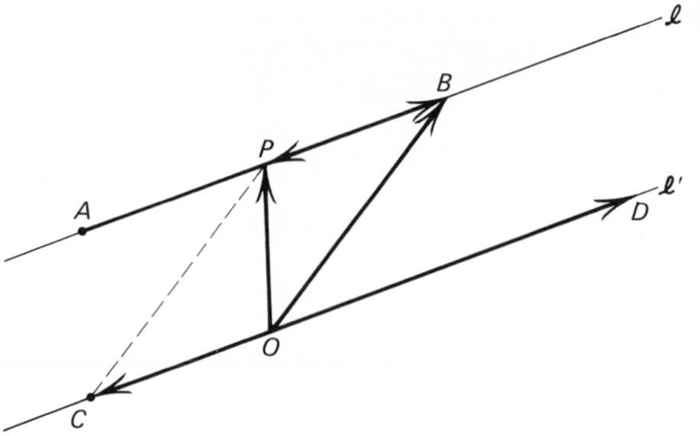

FIGURE 40

EXAMPLE 3 Let ℓ be a line in space through two given points A and B. We will derive the vector equation of ℓ. Let P be any point on ℓ. Let ℓ' be a line through the origin O parallel to ℓ and let $\overrightarrow{OD}$ and $\overrightarrow{OC}$ be vectors on this line with the same lengths and directions as the directed line segments $\overrightarrow{AB}$ and $\overrightarrow{BP}$, respectively (see Figure 40). Then $\overrightarrow{OC} = \overrightarrow{OP} - \overrightarrow{OB}$ and $\overrightarrow{OC} = k \cdot \overrightarrow{OD}$ for some real number k. Therefore,

$$\overrightarrow{OP} = \overrightarrow{OB} + \overrightarrow{OC} = \overrightarrow{OB} + k \cdot \overrightarrow{OD}. \qquad (8\text{–}1)$$

This is the vector equation of ℓ. Indeed, if P is any point on ℓ, then there is a real number k such that $\overrightarrow{OP}$ satisfies equation (8–1), and conversely, if $\overrightarrow{OP}$ is any vector which satisfies (8–1) for some real number k, then the point P lies on ℓ.

EXAMPLE 4 In Example 3, let the coordinates of A and B be (x_A, y_A, z_A) and (x_B, y_B, z_B), respectively. Let P be any point on ℓ with coordinates (x, y, z). Let $\overrightarrow{OD}$ and $\overrightarrow{OC}$ be the vectors described in Example 3. Then the coordinates of $\overrightarrow{OD}$ are $(x_B - x_A, y_B - y_A, z_B - z_A)$ [see Example 3, §7]. By equation (8–1), the coordinates of $\overrightarrow{OP}$ are $(x_B + k(x_B - x_A), y_B + k(y_B - y_A), z_B + k(z_B - z_A))$. Therefore,

$$x = x_B + k(x_B - x_A),\ y = y_B + k(y_B - y_A),\ \ z = z_B + k(z_B - z_A). \quad (8\text{–}2)$$

Equations (8–2) are the *parametric equations* of the line ℓ. These equations give the coordinates of any point P on ℓ in terms of the coordinates of two given points, A and B, on ℓ.

EXAMPLE 5 Let π be a plane determined by three noncollinear points A, B, and C on π. Let π' be a plane through O parallel to π and let $\overrightarrow{OU}$, $\overrightarrow{OV}$, and $\overrightarrow{OW}$ be vectors in the plane π' which have the same length and directions as the directed line segments $\overrightarrow{AP}$, $\overrightarrow{AB}$, and $\overrightarrow{AC}$, respectively (see

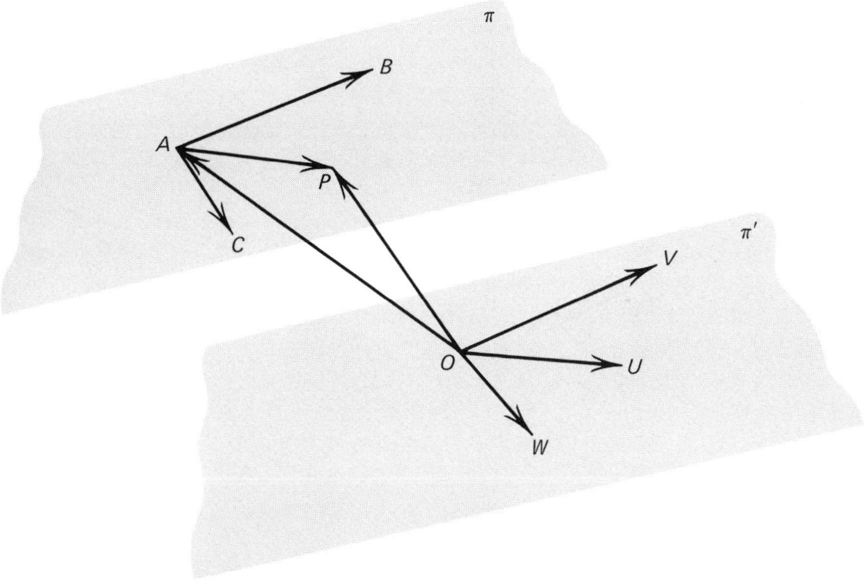

FIGURE 41

Figure 41). Since the vector $\overrightarrow{OU}$ is coplanar with the noncollinear vectors $\overrightarrow{OV}$ and $\overrightarrow{OW}$, it follows from Theorem 3.5 that

$$\overrightarrow{OU} = k_1 \cdot \overrightarrow{OV} + k_2 \cdot \overrightarrow{OW}$$

for some real numbers k_1 and k_2. Moreover, $\overrightarrow{OU} = \overrightarrow{OP} - \overrightarrow{OA}$, or $\overrightarrow{OP} = \overrightarrow{OA} + \overrightarrow{OU}$. Therefore,

$$\overrightarrow{OP} = \overrightarrow{OA} + k_1 \cdot \overrightarrow{OV} + k_2 \cdot \overrightarrow{OW}. \tag{8-3}$$

Equation (8–3) is the vector equation of the plane π. Let the points A, B, and C have coordinates (x_A, y_A, z_A), (x_B, y_B, z_B), and (x_C, y_C, z_C), respectively, and the point P have coordinates (x, y, z). Then the parametric equations of π which give the coordinates of P in terms of the coordinates of A, B, and C can be found directly from equation (8–3) [see Exercise 8.7].

EXAMPLE 6 Let ℓ_1 be a line through the points P and Q with coordinates $(1, 1, 1)$ and $(2, -3, 4)$ respectively, and let ℓ_2 be a line through P and a point R with coordinates $(-1, 6, 2)$. We will determine the acute angle between the lines ℓ_1 and ℓ_2. This angle is either the angle θ between the vectors $\overrightarrow{OQ} - \overrightarrow{OP}$ and $\overrightarrow{OR} - \overrightarrow{OP}$, or the supplement of θ (see Figure 42). The coordinates of $\overrightarrow{OQ} - \overrightarrow{OP}$ are $(1, -4, 3)$ and the coordinates of $\overrightarrow{OR} - \overrightarrow{OP}$ are $(-2, 5, 1)$. The length of $\overrightarrow{OQ} - \overrightarrow{OP}$ is $\sqrt{1 + 16 + 9} = \sqrt{26}$, the length of $\overrightarrow{OR} - \overrightarrow{OP}$ is $\sqrt{4 + 25 + 1} = \sqrt{30}$, and $(\overrightarrow{OQ} - \overrightarrow{OP}) \circ (\overrightarrow{OR} - \overrightarrow{OP}) = -2 - 20 + 3 = -19$. Therefore,

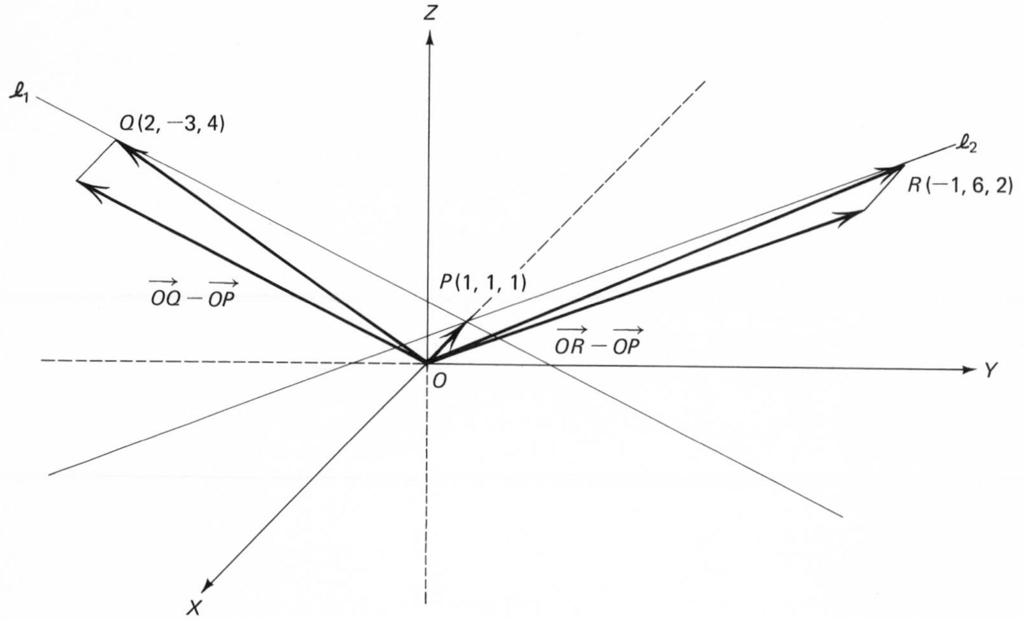

FIGURE 42

$$\cos\theta = \frac{(\overrightarrow{OQ} - \overrightarrow{OP}) \circ (\overrightarrow{OR} - \overrightarrow{OP})}{|\overrightarrow{OQ} - \overrightarrow{OP}|\,|\overrightarrow{OR} - \overrightarrow{OP}|} = \frac{-19}{\sqrt{26}\sqrt{30}}.$$

Thus, the acute angle between ℓ_1 and ℓ_2 is the angle whose cosine is $19/\sqrt{26}\,\sqrt{30}$.

EXAMPLE 7 If the point of application of a force F moves through a displacement S, then the work done by the force is equal to the product of the component of F in the direction of S and the magnitude of S. Let F be represented by a vector $\overrightarrow{OP}$ and S be represented by a vector $\overrightarrow{OQ}$ (see Figure 43). Thus, the work done by F is the component of $\overrightarrow{OP}$ on $\overrightarrow{OQ}$ times the length of $\overrightarrow{OQ}$, which according to Example 3, §4, is precisely $\overrightarrow{OP} \circ \overrightarrow{OQ}$.

EXAMPLE 8 If F and G are two forces which act simultaneously through a displacement S, then the work done by the resultant force $F + G$ is the

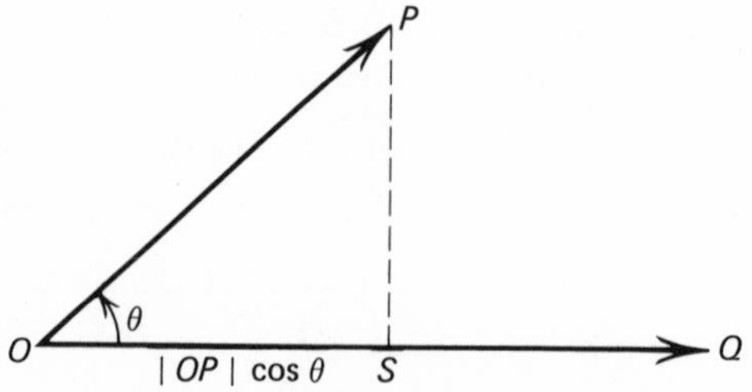

FIGURE 43

sum of the works done by F and G separately. This follows from the observation that if F and G are represented by vectors $\overrightarrow{OP}$ and $\overrightarrow{OR}$, respectively, and S is represented by the vector $\overrightarrow{OQ}$, then by Example 7 above and Corollary 7.3, the work done is

$$(\overrightarrow{OP} + \overrightarrow{OR}) \circ \overrightarrow{OQ} = \overrightarrow{OP} \circ \overrightarrow{OQ} + \overrightarrow{OR} \circ \overrightarrow{OQ}.$$

EXAMPLE 9 In this example we find the equations of the line ℓ through the origin which is perpendicular to the vectors $\overrightarrow{OA} = a_1 \cdot \mathbf{I} + a_2 \cdot \mathbf{J} + a_3 \cdot \mathbf{K}$ and $\overrightarrow{OB} = b_1 \cdot \mathbf{I} + b_2 \cdot \mathbf{J} + b_3 \cdot \mathbf{K}$. Let P be any point on ℓ. Then the vector $\overrightarrow{OP}$ is collinear with $\overrightarrow{OA} \times \overrightarrow{OB}$. Thus, $\overrightarrow{OP} = k \cdot (\overrightarrow{OA} \times \overrightarrow{OB})$ for some real number k. By Theorem 7.5, the coordinates of $\overrightarrow{OA} \times \overrightarrow{OB}$ are $(a_2b_3 - b_2a_3, b_1a_3 - a_1b_3, a_1b_2 - b_1a_2)$. If the coordinates of P are denoted by (x, y, z), we have

$$x = k(a_2b_3 - b_2a_3), \ y = k(b_1a_3 - a_1b_3), \ z = k(a_1b_2 - b_1a_2).$$

EXAMPLE 10 Let $\overrightarrow{OA}$, $\overrightarrow{OB}$, and $\overrightarrow{OC}$ be noncoplanar vectors. We can show that the volume of the parallelepiped determined by these vectors is $\pm (\overrightarrow{OA} \times \overrightarrow{OB}) \circ \overrightarrow{OC}$. Let $\overrightarrow{OP} = \overrightarrow{OA} \times \overrightarrow{OB}$. Then $(\overrightarrow{OA} \times \overrightarrow{OB}) \circ \overrightarrow{OC} = |OP|\,|OC| \cos\theta = |OA|\,|OB|\,|OC| \cos\theta \sin\varphi$, where θ is the angle between $\overrightarrow{OC}$ and $\overrightarrow{OP}$ and φ is the angle between $\overrightarrow{OA}$ and $\overrightarrow{OB}$ (see Figure 44). Indeed, $|OA|\,|OB| \sin\varphi$ is the area of the parallelogram determined by $\overrightarrow{OA}$ and $\overrightarrow{OB}$ (which is the base of the given parallelepiped), and the altitude of the parallelepiped is $\pm |OC| \cos\theta$, according as θ is acute or obtuse.

If the coordinates of A, B, and C are (a_1, a_2, a_3), (b_1, b_2, b_3), and

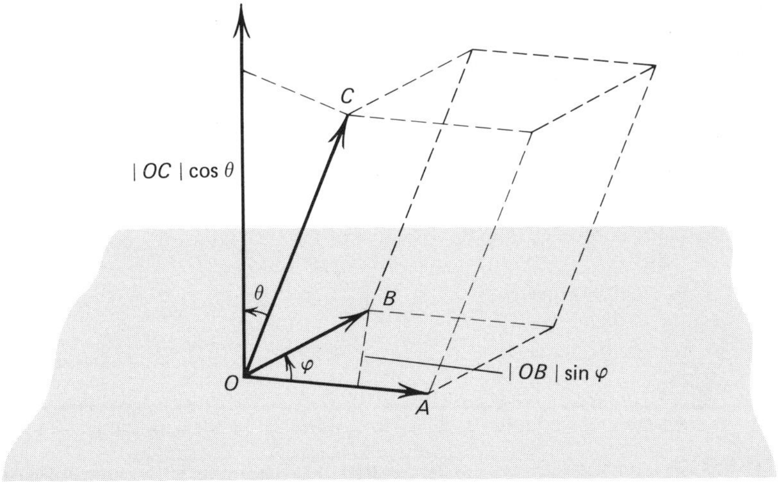

FIGURE 44

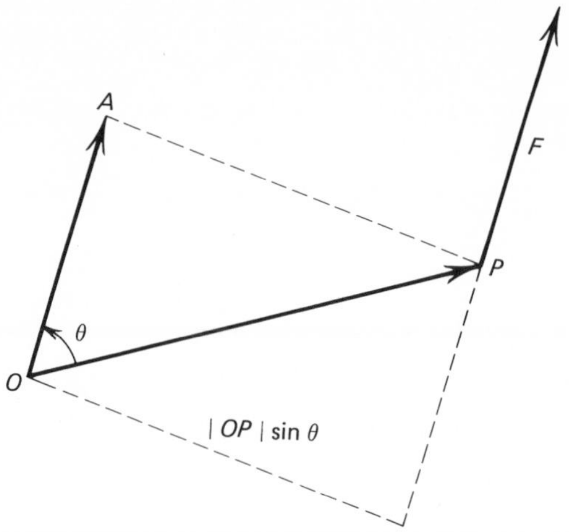

FIGURE 45

(c_1, c_2, c_3), then by Theorems 7.5 and 7.2, the volume of the parallelepiped is

$$\pm [c_1(a_2b_3 - b_2a_3) + c_2(b_1a_3 - a_1b_3) + c_3(a_1b_2 - b_1a_2)].$$

EXAMPLE 11 Let O and P be points in a rigid body and let F be a force acting at P. The force F tends to cause a rotation of the body about an axis through O perpendicular to the plane determined by $\overrightarrow{OP}$ and the line of F. This tendency is called the *moment of the force F* about the axis through O. Let $\overrightarrow{OA}$ be the vector which represents the force F (see Figure 45). Then this moment can be represented by the vector $\overrightarrow{OP} \times \overrightarrow{OA}$. Note that the length $|OP|\,|OA| \sin\theta$ of $\overrightarrow{OP} \times \overrightarrow{OA}$ is the product of the magnitude $|OA|$ of the force F and the length $|OP| \sin\theta$ of the lever arm of the force (the perpendicular distance from O to the line of F). Moreover, $\overrightarrow{OP} \times \overrightarrow{OA}$ is along the axis of rotation and the rotation is in the direction that carries $\overrightarrow{OP}$ into $\overrightarrow{OA}$.

exercises

8.1 Let P be a point with coordinates $(\frac{1}{2}, \frac{1}{3}, \frac{1}{4})$ and Q be a point with coordinates $(-4, 6, 8)$. Find the coordinates of the point R on the line segment QP such that

(a) $|QR|/|QP| = \frac{1}{2}$
(b) $|QR|/|QP| = \frac{4}{5}$
(c) $|QR|/|QP| = 1/\sqrt{2}$.

8.2 Find the coordinates of the point that is one-third of the way from a point P with coordinates $(1, 1, 1)$ to a point Q with coordinates $(-1, -1, -1)$.

8.3 Use vector methods to prove that the diagonals of a rhombus are perpendicular.

8.4 Use vector methods to prove that any two medians of a triangle intersect at a point two-thirds of the way along either median from the vertex through which it passes to the opposite side.

8.5 Find the parametric equations of the lines through the following pairs of points:
(a) $(2, -1, 3)$ $(-1, 5, 6)$
(b) $(-5, 2, 10)$ $(3, 4, 2)$
(c) $(0, 0, 0)$ $(6, 1, 7)$

8.6 Show that the point $(3/2, 1, -2)$ is on the line through the points $(1, -5, 6)$ and $(2, 7, -10)$.

8.7 Let ℓ be a line through a point B parallel to a vector $\overrightarrow{OA}$. Show that the vector equation of ℓ is $\overrightarrow{OP} = \overrightarrow{OB} + k \cdot \overrightarrow{OA}$, where P is any point on ℓ.

8.8 Derive the parametric equations of a plane π through the noncollinear points A, B, and C from the vector equation of π given in Example 5.

8.9 Find the parametric equations of the planes π through the following noncollinear points:
(a) $(3, 1, 7)$, $(2, -1, 4)$, $(-6, 4, -2)$
(b) $(1/2, 3, 1/4)$, $(2/3, 0, 1/3)$, $(0, 0, 1)$
(c) $(2, 3, 5)$, $(-6, \sqrt{7}, 1)$, $(\sqrt{11}, 4, 0)$

8.10 Let A be a fixed point in space. (a) Show that the set of all points P such that $(\overrightarrow{OP} - \overrightarrow{OA}) \circ \overrightarrow{OA} = 0$ is a plane. (b) Show that the set of all points P such that $(\overrightarrow{OP} - \overrightarrow{OA}) \circ \overrightarrow{OP} = 0$ is a sphere.

8.11 In Exercise 8.10, let A have coordinates (a, b, c) and P have coordinates (x, y, z). Derive equations of the form $f(x, y, z, a, b, c) = 0$ from the vector equations of the plane and the sphere given in Exercise 8.10.

8.12 Let ℓ_1 be a line through a point P with coordinates $(3, 2, 1)$ and a point Q with coordinates $(-1, -2, -3)$. Let ℓ_2 be a line through P and a point R with coordinates $(1, 1, 0)$. Find the cosine of the acute angle between ℓ_1 and ℓ_2.

8.13 Let π be a plane through the origin O and the points $(-3, 7, 2)$ and $(6, -2, 1)$. Find the parametric equations of a line ℓ through O perpendicular to π.

8.14 Find the volumes of the parallelepipeds determined by the following vectors (see Figure 43):
(a) $\overrightarrow{OA} = 3 \cdot \mathbf{J}$, $\overrightarrow{OB} = 5 \cdot \mathbf{I} + 2 \cdot \mathbf{J}$, $\overrightarrow{OC} = \mathbf{I} + \mathbf{J} + \mathbf{K}$
(b) $\overrightarrow{OA} = 2 \cdot \mathbf{I} - 3 \cdot \mathbf{J} + 2 \cdot \mathbf{K}$, $\overrightarrow{OB} = \mathbf{I} - 3 \cdot \mathbf{K}$, $\overrightarrow{OC} = \mathbf{J} + 5 \cdot \mathbf{K}$

2

real vector spaces

9 real vector spaces and subspaces

The set of geometric vectors in three-dimensional space with addition, scalar multiplication, inner product, and cross product as defined in Chapter 1 will be denoted by $\mathcal{E}_3$. In our future work with vectors in $\mathcal{E}_3$ it will seldom be necessary to indicate the terminal point P of a vector $\overrightarrow{OP}$. Therefore we will denote vectors by boldface capital letters **U**, **V**, **W**, . . . in place of the more cumbersome notation $\overrightarrow{OP}$. We have already adopted this notation for the unit vectors **I**, **J**, and **K** associated with a rectangular Cartesian coordinate system. The zero vector in $\mathcal{E}_3$ will be denoted by **0**.

In Chapter 1, rules for addition and scalar multiplication were given for the vectors in $\mathcal{E}_3$. If **U** and **V** are any two vectors in $\mathcal{E}_3$, then $\mathbf{U} + \mathbf{V}$ is a unique vector in $\mathcal{E}_3$. Also, if r is any real number and **U** is any vector in $\mathcal{E}_3$, then $r \cdot \mathbf{U}$ is again a unique vector in $\mathcal{E}_3$. We showed that the operations of addition and scalar multiplication of vectors satisfy certain algebraic identities (see Theorems 2.2 and 3.2). These results are summarized in the following theorem, using our new notation for vectors.

(9.1) theorem The vectors in $\mathcal{E}_3$ satisfy the following conditions:

(a) If **U** and **V** are in $\mathcal{E}_3$, then $\mathbf{U} + \mathbf{V}$ is a unique vector in $\mathcal{E}_3$;

(b) $(\mathbf{U} + \mathbf{V}) + \mathbf{W} = \mathbf{U} + (\mathbf{V} + \mathbf{W})$ for all **U**, **V**, **W** in $\mathcal{E}_3$;

(c) $\mathbf{U} + \mathbf{V} = \mathbf{V} + \mathbf{U}$ for all **U**, **V** in $\mathcal{E}_3$;

(d) there is a vector **0** in $\mathcal{E}_3$ such that $\mathbf{U} + \mathbf{0} = \mathbf{0} + \mathbf{U} = \mathbf{U}$ for every **U** in $\mathcal{E}_3$;

(e) if **U** is in $\mathcal{E}_3$, then there is a vector $-\mathbf{U}$ in $\mathcal{E}_3$ such that $\mathbf{U} + (-\mathbf{U}) = (-\mathbf{U}) + \mathbf{U} = \mathbf{0}$;

(f) if **U** is in $\mathcal{E}_3$ and r is any real number, then $r \cdot \mathbf{U}$ is a unique vector in $\mathcal{E}_3$;

(g) $c \cdot (d \cdot \mathbf{U}) = d \cdot (c \cdot \mathbf{U}) = (cd) \cdot \mathbf{U}$ for all real numbers c, d and vectors **U** in $\mathcal{E}_3$;

(h) $(c + d) \cdot \mathbf{U} = c \cdot \mathbf{U} + d \cdot \mathbf{U}$ for all real numbers c, d and vectors **U** in $\mathcal{E}_3$;

(i) $c \cdot (\mathbf{U} + \mathbf{V}) = c \cdot \mathbf{U} + c \cdot \mathbf{V}$ for all real numbers c and vectors **U**, **V** in $\mathcal{E}_3$;

(j) $1 \cdot \mathbf{U} = \mathbf{U}$ for every vector **U** in $\mathcal{E}_3$.

The algebraic properties of $\mathcal{E}_3$ given in Theorem 9.1 admit a generalization that is one of the basic notions of modern mathematics.

(9.2) definition Let $\mathcal{V}$ be a set of elements for which operations of addition (+) and scalar multiplication by a real number $(\cdot)$ are defined that satisfy the conditions (a) through (j) of Theorem 9.1 (with $\mathcal{E}_3$ replaced by $\mathcal{V}$). Then $\mathcal{V}$ is called a *real vector space*.

It should be understood that the concept of a real vector space $\mathcal{V}$ is *abstract* in the following sense. The elements of $\mathcal{V}$ can be any objects whatsoever and the rules of addition and scalar multiplication can be any rules of combination as long as $\mathcal{V}$ together with these operations satisfies the conditions (a) through (j) of Theorem 9.1. Nothing else is assumed about the elements of $\mathcal{V}$ or the operations. Of course, in specific examples of real vector spaces, such as $\mathcal{E}_3$, the elements and the operations may have additional properties.

The elements of any real vector space are called *vectors*. The vector **0**, whose existence is asserted in 9.1(d) is called the *zero vector*. By 9.1(e), for every **U** in $\mathcal{V}$, there is a vector $-\mathbf{U}$ such that $\mathbf{U} + (-\mathbf{U}) = \mathbf{0}$. The vector $-\mathbf{U}$ is called the *negative* of **U**. *Subtraction* is defined in $\mathcal{V}$ by $\mathbf{U} - \mathbf{V} = \mathbf{U} + (-\mathbf{V})$.

EXAMPLE 1 The set of all polynomials in a variable x with real coefficients, together with ordinary polynomial addition and multiplication by a real number, is a real vector space. Denote this set of polynomials by $R[x]$. Let

$$f(x) = a_n x^n + a_{n-1} x^{n-1} + \cdots + a_1 x + a_0$$

and

$$g(x) = b_m x^m + b_{m-1} x^{m-1} + \cdots + b_1 x + b_0$$

be in $R[x]$. Suppose that $n \geq m$. Then

$$f(x) + g(x) = a_n x^n + \cdots + a_{m+1} x^{m+1} + (a_m + b_m) x^m + \cdots + (a_1 + b_1) x + (a_0 + b_0)$$

is in $R[x]$. If r is any real number, then

$$r \cdot f(x) = (ra_n) x^n + (ra_{n-1}) x^{n-1} + \cdots + (ra_1) x + ra_0$$

is in $R[x]$.

The zero polynomial is the constant polynomial a_0, where $a_0 = 0$. The negative of the polynomial $f(x)$ is

$$(-1) \cdot f(x) = (-a_n)x^n + (-a_{n-1})x^{n-1} + \cdots + (-a_1)x + (-a_0).$$

Thus $R[x]$ satisfies (a), (f), (d), and (e) of Theorem 9.1 (with $\mathcal{E}_3$ replaced by $R[x]$). The fact that $R[x]$ has the rest of the properties listed in 9.1 should be well known to the reader from elementary algebra.

EXAMPLE 2 Let $\mathfrak{F} = \{f, g, \ldots\}$ be the set of all real-valued functions defined at every point of the closed interval $[a, b]$, with addition and scalar multiplication defined by

$$(f + g)(x) = f(x) + g(x), \qquad (c \cdot f)(x) = cf(x)$$

for all x in $[a, b]$. Then $\mathfrak{F}$ is a real vector space. The subset $\mathcal{S}$ of $\mathfrak{F}$, consisting of the continuous functions, is also a real vector space with respect to the same operations of addition and scalar multiplication.

EXAMPLE 3 Let C be the set of all complex numbers. Then if $a + bi$ and $c + di$ are in C and r is any real number, it follows that

$$(a + bi) + (c + di) = (a + c) + (b + d)i,$$

and

$$r \cdot (a + bi) = (ra) + (rb)i.$$

The reader can verify that C is a real vector space.

EXAMPLE 4 Let $\mathcal{V}_n$, for any positive integer n, be the set of all ordered n-tuples $(a_1, a_2, \ldots, a_n)$ of real numbers. Define

$$(a_1, a_2, \ldots, a_n) = (b_1, b_2, \ldots, b_n)$$

if and only if $a_i = b_i$ for $i = 1, 2, \ldots, n$; furthermore, let

$$(a_1, a_2, \ldots, a_n) + (b_1, b_2, \ldots, b_n) = (a_1 + b_1, a_2 + b_2, \ldots, a_n + b_n)$$
$$r \cdot (a_1, a_2, \ldots, a_n) = (ra_1, ra_2, \ldots, ra_n)$$

where r is any real number. Note that the definitions of addition and scalar multiplication in $\mathcal{V}_n$ are analogous to the descriptions of these operations in $\mathcal{E}_3$, when the vectors in $\mathcal{E}_3$ are given by coordinates. It is easy to verify that $\mathcal{V}_n$ satisfies (a) through (j) of Theorem 9.1.

The real vector spaces $\mathcal{V}_n$ for each positive integer n, which are described in Example 4, are of considerable importance, because we will discover that they serve as models for an important class of real vector spaces. For any vector $(a_1, a_2, \ldots, a_n)$ in $\mathcal{V}_n$, the real numbers $a_1, a_2, \ldots, a_n$ are called the *components* of the vector; a_i is called the ith *component*.

Henceforth, we will refer to a real vector space simply as a *vector space*. The following theorem lists some useful properties of addition and scalar multiplication in a vector space that follow from the basic assumptions.

(9.3) theorem Let $\mathcal{V}$ be a vector space. Let **U** and **V** be any vectors in $\mathcal{V}$, and let r be any real number.

(a) There is one and only one vector **X** in $\mathcal{V}$ such that $\mathbf{U} + \mathbf{X} = \mathbf{V}$.

(b) $r \cdot \mathbf{0} = \mathbf{0}$ and $0 \cdot \mathbf{U} = \mathbf{0}$.

(c) $r \cdot (-\mathbf{U}) = (-r) \cdot \mathbf{U} = -(r \cdot \mathbf{U})$.

(d) If $r \cdot \mathbf{U} = \mathbf{0}$, then either $r = 0$ or $\mathbf{U} = \mathbf{0}$.

PROOF (a) If **X** is a vector such that $\mathbf{U} + \mathbf{X} = \mathbf{V}$, then, by 9.1(d), (e), and (b),

$$\mathbf{X} = \mathbf{0} + \mathbf{X} = (-\mathbf{U} + \mathbf{U}) + \mathbf{X} = -\mathbf{U} + (\mathbf{U} + \mathbf{X}) = -\mathbf{U} + \mathbf{V}.$$

Moreover, by direct substitution, $\mathbf{U} + (-\mathbf{U} + \mathbf{V}) = \mathbf{V}$. Thus, $\mathbf{X} = -\mathbf{U} + \mathbf{V}$ is the unique solution of the equation $\mathbf{U} + \mathbf{X} = \mathbf{V}$.

(b) By 9.1(d) and (i), $r \cdot \mathbf{0} = r \cdot (\mathbf{0} + \mathbf{0}) = r \cdot \mathbf{0} + r \cdot \mathbf{0}$. Also, $r \cdot \mathbf{0} = r \cdot \mathbf{0} + \mathbf{0}$. By part (a), the equation $r \cdot \mathbf{0} = r \cdot \mathbf{0} + \mathbf{X}$ has a unique solution. Hence $r \cdot \mathbf{0} = \mathbf{0}$. By 9.1(h),

$$0 \cdot \mathbf{U} = (0 + 0) \cdot \mathbf{U} = 0 \cdot \mathbf{U} + 0 \cdot \mathbf{U}.$$

Again using part (a), $0 \cdot \mathbf{U} = \mathbf{0}$.

(c) $r \cdot \mathbf{U} + r \cdot (-\mathbf{U}) = r \cdot [\mathbf{U} + (-\mathbf{U})] = r \cdot \mathbf{0} = \mathbf{0}$ by 9.1(e) and (i) and part (b) above. Moreover, $r \cdot \mathbf{U} + [-(r \cdot \mathbf{U})] = \mathbf{0}$. Therefore, $r \cdot (-\mathbf{U}) = -(r \cdot \mathbf{U})$ by (a). A similar argument shows that $(-r) \cdot \mathbf{U} = -(r \cdot \mathbf{U})$.

(d) Suppose that $r \cdot \mathbf{U} = \mathbf{0}$ and $r \neq 0$. Then

$$\mathbf{U} = 1 \cdot \mathbf{U} = \frac{1}{r} \cdot (r \cdot \mathbf{U}) = \frac{1}{r} \cdot \mathbf{0} = \mathbf{0}$$

by 9.1(j) and (g) and by part (b). ∎

If $\mathcal{V}$ is a vector space, there are, in general, many subsets of $\mathcal{V}$ that are also vector spaces with respect to the operations of addition and scalar multiplication defined in $\mathcal{V}$. We have already remarked in Example 2 that the set $\mathcal{S}$ of real-valued continuous functions defined on a closed interval $[a, b]$ is a vector space with respect to the operations given for the set $\mathcal{F}$ of all real-valued functions defined on $[a, b]$. As another example, consider the set $\mathcal{S}$ of vectors in $\mathcal{V}_5$ such that the third component is the sum of the first and fifth components. That is, $\mathbf{A} = (a_1, a_2, a_3, a_4, a_5)$ is in $\mathcal{S}$ if and only if $a_3 = a_1 + a_5$. Suppose that **A** and $\mathbf{B} = (b_1, b_2, b_3, b_4, b_5)$ are in $\mathcal{S}$. Then

$$\mathbf{A} + \mathbf{B} = (a_1 + b_1, a_2 + b_2, a_3 + b_3, a_4 + b_4, a_5 + b_5)$$

and

$$a_3 + b_3 = (a_1 + a_5) + (b_1 + b_5) = (a_1 + b_1) + (a_5 + b_5).$$

Hence, $\mathbf{A} + \mathbf{B}$ is in $\mathcal{S}$, and $\mathcal{S}$ satisfies (a) of Theorem 9.1. If r is any real number, then

$$r \cdot \mathbf{A} = (ra_1, ra_2, ra_3, ra_4, ra_5)$$

and

$$ra_3 = r(a_1 + a_5) = ra_1 + ra_5.$$

Thus, $r \cdot \mathbf{A}$ is in $\mathcal{S}$, so that $\mathcal{S}$ satisfies 9.1(f). Taking $r = 0$ and $r = -1$, we find that $\mathbf{0} = (0, 0, 0, 0, 0)$ is in $\mathcal{S}$ and $-\mathbf{A} = (-a_1, -a_2, -a_3, -a_4, -a_5)$ is in $\mathcal{S}$. Hence, (d) and (e) of Theorem 9.1 are satisfied. The other parts of Theorem 9.1 are automatically satisfied by the vectors in $\mathcal{S}$, since they hold for all vectors in $\mathcal{V}_5$ and $\mathcal{S}$ is a subset of $\mathcal{V}_5$. Thus, $\mathcal{S}$ is a vector space. These examples suggest the following definition.

(9.4) definition A nonempty subset $\mathcal{S}$ of a vector space $\mathcal{V}$ that is a vector space with respect to the operations of addition and scalar multiplication defined in $\mathcal{V}$ is called a *subspace* of $\mathcal{V}$.

(9.5) theorem Let $\mathcal{S}$ be a nonempty subset of a vector space $\mathcal{V}$ such that: (a) If $\mathbf{A}$ and $\mathbf{B}$ are in $\mathcal{S}$, then $\mathbf{A} + \mathbf{B}$ is in $\mathcal{S}$; and (b) if $\mathbf{A}$ is in $\mathcal{S}$ and r is any real number, then $r \cdot \mathbf{A}$ is in $\mathcal{S}$. Then $\mathcal{S}$ is a subspace of $\mathcal{V}$.

PROOF Conditions (a) and (b) of the theorem are just the properties (a) and (f) of Theorem 9.1. As in the example of the subspace of $\mathcal{V}_5$ discussed above, the laws (b), (c), (g), (h), (i), and (j) of Theorem 9.1 are satisfied by the vectors in $\mathcal{S}$, since they are satisfied by all vectors in $\mathcal{V}$ and the operations in $\mathcal{S}$ are those of $\mathcal{V}$. Since $\mathcal{S}$ is nonempty, there is a vector $\mathbf{U}$ in $\mathcal{S}$. Hence, $0 \cdot \mathbf{U}$ is in $\mathcal{S}$. By Theorem 9.3(b), $0 \cdot \mathbf{U} = \mathbf{0}$. Thus, the zero vector of $\mathcal{V}$ is in $\mathcal{S}$, so that 9.1(d) is satisfied. Moreover, for every vector $\mathbf{U}$ in $\mathcal{S}$, $(-1) \cdot \mathbf{U}$ is in $\mathcal{S}$. It follows from Theorem 9.3(c) that $(-1) \cdot \mathbf{U} = -(1 \cdot \mathbf{U}) = -\mathbf{U}$ is in $\mathcal{S}$. Therefore, $\mathcal{S}$ satisfies 9.1(e), which completes the proof of the theorem. ▮

We observe that the set containing only the vector $\mathbf{0}$ is a subspace of $\mathcal{V}$. This subspace is called the *zero subspace* and is denoted by $\{\mathbf{0}\}$. By Definition 9.4, a vector space $\mathcal{V}$ is a subspace of itself. If $\mathcal{S}$ is a subspace of $\mathcal{V}$ that does not contain every vector in $\mathcal{V}$, then $\mathcal{S}$ is called a *proper subspace*.

EXAMPLE 5 Consider the vector space $R[x]$ of Example 1. Let $\mathcal{S}$ be the set of polynomials in $R[x]$ of degree at most k, where $k \geq 0$, including the zero polynomial (the zero polynomial has no degree and a polynomial has degree zero if and only if it is a constant polynomial $a_0 \neq 0$). Clearly, $\mathcal{S}$ is nonempty. Since the sum of two polynomials of degree at most k is either the zero polynomial or is a polynomial of degree at most k, condition (a) of Theorem 9.5 is satisfied. Also $r \cdot f(x)$, where r is a real number, is either 0 or is a polynomial with the same degree as $f(x)$. Hence, condition (b) of Theorem 9.5 is also satisfied. Therefore, $\mathcal{S}$ is a subspace of $R[x]$.

EXAMPLE 6 Let $\mathbf{U}$ be any nonzero vector in $\mathcal{E}_3$. Let $\mathcal{S}$ be the set of all vectors in $\mathcal{E}_3$ that are collinear with $\mathbf{U}$. Since $\mathbf{U}$ is in $\mathcal{S}$, $\mathcal{S}$ is nonempty. If $\mathbf{V}$ is any vector in $\mathcal{E}_3$ that is collinear with $\mathbf{U}$, then by Theorem 3.3, $\mathbf{V} = r \cdot \mathbf{U}$ for some real number r. Let $\mathbf{V}$ and $\mathbf{W}$ be vectors in $\mathcal{S}$. Then $\mathbf{U}$ and $\mathbf{V}$ are collinear and **U** and **W** are collinear. Thus, $\mathbf{V} = r \cdot \mathbf{U}$ and $\mathbf{W} = s \cdot \mathbf{U}$ for real numbers r and s. Hence,

$$\mathbf{V} + \mathbf{W} = r \cdot \mathbf{U} + s \cdot \mathbf{U} = (r + s) \cdot \mathbf{U}.$$

Therefore, $\mathbf{V} + \mathbf{W}$ and **U** are collinear; that is, $\mathbf{V} + \mathbf{W}$ is in $\mathcal{S}$, and Theorem 9.5(a) is satisfied. If t is any real number, then $t \cdot \mathbf{V} = t \cdot (r \cdot \mathbf{U}) = (tr) \cdot \mathbf{U}$. Thus, $t \cdot \mathbf{V}$ and **U** are collinear, so that $t \cdot \mathbf{V}$ is in $\mathcal{S}$. Therefore, Theorem 9.5(b) is satisfied, and $\mathcal{S}$ is a subspace of $\mathcal{E}_3$.

EXAMPLE 7 Let π be a plane through the origin O in three-dimensional space. Denote by $\mathcal{E}_\pi$ the set of all vectors in $\mathcal{E}_3$ that lie in the plane π. Suppose that **U** and **V** are in $\mathcal{E}_\pi$. It follows from Theorem 3.5 that $r \cdot \mathbf{U} + s \cdot \mathbf{V}$, for any real numbers r and s, is coplanar with **U** and **V**. That is, $r \cdot \mathbf{U} + s \cdot \mathbf{V}$ lies in the plane π, or in other words $r \cdot \mathbf{U} + s \cdot \mathbf{V}$ is in the set $\mathcal{E}_\pi$. In particular, $\mathbf{U} + \mathbf{V} = 1 \cdot \mathbf{U} + 1 \cdot \mathbf{V}$ is in $\mathcal{E}_\pi$, and $r \cdot \mathbf{U} = r \cdot \mathbf{U} + 0 \cdot \mathbf{V}$ is in $\mathcal{E}_\pi$, for any real number r. Therefore, by Theorem 9.5 $\mathcal{E}_\pi$ is a subspace of $\mathcal{E}_3$. If, for example, π is the XY plane in a given coordinate system, then $\mathcal{E}_\pi$ consists of all vectors **U** in $\mathcal{E}_3$ with coordinates $(x, y, 0)$. That is, if **U** is in $\mathcal{E}_\pi$, then $\mathbf{U} = x \cdot \mathbf{I} + y \cdot \mathbf{J}$. We will denote this particular subspace of $\mathcal{E}_3$, consisting of all vectors in the XY plane, by $\mathcal{E}_2$.

Let $\mathcal{E}_\pi$ be the subspace of $\mathcal{E}_3$ discussed in Example 7. Suppose that $S = \{\mathbf{U}_1, \mathbf{U}_2, \mathbf{U}_3\}$ is a set of vectors in $\mathcal{E}_\pi$ such that $\mathbf{U}_1$ and $\mathbf{U}_2$ are noncollinear and $\mathbf{U}_3 = 2 \cdot \mathbf{U}_1 + 3 \cdot \mathbf{U}_2$. If **V** is any vector in $\mathcal{E}_\pi$, it follows from Theorem 3.5 that $\mathbf{V} = r \cdot \mathbf{U}_1 + s \cdot \mathbf{U}_2$ for some real numbers r and s. Let t be any real number. Then

$$\begin{aligned}
&(r - 2t) \cdot \mathbf{U}_1 + (s - 3t) \cdot \mathbf{U}_2 + t \cdot \mathbf{U}_3 \\
&\quad = (r - 2t) \cdot \mathbf{U}_1 + (s - 3t) \cdot \mathbf{U}_2 + t \cdot (2 \cdot \mathbf{U}_1 + 3 \cdot \mathbf{U}_2) \\
&\quad = r \cdot \mathbf{U}_1 - (2t) \cdot \mathbf{U}_1 + s \cdot \mathbf{U}_2 - (3t) \cdot \mathbf{U}_2 + (2t) \cdot \mathbf{U}_1 + (3t) \cdot \mathbf{U}_2 \\
&\quad = r \cdot \mathbf{U}_1 + s \cdot \mathbf{U}_2 = \mathbf{V}.
\end{aligned}$$

Thus, each vector **V** in $\mathcal{E}_\pi$ can be expressed as a linear combination of the vectors in the set $S = \{\mathbf{U}_1, \mathbf{U}_2, \mathbf{U}_3\}$. Since t is an arbitrary real number, **V** can be written in infinitely many different ways as a linear combination of the vectors in S.

In $\mathcal{E}_3$, each vector $\mathbf{V} = x \cdot \mathbf{I} + y \cdot \mathbf{J} + z \cdot \mathbf{K}$ is expressed as a linear combination of the vectors in the coordinate system $\{\mathbf{I}, \mathbf{J}, \mathbf{K}\}$. In this case, it follows from Theorem 6.2, that a vector $\mathbf{V} \in \mathcal{E}_3$ can be written in only one way as a linear combination of the vectors in the set $\{\mathbf{I}, \mathbf{J}, \mathbf{K}\}$.

The two examples discussed above suggest the following definition.

(9.6) definition Let $S = \{\mathbf{U}_1, \mathbf{U}_2, \ldots, \mathbf{U}_n\}$ be a set of vectors contained in a subspace $\mathcal{S}$ of a vector space $\mathcal{V}$. The subspace $\mathcal{S}$ *is spanned by the set* S if every vector $\mathbf{V} \in \mathcal{S}$ can be expressed as a linear combination of the vectors in S. That is,

$$\mathbf{V} = r_1 \cdot \mathbf{U}_1 + r_2 \cdot \mathbf{U}_2 + \cdots + r_n \cdot \mathbf{U}_n$$

for real numbers $r_1, r_2, \ldots, r_n$.

In the examples preceding Definition 9.6, $\mathcal{E}_\pi$ is spanned by the set $S = \{\mathbf{U}_1, \mathbf{U}_2, \mathbf{U}_3\}$, and $\mathcal{E}_3$ is spanned by the set $\{\mathbf{I}, \mathbf{J}, \mathbf{K}\}$. In Example 6, the subspace $\mathcal{S}$ of $\mathcal{E}_3$ is spanned by the single vector $\mathbf{U}$.

Suppose that the subspace $\mathcal{S}$ of a vector space $\mathcal{V}$ is spanned by the set $S = \{\mathbf{U}_1, \mathbf{U}_2, \ldots, \mathbf{U}_n\}$, and that $\mathcal{T}$ is any subspace of $\mathcal{V}$ which contains S. We can show that $\mathcal{S}$ is contained in $\mathcal{T}$. Suppose $\mathcal{T}$ contains $\mathbf{U}_1, \mathbf{U}_2, \ldots, \mathbf{U}_n$; then it is a consequence of Definition 9.4 that $\mathcal{T}$ contains every linear combination $r_1 \cdot \mathbf{U}_1 + r_2 \cdot \mathbf{U}_2 + \cdots + r_n \cdot \mathbf{U}_n$ of the vectors $\mathbf{U}_1, \mathbf{U}_2, \ldots, \mathbf{U}_n$ (see Exercise 9.13). Since each vector in $\mathcal{S}$ is a linear combination of the vectors $\mathbf{U}_1, \mathbf{U}_2, \ldots, \mathbf{U}_n$, it follows that $\mathcal{S}$ is contained in $\mathcal{T}$. Therefore, $\mathcal{S}$ is the "smallest" subspace of $\mathcal{V}$ that contains the set S.

If one begins with a finite set of n vectors, $S = \{\mathbf{U}_1, \mathbf{U}_2, \ldots, \mathbf{U}_n\}$ in a vector space $\mathcal{V}$, then the set $\mathcal{S}$ of all linear combinations, $r_1 \cdot \mathbf{U}_1 + r_2 \cdot \mathbf{U}_2 + \cdots + r_n \cdot \mathbf{U}_n$, of the vectors $\mathbf{U}_1, \mathbf{U}_2, \ldots, \mathbf{U}_n$ is a subspace of $\mathcal{V}$. In fact, if $r_1 \cdot \mathbf{U}_1 + r_2 \cdot \mathbf{U}_2 + \cdots + r_n \cdot \mathbf{U}_n$ and $s_1 \cdot \mathbf{U}_1 + s_2 \cdot \mathbf{U}_2 + \cdots + s_n \cdot \mathbf{U}_n$ are any two vectors in $\mathcal{S}$, then using the properties of addition and scalar multiplication in $\mathcal{V}$, we obtain

$$(r_1 \cdot \mathbf{U}_1 + r_2 \cdot \mathbf{U}_2 + \cdots + r_n \cdot \mathbf{U}_n) + (s_1 \cdot \mathbf{U}_1 + s_2 \cdot \mathbf{U}_2 + \cdots + s_n \cdot \mathbf{U}_n)$$
$$= (r_1 + s_1) \cdot \mathbf{U}_1 + (r_2 + s_2) \cdot \mathbf{U}_2 + \cdots + (r_n + s_n) \cdot \mathbf{U}_n.$$

Since $(r_1 + s_1) \cdot \mathbf{U}_1 + (r_2 + s_2) \cdot \mathbf{U}_2 + \cdots + (r_n + s_n) \cdot \mathbf{U}_n$ is also a linear combination of $\mathbf{U}_1, \mathbf{U}_2, \ldots, \mathbf{U}_n$, we have shown that the sum of any two vectors in $\mathcal{S}$ is again a vector in $\mathcal{S}$. Furthermore, if t is any real number, then

$$t \cdot (r_1 \cdot \mathbf{U}_1 + r_2 \cdot \mathbf{U}_2 + \cdots + r_n \cdot \mathbf{U}_n)$$
$$= (tr_1) \cdot \mathbf{U}_1 + (tr_2) \cdot \mathbf{U}_2 + \cdots + (tr_n) \cdot \mathbf{U}_n$$

is a linear combination of $\mathbf{U}_1, \mathbf{U}_2, \ldots, \mathbf{U}_n$. Thus, a scalar multiple of a vector in $\mathcal{S}$ is again a vector in $\mathcal{S}$. By Theorem 9.5, $\mathcal{S}$ is a subspace of $\mathcal{V}$. Since

$$\mathbf{U}_i = 0 \cdot \mathbf{U}_1 + \cdots + 0 \cdot \mathbf{U}_{i-1} + 1 \cdot \mathbf{U}_i + 0 \cdot \mathbf{U}_{i+1} + \cdots + 0 \cdot \mathbf{U}_n$$

is in $\mathcal{S}$ for $i = 1, 2, \ldots, n$, $\mathcal{S}$ contains the set $S = \{\mathbf{U}_1, \mathbf{U}_2, \ldots, \mathbf{U}_n\}$. Moreover, every vector in $\mathcal{S}$ is a linear combination of the vectors in S. Therefore, by definition 9.6, the subspace $\mathcal{S}$ is spanned by S. We have shown that any finite set S of vectors in a vector space $\mathcal{V}$ spans a subspace of $\mathcal{V}$, namely, the subspace $\mathcal{S}$ consisting of all linear combinations of the vectors in S.

EXAMPLE 8 Consider the set of vectors

$$S = \{(2, 1, -3, 5), (4, 1, 1, -1), (1, 0, 2, -3), (12, 5, -11, 19)\}$$

contained in the vector space $\mathcal{V}_4$. Then the subspace $\mathcal{S}$ of $\mathcal{V}_4$ spanned by S consists of all vectors of the form

$$r_1 \cdot (2, 1, -3, 5) + r_2 \cdot (4, 1, 1, -1) + r_3 \cdot (1, 0, 2, -3)$$
$$+ r_4 \cdot (12, 5, -11, 19) = (2r_1 + 4r_2 + r_3 + 12r_4,$$
$$r_1 + r_2 + 5r_4, -3r_1 + r_2 + 2r_3 - 11r_4, 5r_1 - r_2 - 3r_3 + 19r_4)$$

where r_1, r_2, r_3, and r_4 are real numbers. Now, $(4, 1, 1, -1) = 1 \cdot (2, 1, -3, 5) + 2 \cdot (1, 0, 2, -3)$, and $(12, 5, -11, 19) = 5 \cdot (2, 1, -3, 5) + 2 \cdot (1, 0, 2, -3)$. Hence, every vector in $\mathcal{S}$ can be expressed

$$\begin{aligned} r_1 \cdot (2, 1, -3, 5) &+ r_2 \cdot [1 \cdot (2, 1, -3, 5) + 2 \cdot (1, 0, 2, -3)] \\ &+ r_3 \cdot (1, 0, 2, -3) + r_4 \cdot [5 \cdot (2, 1, -3, 5) + 2 \cdot (1, 0, 2, -3)] \\ &= [r_1 + r_2 + 5r_4] \cdot (2, 1, -3, 5) + [2r_2 + r_3 + 2r_4] \cdot (1, 0, 2, -3). \end{aligned}$$

Thus, every vector in $\mathcal{S}$ is a linear combination of $(2, 1, -3, 5)$ and $(1, 0, 2, -3)$. Hence, the set $T = \{(2, 1, -3, 5), (1, 0, 2, -3)\}$ also spans $\mathcal{S}$.

Example 8 suggests the following theorems concerning sets that span a subspace $\mathcal{S}$ of a vector space $\mathcal{V}$.

(9.7) theorem Let $\mathcal{S}$ be a subspace of a vector space $\mathcal{V}$ spanned by $S = \{\mathbf{U}_1, \mathbf{U}_2, \ldots, \mathbf{U}_n\}$. If some $\mathbf{U}_i$ is a linear combination of the remaining vectors in S, then $\mathcal{S}$ is spanned by $\{\mathbf{U}_1, \mathbf{U}_2, \ldots, \mathbf{U}_{i-1}, \mathbf{U}_{i+1}, \ldots, \mathbf{U}_n\}$.

PROOF Each vector $\mathbf{V}$ in $\mathcal{S}$ has the form

$$\mathbf{V} = r_1 \cdot \mathbf{U}_1 + r_2 \cdot \mathbf{U}_2 + \cdots + r_i \cdot \mathbf{U}_i + \cdots + r_n \cdot \mathbf{U}_n.$$

Moreover,

$$\mathbf{U}_i = s_1 \cdot \mathbf{U}_1 + s_2 \cdot \mathbf{U}_2 + \cdots + s_{i-1} \cdot \mathbf{U}_{i-1} + s_{i+1} \cdot \mathbf{U}_{i+1} + \cdots + s_n \cdot \mathbf{U}_n.$$

Therefore,

$$\begin{aligned} \mathbf{V} = (r_1 + r_i s_1) \cdot \mathbf{U}_1 + \cdots + (r_{i-1} + r_i s_{i-1}) \cdot \mathbf{U}_{i-1} \\ + (r_{i+1} + r_i s_{i+1}) \cdot \mathbf{U}_{i+1} + \cdots + (r_n + r_i s_n) \cdot \mathbf{U}_n. \end{aligned}$$

Hence, $\mathcal{S}$ is spanned by $\{\mathbf{U}_1, \mathbf{U}_2, \ldots, \mathbf{U}_{i-1}, \mathbf{U}_{i+1}, \ldots, \mathbf{U}_n\}$. ∎

In Example 8, let $\mathbf{U}_1 = (2, 1, -3, 5)$, $\mathbf{U}_2 = (4, 1, 1, -1)$, $\mathbf{U}_3 = (1, 0, 2, -3)$, and $\mathbf{U}_4 = (12, 5, -11, 19)$. Then in this example $\mathcal{S}$ is the subspace spanned by $S = \{\mathbf{U}_1, \mathbf{U}_2, \mathbf{U}_3, \mathbf{U}_4\}$. Since $\mathbf{U}_2 = 1 \cdot \mathbf{U}_1 + 2 \cdot \mathbf{U}_3 + 0 \cdot \mathbf{U}_4$, it follows that $\mathbf{U}_2$ is a linear combination of the remaining vectors in S. By Theorem 9.7, $\mathcal{S}$ is spanned by $\{\mathbf{U}_1, \mathbf{U}_3, \mathbf{U}_4\}$. However, we also have $\mathbf{U}_4 = 5 \cdot \mathbf{U}_1 + 2 \cdot \mathbf{U}_3$, so using Theorem 9.7 again, $\mathcal{S}$ is spanned by the set $T = \{\mathbf{U}_1, \mathbf{U}_3\}$ as we showed in Example 8.

(9.8) theorem Let $S = \{\mathbf{U}_1, \ldots, \mathbf{U}_m\}$ and $T = \{\mathbf{V}_1, \ldots, \mathbf{V}_n\}$ be subsets of a vector space $\mathcal{V}$. If each $\mathbf{U}_i$, $i = 1, 2, \ldots, m$, is a linear combination of the vectors in T and each $\mathbf{V}_j$, $j = 1, 2, \ldots, n$, is a linear combination of the vectors in S, then S and T span the same subspace of $\mathcal{V}$.

PROOF Let $\mathcal{S}$ and $\mathcal{T}$ be the subspaces of $\mathcal{V}$ spanned by S and T, respectively. Every element of $\mathcal{S}$ is a linear combination of the $\mathbf{U}_i$, $i = 1, 2, \ldots, m$, and each $\mathbf{U}_i$ is a linear combination of the $\mathbf{V}_j$, $j = 1, 2, \ldots, n$. Hence, every element of $\mathcal{S}$ is a linear combination of the $\mathbf{V}_j$. Thus, $\mathcal{S} \subseteq \mathcal{T}$. A similar argument shows that $\mathcal{T} \subseteq \mathcal{S}$. Therefore, $\mathcal{S} = \mathcal{T}$. ∎

The concept of a set S of vectors spanning a subspace $\mathcal{S}$ of a vector space $\mathcal{V}$ can be generalized to include infinite as well as finite sets. We simply say that $\mathcal{S}$ is spanned by S if $\mathcal{S}$ contains S, and every vector $\mathbf{V} \in \mathcal{S}$ can be expressed as a (finite) linear combination of vectors in S. For the subspace $\mathcal{E}_\pi$ of $\mathcal{E}_3$, discussed in Example 7, let S consist of all vectors on a line ℓ through O in the plane π and a vector $\mathbf{U}$ in π not on ℓ. Then S is an infinite set of vectors that spans $\mathcal{E}_\pi$. For if $\mathbf{V} \in \mathcal{E}_\pi$, then $\mathbf{V} = r \cdot \mathbf{U} + s \cdot \mathbf{W}$, where $\mathbf{W}$ is any nonzero vector on ℓ.

As in the earlier case where S was restricted to be a finite set, a subspace $\mathcal{S}$ of $\mathcal{V}$ spanned by S is contained in every subspace $\mathcal{T}$ of $\mathcal{V}$ that contains S.

If S is any nonempty subset of a vector space $\mathcal{V}$, then the collection of all (finite) linear combinations of vectors in S is a subspace $\mathcal{S}$ of $\mathcal{V}$ (see Exercise 9.14). It follows as before that $\mathcal{S}$ is spanned by S.

EXAMPLE 9 Let $\mathcal{V}$ be the vector space $R[x]$ of Example 1. Let $S = \{1, x^2, x^4, \ldots, x^{2n}, \ldots\}$ be the infinite subset of $R[x]$ containing 1 and all even powers of x. The subspace $\mathcal{S}$ spanned by S consists of all polynomials in $R[x]$ which involve only even powers of x. For example, $2 + 3x^2 + 5x^6$, $\sqrt{5} - \sqrt{3}\,x^4$, $x^{10} + x^{22}$, and 10 are elements in $\mathcal{S}$. Note that $\mathcal{S}$ is not spanned by any finite subset of $R[x]$ and, in fact, $\mathcal{S}$ is not spanned by any proper subset of S.

Let $\mathcal{S}$ and $\mathcal{T}$ be subspaces of a vector space $\mathcal{V}$. Other subspaces of $\mathcal{V}$ can be obtained from these given subspaces. The ordinary set intersection, $\mathcal{S} \cap \mathcal{T}$, of $\mathcal{S}$ and $\mathcal{T}$ contains the vectors that are in both $\mathcal{S}$ and $\mathcal{T}$. The reader can easily show that $\mathcal{S} \cap \mathcal{T}$ is a subspace of $\mathcal{V}$. The subset of $\mathcal{V}$ consisting of all vectors $\mathbf{U} + \mathbf{V}$, where $\mathbf{U} \in \mathcal{S}$, $\mathbf{V} \in \mathcal{T}$, is called the *sum* of $\mathcal{S}$ and $\mathcal{T}$ and is denoted by $\mathcal{S} + \mathcal{T}$. [It is left as an exercise (Exercise 9.23) to prove that $\mathcal{S} + \mathcal{T}$ is a subspace of $\mathcal{V}$.]

If $\mathcal{V} = \mathcal{S} + \mathcal{T}$, then $\mathcal{V}$ is said to be the *sum of its subspaces* $\mathcal{S}$ *and* $\mathcal{T}$. In the particular case where $\mathcal{V} = \mathcal{S} + \mathcal{T}$ and $\mathcal{S} \cap \mathcal{T} = \{\mathbf{0}\}$, the zero subspace, $\mathcal{V}$ is called the *direct sum* of $\mathcal{S}$ *and* $\mathcal{T}$ and is written $\mathcal{V} = \mathcal{S} \oplus \mathcal{T}$.

For example, if $\mathbf{U}$, $\mathbf{V}$, and $\mathbf{W}$ are noncoplanar vectors in $\mathcal{E}_3$, then $\mathcal{E}_3$ is the sum of the subspace $\mathcal{S}$ spanned by $\mathbf{U}$ and $\mathbf{V}$ and the subspace $\mathcal{T}$ spanned by $\mathbf{U}$ and $\mathbf{W}$, but $\mathcal{E}_3$ is not the direct sum of $\mathcal{S}$ and $\mathcal{T}$. However, $\mathcal{E}_3$ is the direct sum of the subspace $\mathcal{S}$ spanned by $\mathbf{U}$ and $\mathbf{V}$ and the subspace $\mathcal{W}$ spanned by $\mathbf{W}$.

exercises

9.1 Show that the set $\mathcal{F}$ of functions described in Example 2 is a real vector space. Show that the subset of continuous functions is also a real vector space.

9.2 Verify that the set of all complex numbers C is a real vector space with respect to the operations of ordinary addition and multiplication of complex numbers (see Example 3). Show that the set of all real numbers R is a real vector space with respect to ordinary addition and multiplication.

9.3 Let $\mathbf{U} = (-3, 0, \frac{1}{2}, \sqrt{2})$, $\mathbf{V} = (0, \sqrt{3}, -\frac{2}{9}, 1)$, and $\mathbf{W} = (1, 1, -1, -1)$ be vectors in $\mathcal{V}_4$ (see Example 4). Compute the following vectors:
(a) $2 \cdot \mathbf{U} - 3 \cdot (\mathbf{V} - \frac{1}{2} \cdot \mathbf{W})$
(b) $\sqrt{3} \cdot \mathbf{U} + \sqrt{2} \cdot \mathbf{V} + \sqrt{6} \cdot \mathbf{W}$
(c) $\mathbf{U} - \mathbf{V} - \mathbf{W}$

9.4 Let $\mathfrak{F}_\infty$ be the set of all real-valued functions defined for every real number x, with addition and scalar multiplication defined as in Example 2. Show that $\mathfrak{F}_\infty$ is a real vector space. Which of the following functions f, defined by the given expressions for $f(x)$, are in $\mathfrak{F}_\infty$? (a) $f(x) = e^{-x}$; (b) $f(x) = 1/x$; (c) $f(x) = \sin x$; (d) $f(x) = \log_{10} x$; (e) $f(x) = 1/x^4 + x^2 + 1$; (f) $f(x) = 2 + x^3 - x^7$; (g) $f(x) = \sqrt{x^2 + 1}$; (h) $f(x) = \sqrt{x - 1}$.

9.5 Let f, g, h, and k be the vectors (functions) in $\mathfrak{F}_\infty$ (see Exercise 9.4) defined by $f(x) = e^x$, $g(x) = e^{-x}$, $h(x) = x^3 + x + 1$, and $k(x) = 2 - x + x^3$, respectively. Write the expressions for $u(x)$ where u is the element of $\mathfrak{F}_\infty$ defined by
(a) $u = \frac{1}{2} \cdot (f - g)$
(b) $u = 4 \cdot h - 3 \cdot k$
(c) $u = 2 \cdot [\sqrt{2} \cdot f + \frac{1}{3} \cdot h]$
(d) $u = f + g + h + k$

9.6 For the functions u in $\mathfrak{F}_\infty$ given in (a) through (d) of Exercise 9.5, find $u(0)$, $u(-1)$, and $u(\frac{1}{2})$.

9.7 Solve the vector equation $\mathbf{U} + \mathbf{X} = \mathbf{V}$ for the following pairs of vectors $\mathbf{U}$ and $\mathbf{V}$ in $\mathcal{V}_5$:
(a) $\mathbf{U} = (2, -1, 0, 3, 6)$, $\mathbf{V} = (0, 1, 2, -1, -2)$
(b) $\mathbf{U} = (\frac{1}{2}, 3, \frac{1}{3}, -6, 1)$, $\mathbf{V} = (\frac{1}{2}, 4, -1, \frac{1}{3}, 1)$
(c) $\mathbf{U} = (1, 2, 3, 4, 5)$, $\mathbf{V} = (5, 4, 3, 2, 1)$

9.8 For any positive integer n, show that the set of all vectors in $\mathcal{V}_n$ with first component 0 is a subspace of $\mathcal{V}_n$.

9.9 Which of the following are subspaces of $\mathcal{V}_3$? (a) All vectors (a_1, a_2, a_3) with $a_2 = 6a_3$; (b) all vectors (a_1, a_2, a_3) with $a_1^2 = a_2^2$; (c) all vectors (a_1, a_2, a_3) with $a_2 - a_3 = a_1$; (d) all vectors (a_1, a_2, a_3) with $a_1 + a_2 + a_3 = 1$.

9.10 Which of the following are subspaces of $\mathfrak{F}_\infty$ (see Exercise 9.4)? (a) The set of all real-valued functions defined for every real number in the closed interval $[-2, 2]$; (b) the set of all polynomial functions (a polynomial function is a function f defined by a polynomial $f(x) = a_n x^n + a_{n-1}x^{n-1} + \cdots + a_1 x + a_0$); (c) the set of all functions $f \in \mathfrak{F}_\infty$ such that $f(0) = 1$; (d) the set of all constant functions (a constant function is a function f such that $f(x) = a$ for all x, where a is a real number); (e) the set of all functions f of the form $f(x) = ae^{bx}$ for real numbers a and b; (f) the set of all functions $f \in \mathfrak{F}_\infty$ such that $f(1) = 0$; (g) the set of all differentiable functions in $\mathfrak{F}_\infty$.

9.11 Describe the subspaces of $\mathcal{E}_3$ geometrically.

9.12 Prove that $\mathcal{S}$ is a subspace of a vector space $\mathcal{V}$ if and only if $\mathcal{S}$ is nonempty and $r \cdot \mathbf{A} + s \cdot \mathbf{B}$ is in $\mathcal{S}$ for all $\mathbf{A}$, $\mathbf{B}$ in $\mathcal{S}$ and all real numbers r, s.

9.13 Prove that if $\mathcal{T}$ is a subspace of a vector space $\mathcal{V}$ and $\mathcal{T}$ contains the vectors $\mathbf{U}_1, \mathbf{U}_2, \ldots, \mathbf{U}_n$, then $\mathcal{T}$ contains every linear combination $r_1 \cdot \mathbf{U}_1 + r_2 \cdot \mathbf{U}_2 + \cdots + r_n \cdot \mathbf{U}_n$ of $\mathbf{U}_1, \mathbf{U}_2, \ldots, \mathbf{U}_n$.

***9.14** Let S be any nonempty subset of a vector space $\mathcal{V}$. Prove that the set of all linear combinations of vectors in S is a subspace $\mathcal{S}$ of $\mathcal{V}$. Prove that $\mathcal{S}$ is contained in every subspace of $\mathcal{V}$ that contains the set S.

9.15 Describe geometrically the subspace of $\mathcal{E}_3$ spanned by the following sets of vectors:
(a) $\{\mathbf{I}, \mathbf{J}\}$
(b) $\{\mathbf{K}, \mathbf{J}\}$
(c) $\{\mathbf{I} + \mathbf{J}, \mathbf{K}\}$
(d) $\{\mathbf{I} - \mathbf{J}\}$
(e) $\{\mathbf{I}, \mathbf{J}, \mathbf{K}\}$
(f) $\{\mathbf{K}\}$

9.16 Let $\mathcal{S}$ be the subspace of $\mathcal{V}_4$ spanned by the set $S = \{(2, -1, 3, 0), (5/2, 7, 3/2, -3), (1, 5, 0, -2)\}$. Find a set of two vectors which spans $\mathcal{S}$.

9.17 Find a finite set of vectors in $\mathcal{V}_5$ which spans $\mathcal{V}_5$.

9.18 Find a finite set of polynomials which spans the subspace of $R[x]$ described in Example 5.

9.19 Find a set of polynomials which spans $R[x]$.

9.20 Let $\mathcal{S}$ be the subspace of $\mathcal{V}_5$ consisting of all vectors $(a_1, a_2, a_3, a_4, a_5)$ such that $a_1 = 2a_5$ and $a_3 - a_4 = 0$. Find a set of three vectors in $\mathcal{V}_5$ that spans $\mathcal{S}$.

9.21 Prove the converse of Theorem 9.8.

9.22 Let $\mathcal{S}_1, \mathcal{S}_2, \mathcal{S}_3, \mathcal{S}_4, \mathcal{S}_5$, and $\mathcal{S}_6$ be the subspaces of $\mathcal{E}_3$ described by (a), (b), (c), (d), (e), and (f) of Exercise 9.15, respectively. Describe geometrically the following subspaces of $\mathcal{E}_3$:
(a) $\mathcal{S}_1 \cap \mathcal{S}_2$
(b) $\mathcal{S}_1 + \mathcal{S}_2$
(c) $\mathcal{S}_1 \cap \mathcal{S}_4$
(d) $\mathcal{S}_1 + \mathcal{S}_4$
(e) $\mathcal{S}_1 \cap \mathcal{S}_6$
(f) $\mathcal{S}_1 + \mathcal{S}_6$
(g) $\mathcal{S}_3 \cap \mathcal{S}_6$
(h) $\mathcal{S}_3 + \mathcal{S}_6$
(i) $(\mathcal{S}_1 + \mathcal{S}_4) \cap \mathcal{S}_3$

9.23 Let $\mathcal{S}$ and $\mathcal{T}$ be subspaces of a vector space $\mathcal{V}$. Prove that $\mathcal{S} \cap \mathcal{T}$ and $\mathcal{S} + \mathcal{T}$ are subspaces of $\mathcal{V}$.

***9.24** Let $\mathcal{S}_1, \mathcal{S}_2, \ldots, \mathcal{S}_k$ be subspaces of a vector space $\mathcal{V}$. Prove that the set of all vectors $\mathbf{U}_1 + \mathbf{U}_2 + \cdots + \mathbf{U}_k$, where $\mathbf{U}_i \in \mathcal{S}_i$ for $i = 1, 2, \ldots, k$, is a subspace of $\mathcal{V}$. This subspace is denoted by $\mathcal{S}_1 + \mathcal{S}_2 + \cdots + \mathcal{S}_k$ and is called the *sum of the subspaces* $\mathcal{S}_1, \mathcal{S}_2, \ldots, \mathcal{S}_k$. If

$$\mathcal{S}_i \cap (\mathcal{S}_1 + \mathcal{S}_2 + \cdots + \mathcal{S}_{i-1} + \mathcal{S}_{i+1} + \cdots + \mathcal{S}_k) = \{\mathbf{0}\}$$

for $i = 1, 2, \ldots, k$, then $\mathcal{S}_1 + \mathcal{S}_2 + \cdots + \mathcal{S}_k$ is the *direct sum of the subspaces* $\mathcal{S}_1, \mathcal{S}_2, \ldots, \mathcal{S}_k$, and is written $\mathcal{S}_1 \oplus \mathcal{S}_2 \oplus \cdots \oplus \mathcal{S}_k$.

9.25 Express $\mathcal{V}_4$ as a direct sum: (a) of four nonzero subspaces; (b) of two nonzero subspaces.

9.26 Prove that the set of all vectors in $\mathcal{E}_3$ that are orthogonal to a fixed vector $\mathbf{U} \in \mathcal{E}_3$ is a subspace $\mathcal{S}$ of $\mathcal{E}_3$. Prove that $\mathcal{E}_3$ is the direct sum of $\mathcal{S}$ and the subspace spanned by $\mathbf{U}$.

10 dimension of a vector space

In our study of geometric vectors in Chapter 1, we showed that two vectors $\mathbf{U}$ and $\mathbf{V}$ in $\mathcal{E}_3$ are collinear if and only if there exist real numbers k_1 and k_2, not both zero, such that $k_1 \cdot \mathbf{U} + k_2 \cdot \mathbf{V} = \mathbf{0}$ (see Corollary 3.4). Thus, collinear vectors in $\mathcal{E}_3$ are related or *dependent*. Similarly, by Corollary 3.6, three vectors $\mathbf{U}$, $\mathbf{V}$, and $\mathbf{W}$ in $\mathcal{E}_3$ are coplanar if and only if there exist real numbers k_1, k_2, and k_3, not all zero, such that $k_1 \cdot \mathbf{U} + k_2 \cdot \mathbf{V} + k_3 \cdot \mathbf{W} = \mathbf{0}$. That is, there is a similar dependency relating any three coplanar vectors in $\mathcal{E}_3$. This concept of dependent vectors is valuable not only in the vector space $\mathcal{E}_3$, but is one of the central ideas in our study of real vector spaces.

(10.1) definition Let S be a subset of a vector space $\mathcal{V}$. The subset S is called a *linearly dependent set* if there is a finite subset $\{\mathbf{U}_1, \mathbf{U}_2, \ldots, \mathbf{U}_k\}$ of S such that $r_1 \cdot \mathbf{U}_1 + r_2 \cdot \mathbf{U}_2 + \cdots + r_k \cdot \mathbf{U}_k = \mathbf{0}$, where at least one of the real numbers $r_1, r_2, \ldots, r_k$ is not zero.

It follows from Definition 10.1 that any subset S of $\mathcal{V}$ that contains the zero vector is a linearly dependent set since $k \cdot \mathbf{0} = \mathbf{0}$ for $k \neq 0$. Furthermore, if $T \supseteq S$ and S is a linearly dependent set, then T is a linearly dependent set (see Exercise 10.4). A set S of vectors in $\mathcal{V}$ is *linearly independent* if S is not a linearly dependent set. Thus, S is a linearly independent set in $\mathcal{V}$ if for every finite subset $\{\mathbf{U}_1, \mathbf{U}_2, \ldots, \mathbf{U}_k\}$ of S, $r \cdot \mathbf{U}_1 + r_2 \cdot \mathbf{U}_2 + \cdots + r_k \cdot \mathbf{U}_k = \mathbf{0}$ implies that $r_1 = r_2 = \cdots = r_k = 0$.

EXAMPLE 1 In $\mathcal{E}_3$ linear dependence can, of course, be interpreted geometrically. In fact, according to the introductory paragraph in this section, a set of two vectors $\{\mathbf{V}_1, \mathbf{V}_2\}$ in $\mathcal{E}_3$ is linearly dependent if and only if $\mathbf{V}_1$ and $\mathbf{V}_2$ are collinear, and a set of three vectors $\{\mathbf{V}_1, \mathbf{V}_2, \mathbf{V}_3\}$ in $\mathcal{E}_3$ is linearly dependent if and only if $\mathbf{V}_1$, $\mathbf{V}_2$, and $\mathbf{V}_3$ are coplanar. To complete the study of linear dependence in $\mathcal{E}_3$, we note that: (a) A set $\{\mathbf{V}\}$, which consists of the single vector $\mathbf{V}$, is linearly dependent if and only if $\mathbf{V} = \mathbf{0}$; and (b) any four or more vectors in $\mathcal{E}_3$ form a linearly dependent set.

To prove (a), we observe that by Definition 10.1, $\{\mathbf{V}\}$ is a linearly dependent set if and only if $k \cdot \mathbf{V} = \mathbf{0}$, where $k \neq 0$. By Theorem 9.3 (b) and (d), it follows that $k \cdot \mathbf{V} = \mathbf{0}$ if and only if $\mathbf{V} = \mathbf{0}$.

Suppose that $\{\mathbf{V}_1, \mathbf{V}_2, \mathbf{V}_3, \ldots, \mathbf{V}_m\}$ is a set of vectors in $\mathcal{E}_3$, where $m \geq 4$. If any three of these vectors are coplanar, say $\mathbf{V}_1$, $\mathbf{V}_2$, $\mathbf{V}_3$, then by the above remarks, they are linearly dependent. Thus, $\{\mathbf{V}_1, \mathbf{V}_2, \mathbf{V}_3, \ldots, \mathbf{V}_m\}$ contains a linearly dependent set $\{\mathbf{V}_1, \mathbf{V}_2, \mathbf{V}_3\}$. Therefore, by the remark above, $\{\mathbf{V}_1, \mathbf{V}_2, \mathbf{V}_3, \ldots, \mathbf{V}_m\}$ is a linearly dependent set. If no three of the vectors in the given set are coplanar, then it follows from the result of Exercise 6.5 that $\mathbf{V}_4 = k_1 \cdot \mathbf{V}_1 + k_2 \cdot \mathbf{V}_2 + k_3 \cdot \mathbf{V}_3$. Choose $k_4 = -1$, and $k_5 = \cdots = k_m = 0$ if $m > 4$. Then $k_1 \cdot \mathbf{V}_1 + k_2 \cdot \mathbf{V}_2 + k_3 \cdot \mathbf{V}_3 + k_4 \cdot \mathbf{V}_4 + k_5 \cdot \mathbf{V}_5 + \cdots + k_m \cdot \mathbf{V}_m = \mathbf{0}$, where $k_4 = -1 \neq 0$. Thus, $k_1, k_2, k_3, k_4, \ldots, k_m$ are not all zero, and by Definition 10.1, $\{\mathbf{V}_1, \mathbf{V}_2, \mathbf{V}_3, \ldots, \mathbf{V}_m\}$ is a linearly dependent set. This completes the proof of statement (b).

The result in Example 1 that a set $\{\mathbf{V}\}$ containing a single vector in $\mathcal{E}_3$ is linearly dependent if and only if $\mathbf{V} = \mathbf{0}$ is not peculiar to $\mathcal{E}_3$. Indeed, the result followed from Theorem 9.3 which is valid in any vector space $\mathcal{V}$. As a consequence of this remark, we note that if $\mathbf{V}$ is a nonzero vector in any vector space $\mathcal{V}$, then $\{\mathbf{V}\}$ is a linearly independent set.

EXAMPLE 2 Let $\mathcal{V}$ be the vector space of Example 1, §9. Let S be the set $\{2x, x + 5, 3x - 4\}$ in $\mathcal{V} = R[x]$. Then S is linearly dependent, since

$$19(2x) + (-8)(x + 5) + (-10)(3x - 4) = 0.$$

Let T be the set $\{1, x, x^2, \ldots, x^n, \ldots\}$ consisting of all powers of x. Let $\{x^{n_1}, x^{n_2}, \ldots, x^{n_k}\}$ be any finite subset of T (where $0 \leq n_1 < n_2 < \cdots < n_k$). If

$$a_{n_k}x^{n_k} + a_{n_{k-1}}x^{n_{k-1}} + \cdots + a_{n_2}x^{n_2} + a_{n_1}x^{n_1}$$

is the zero polynomial for real numbers $a_{n_1}, a_{n_2}, \ldots, a_{n_k}$, then $a_{n_1} = a_{n_2} = \cdots = a_{n_k} = 0$. Hence, T is a linearly independent set.

EXAMPLE 3 In $\mathcal{V}_n$, denote the vector whose ith component is equal to 1 and all other components 0 by $\mathbf{E}_i^{(n)}$. That is, $\mathbf{E}_1^{(n)} = (1, 0, 0, \ldots, 0)$, $\mathbf{E}_2^{(n)} = (0, 1, 0, \ldots, 0)$, and so forth. Let $E^{(n)} = \{\mathbf{E}_1^{(n)}, \mathbf{E}_2^{(n)}, \ldots, \mathbf{E}_n^{(n)}\}$. If $r_1, r_2, \ldots, r_n$ are real numbers such that

$$r_1 \cdot \mathbf{E}_1^{(n)} + r_2 \cdot \mathbf{E}_2^{(n)} + \cdots + r_n \cdot \mathbf{E}_n^{(n)} = \mathbf{0},$$

then

$$(r_1, r_2, \ldots, r_n) = (r_1, 0, 0, \ldots, 0) + (0, r_2, 0, \ldots, 0) + \cdots + (0, 0, \ldots, 0, r_n) = (0, 0, \ldots, 0).$$

This implies that $r_1 = r_2 = \cdots = r_n = 0$. Hence, $E^{(n)}$ is a linearly independent set. We also note that the set $E^{(n)}$ spans $\mathcal{V}_n$. Indeed, if $\mathbf{U} = (a_1, a_2, \ldots, a_n)$ is any vector in $\mathcal{V}_n$, then

$$\mathbf{U} = a_1 \cdot \mathbf{E}_1^{(n)} + a_2 \cdot \mathbf{E}_2^{(n)} + \cdots + a_n \cdot \mathbf{E}_n^{(n)}.$$

EXAMPLE 4 Let $f_1, f_2, \ldots, f_k$ be functions that are elements of the vector space $\mathcal{F}_\infty$ consisting of all real-valued functions defined for every real number x (see Exercise 9.4). The zero vector in $\mathfrak{F}_\infty$ is the function O such that $O(x) = 0$ for all x. Thus, the condition

$$r_1 \cdot f_1 + r_2 \cdot f_2 + \cdots + r_k \cdot f_k = O$$

for real numbers $r_1, r_2, \ldots, r_k$, means that

$$(r_1 \cdot f_1 + r_2 \cdot f_2 + \cdots + r_k \cdot f_k)(x) = r_1 f_1(x) + r_2 f_2(x) + \cdots + r_k f_k(x) = 0$$

for all x.

Consider the functions f_1 and f_2, where $f_1(x) = \sin x$ and $f_2(x) = \cos x$. That is, f_1 and f_2 are the ordinary trigonometric sine and cosine functions. Suppose that there are real numbers r_1 and r_2 such that $r_1 \cdot f_1 + r_2 \cdot f_2 = O$. Then $r_1 \sin x + r_2 \cos x = 0$, for all x. Since the sine and cosine are differentiable functions, we obtain $r_1 \cos x - r_2 \sin x = 0$, for all x, by differentiating the latter equation. Multiplying these two equations by r_2 and r_1 respectively, we have

$$r_1 r_2 \sin x + r_2^2 \cos x = 0$$
$$r_1^2 \cos x - r_1 r_2 \sin x = 0$$

for all x. Adding these equations yields

$$(r_1^2 + r_2^2) \cos x = 0$$

for all x. In particular, for $x = 0$, we obtain

$$(r_1^2 + r_2^2) \cos 0 = r_1^2 + r_2^2 = 0$$

which implies $r_1 = r_2 = 0$. Hence, we have shown that the sine and cosine functions are linearly independent in $\mathfrak{F}_\infty$.

The following theorem gives a useful characterization of linear dependence of finite sets.

(10.2) theorem Let $S = \{\mathbf{U}_1, \mathbf{U}_2, \ldots, \mathbf{U}_k\}$ be a subset of a vector space $\mathcal{V}$. Then S is a linearly dependent set if and only if either $\mathbf{U}_1 = \mathbf{0}$ or some $\mathbf{U}_j$ for $j \geq 2$ is contained in the subspace of $\mathcal{V}$ spanned by $\{\mathbf{U}_1, \mathbf{U}_2, \ldots, \mathbf{U}_{j-1}\}$.

PROOF Suppose first that S is a linearly dependent set. If $\mathbf{U}_1 = \mathbf{0}$, there is nothing to prove, since this is one of the alternatives of the conclusion. Thus, it is sufficient to assume that $\mathbf{U}_1 \neq \mathbf{0}$ and show that the other alternative of the conclusion holds. If $\mathbf{U}_1 \neq \mathbf{0}$, then $\{\mathbf{U}_1\}$ is a linearly independent set by the remark following Example 1, above. Therefore, if j is the least positive integer such that $\{\mathbf{U}_1, \mathbf{U}_2, \ldots, \mathbf{U}_j\}$ is linearly dependent, it follows that $2 \leq j \leq k$. By Definition 10.1, there are real numbers $r_1, r_2, \ldots, r_j$, not all zero, such that $r_1 \cdot \mathbf{U}_1 + r_2 \cdot \mathbf{U}_2 + \cdots + r_j \cdot \mathbf{U}_j = \mathbf{0}$. Moreover, $r_j \neq 0$, for if $r_j = 0$, then some $r_i \neq 0$ with $1 \leq i \leq j - 1$ which means $\{\mathbf{U}_1, \mathbf{U}_2, \ldots, \mathbf{U}_{j-1}\}$ is a linearly dependent set, contrary to the choice of j. Since $r_j \neq 0$, we have

$$\mathbf{0} = \frac{1}{r_j} \cdot \mathbf{0} = \frac{1}{r_j} \cdot (r_1 \cdot \mathbf{U}_1 + r_2 \cdot \mathbf{U}_2 + \cdots + r_j \cdot \mathbf{U}_j)$$
$$= \frac{r_1}{r_j} \cdot \mathbf{U}_1 + \frac{r_2}{r_j} \cdot \mathbf{U}_2 + \cdots + \frac{r_{j-1}}{r_j} \cdot \mathbf{U}_{j-1} + \mathbf{U}_j$$

Therefore,

$$\mathbf{U}_j = \frac{-r_1}{r_j} \cdot \mathbf{U}_1 + \frac{-r_2}{r_j} \cdot \mathbf{U}_2 + \cdots + \frac{-r_{j-1}}{r_j} \cdot \mathbf{U}_{j-1}$$

is in the subspace spanned by $\{\mathbf{U}_1, \mathbf{U}_2, \ldots, \mathbf{U}_{j-1}\}$.

Conversely, suppose that $\mathbf{U}_1 = \mathbf{0}$, or $\mathbf{U}_j$ for some $j \geq 2$, is in the subspace spanned by $\{\mathbf{U}_1, \mathbf{U}_2, \ldots, \mathbf{U}_{j-1}\}$. If $\mathbf{U}_1 = \mathbf{0}$, then S contains the zero vector and is a linearly dependent set. If

$$\mathbf{U}_j = r_1 \cdot \mathbf{U}_1 + r_2 \cdot \mathbf{U}_2 + \cdots + r_{j-1} \cdot \mathbf{U}_{j-1},$$

then

$$r_1 \cdot \mathbf{U}_1 + r_2 \cdot \mathbf{U}_2 + \cdots + r_{j-1} \cdot \mathbf{U}_{j-1} + r_j \cdot \mathbf{U}_j = \mathbf{0},$$

with $r_j = -1 \neq 0$. Hence, $\{\mathbf{U}_1, \mathbf{U}_2, \ldots, \mathbf{U}_j\}$ is a linearly dependent set. Since $S \supseteq \{\mathbf{U}_1, \mathbf{U}_2, \ldots, \mathbf{U}_j\}$ it follows that S is linearly dependent. ∎

By Theorem 6.2, the vector space $\mathcal{E}_3$ is spanned by the set of three vectors $\{\mathbf{I}, \mathbf{J}, \mathbf{K}\}$. It follows from our study of linear dependence in $\mathcal{E}_3$ in Example 1, that $\{\mathbf{I}, \mathbf{J}, \mathbf{K}\}$ is a linearly independent set since $\mathbf{I}$, $\mathbf{J}$, and $\mathbf{K}$ are noncoplanar. Moreover, we showed in Example 1 that any four vectors in $\mathcal{E}_3$ are linearly dependent. Thus, $\mathcal{E}_3$ is spanned by three linearly independent vectors, and any four vectors in $\mathcal{E}_3$ are linearly dependent. We will be principally concerned with vector spaces that are spanned by a finite number of linearly independent vectors. The following theorem, which is a generalization of the result in $\mathcal{E}_3$ that we have just mentioned, is fundamental in the study of such vector spaces.

(10.3) theorem Let $\mathcal{V}$ be a vector space that is spanned by a finite set $\{\mathbf{U}_1, \mathbf{U}_2, \ldots, \mathbf{U}_k\}$ of k vectors that is linearly independent. Then any set of $k + 1$ vectors in $\mathcal{V}$ is linearly dependent.

PROOF Assume that there is some set of $k + 1$ vectors $\{\mathbf{V}_1, \mathbf{V}_2, \ldots, \mathbf{V}_k, \mathbf{V}_{k+1}\}$ in $\mathcal{V}$ that is linearly independent. Since $\{\mathbf{U}_1, \mathbf{U}_2, \ldots, \mathbf{U}_k\}$ spans $\mathcal{V}$, it follows that

$$\mathbf{V}_1 = r_1 \cdot \mathbf{U}_1 + r_2 \cdot \mathbf{U}_2 + \cdots + r_k \cdot \mathbf{U}_k.$$

Hence, $\{\mathbf{V}_1, \mathbf{U}_1, \mathbf{U}_2, \ldots, \mathbf{U}_k\}$ is a linearly dependent set. Moreover, $\mathbf{V}_1 \neq \mathbf{0}$ since $\mathbf{V}_1$ is a vector in a linearly independent set. By Theorem 10.2, there is some i such that $\mathbf{U}_i$ is in the subspace spanned by $\{\mathbf{V}_1, \mathbf{U}_1, \ldots, \mathbf{U}_{i-1}\}$ (here i may be equal to 1, in which case $\mathbf{U}_1$ is in the subspace spanned by $\mathbf{V}_1$). Clearly, $\{\mathbf{V}_1, \mathbf{U}_1, \ldots, \mathbf{U}_{i-1}, \mathbf{U}_i, \mathbf{U}_{i+1}, \ldots, \mathbf{U}_k\}$ spans $\mathcal{V}$. Therefore, by Theorem 9.7, $\{\mathbf{V}_1, \mathbf{U}_1, \ldots, \mathbf{U}_{i-1}, \mathbf{U}_{i+1}, \ldots, \mathbf{U}_k\}$

spans $\mathcal{V}$. Repeating the argument above, the set $\{\mathbf{V}_1, \mathbf{V}_2, \mathbf{U}_1, \ldots, \mathbf{U}_{i-1}, \mathbf{U}_{i+1}, \ldots, \mathbf{U}_k\}$ spans $\mathcal{V}$ and is linearly dependent. Since $\{\mathbf{V}_1, \mathbf{V}_2\}$ is linearly independent, again by Theorem 10.2 it follows that some $\mathbf{U}_j$ is a linear combination of preceding vectors. By Theorem 9.7, $\{\mathbf{V}_1, \mathbf{V}_2, \mathbf{U}_1, \ldots, \mathbf{U}_{j-1}, \mathbf{U}_{j+1}, \ldots, \mathbf{U}_{i-1}, \mathbf{U}_{i+1}, \ldots, \mathbf{U}_k\}$ spans $\mathcal{V}$.

This replacement process can be continued. Each time a **V** is added, a **U** can be deleted so that the new set spans $\mathcal{V}$. After k steps all of the **U**'s have been deleted and $\{\mathbf{V}_1, \mathbf{V}_2, \ldots, \mathbf{V}_k\}$ spans $\mathcal{V}$. Then

$$\mathbf{V}_{k+1} = s_1 \cdot \mathbf{V}_1 + s_2 \cdot \mathbf{V}_2 + \cdots + s_k \cdot \mathbf{V}_k$$

so that $\{\mathbf{V}_1, \mathbf{V}_2, \ldots, \mathbf{V}_k, \mathbf{V}_{k+1}\}$ is a linearly dependent set. But this is a contradiction, since it was assumed that this set was linearly independent. Therefore, there is no set of $k + 1$ vectors in $\mathcal{V}$ that is linearly independent. That is, every set of $k + 1$ vectors is linearly dependent. ∎

We observed in the discussion preceding Theorem 10.3 that $\mathcal{E}_3$ is spanned by the linearly independent set $\{\mathbf{I}, \mathbf{J}, \mathbf{K}\}$. Similarly in Example 3 of this section, it was remarked that the set $\{\mathbf{E}_1^{(n)}, \mathbf{E}_2^{(n)}, \ldots, \mathbf{E}_n^{(n)}\}$ is linearly independent and spans $\mathcal{V}_n$. These are examples of the following concept.

(10.4) definition A linearly independent set of vectors that spans a vector space $\mathcal{V}$ is called a *basis* of $\mathcal{V}$.

The discussion preceding Definition 10.4 shows that $\{\mathbf{I}, \mathbf{J}, \mathbf{K}\}$ is a basis of $\mathcal{E}_3$, and that for each positive integer n, $\{\mathbf{E}_1^{(n)}, \mathbf{E}_2^{(n)}, \ldots, \mathbf{E}_n^{(n)}\}$ is a basis of $\mathcal{V}_n$. Referring to Example 4 above, the functions f_1 and f_2, defined by $f_1(x) = \sin x$ and $f_2(x) = \cos x$, form a basis of the subspace $\mathcal{S}$ of $\mathcal{F}_\infty$ spanned by f_1 and f_2. Indeed, by the definition of $\mathcal{S}$, the set $\{f_1, f_2\}$ spans $\mathcal{S}$, and by the result of Example 4, this set is linearly independent.

Although we will not prove it here, every real vector space, except the space consisting of the zero vector alone, possesses a basis.

The following theorem gives a useful characterization of a basis of a vector space.

(10.5) theorem The subset S of a vector space $\mathcal{V}$ is a basis of $\mathcal{V}$ if and only if each vector **V** in $\mathcal{V}$ has a unique expression as a linear combination of elements of S.

PROOF To say that a vector **V** has a unique expression as a linear combination of elements of S amounts to two statements. First,

$$\mathbf{V} = r_1 \cdot \mathbf{U}_{i_1} + r_2 \cdot \mathbf{U}_{i_2} + \cdots + r_k \cdot \mathbf{U}_{i_k} \tag{10-1}$$

where $\mathbf{U}_{i_1}, \mathbf{U}_{i_2}, \ldots, \mathbf{U}_{i_k}$ are distinct vectors in S. Second, if

$$\mathbf{V} = s_1 \cdot \mathbf{U}_{i_1} + s_2 \cdot \mathbf{U}_{i_2} + \cdots + s_k \cdot \mathbf{U}_{i_k} + t_1 \cdot \mathbf{U}_{i_{k+1}} + \cdots + t_l \cdot \mathbf{U}_{i_{k+l}}$$

is any other expression for **V** as a linear combination of elements of S, then $s_j = r_j$ for $j = 1, 2, \ldots, k$, and $t_i = 0$ for $i = 1, 2, \ldots, l$.

If S is a basis of $\mathcal{V}$, then S spans $\mathcal{V}$. Hence, if $\mathbf{V}$ is any vector in $\mathcal{V}$, $\mathbf{V}$ can be expressed as in (10–1). Assume that $\mathbf{V}$ has a different expression as a linear combination of elements in S. Then

$$\begin{aligned}\mathbf{V} = s_1 \cdot \mathbf{U}_{i_1} + s_2 \cdot \mathbf{U}_{i_2} + \cdots + s_k \cdot \mathbf{U}_{i_k} + t_1 \cdot \mathbf{U}_{i_{k+1}} \\ + \cdots + t_l \cdot \mathbf{U}_{i_{k+l}}\end{aligned} \qquad (10\text{–}2)$$

Where either some $s_j \neq r_j$ or some $t_j \neq 0$. Here the vectors $\mathbf{U}_{i_{k+1}}, \ldots, \mathbf{U}_{i_{k+l}}$ are distinct vectors in S and none of them is in the set $\{\mathbf{U}_{i_1}, \mathbf{U}_{i_2}, \ldots, \mathbf{U}_{i_k}\}$. Subtracting (10–1) from (10–2), we obtain

$$\begin{aligned}(s_1 - r_1) \cdot \mathbf{U}_{i_1} + (s_2 - r_2) \cdot \mathbf{U}_{i_2} + \cdots \\ + (s_k - r_k) \cdot \mathbf{U}_{i_k} + t_1 \cdot \mathbf{U}_{i_{k+1}} + \cdots + t_l \cdot \mathbf{U}_{i_{k+l}} = \mathbf{0}.\end{aligned}$$

Since S is a basis of $\mathcal{V}$, S is a linearly independent set. Hence, the set $\{\mathbf{U}_{i_1}, \mathbf{U}_{i_2}, \ldots, \mathbf{U}_{i_k}, \mathbf{U}_{i_{k+1}}, \ldots, \mathbf{U}_{i_{k+l}}\}$ is linearly independent. Therefore, $s_j - r_j = 0$ for $j = 1, 2, \ldots, k$ and $t_j = 0$ for $j = 1, 2, \ldots, l$. This contradicts our assumption concerning the expression (10–2). Thus, (10–1) is the unique expression for $\mathbf{V}$.

Conversely, if each vector in $\mathcal{V}$ has a unique expression as a linear combination of elements in S, then S spans $\mathcal{V}$. We complete the proof by showing that S is a linearly independent set. Let $\{\mathbf{U}_{i_1}, \mathbf{U}_{i_2}, \ldots, \mathbf{U}_{i_k}\}$ be any finite subset of S. Assume that

$$r_1 \cdot \mathbf{U}_{i_1} + r_2 \cdot \mathbf{U}_{i_2} + \cdots + r_k \cdot \mathbf{U}_{i_k} = \mathbf{0}.$$

In particular, the vector $\mathbf{0}$ has a unique expression as a linear combination of elements in S, and, certainly,

$$0 \cdot \mathbf{U}_{i_1} + 0 \cdot \mathbf{U}_{i_2} + \cdots + 0 \cdot \mathbf{U}_{i_k} = \mathbf{0}.$$

Therefore, $r_1 = r_2 = \cdots = r_k = 0$. This proves that S is a linearly independent set (see the discussion following Definition 10.1). ∎

It follows from Theorem 10.3 that if a vector space $\mathcal{V}$ has a finite basis, then any basis of $\mathcal{V}$ is finite and any two bases have the same number of elements. Suppose that $\{\mathbf{U}_1, \mathbf{U}_2, \ldots, \mathbf{U}_k\}$ is a basis of $\mathcal{V}$. Then by Theorem 10.3, any set of $k + 1$ vectors in $\mathcal{V}$ is linearly dependent. Hence, any other basis of $\mathcal{V}$ has h elements, where $h \leq k$. However, if $h < k$, then again by Theorem 10.3, any set of $(h + 1) \leq k$ vectors is linearly dependent. This is a contradiction, since $\{\mathbf{U}_1, \mathbf{U}_2, \ldots, \mathbf{U}_k\}$ is a linearly independent set. Therefore, $h = k$. In particular, every basis of $\mathcal{V}$ is finite. We are now prepared to make the following definition.

(10.6) definition A vector space $\mathcal{V} \neq \{\mathbf{0}\}$ is *finite dimensional* if $\mathcal{V}$ has a finite basis. The *dimension* of a finite dimensional vector space, *dim* $\mathcal{V}$, is the number of elements in a basis. The vector space $\mathcal{V} = \{\mathbf{0}\}$ is finite dimensional with dimension 0.

Thus, $\mathcal{E}_3$ is a finite dimensional vector space of dimension three since $\{\mathbf{I}, \mathbf{J}, \mathbf{K}\}$ is a basis. $\mathcal{E}_2$, the subspace of $\mathcal{E}_3$, consisting of the vectors in the XY plane, has $\{\mathbf{I}, \mathbf{J}\}$ as a basis, so that $\mathcal{E}_2$ has dimension two. Since $\{\mathbf{E}_1^{(n)}, \mathbf{E}_2^{(n)}, \ldots, \mathbf{E}_n^{(n)}\}$ is a basis of $\mathcal{V}_n$, each $\mathcal{V}_n$ is finite dimensional with dimension n. In Example 2 of this section, we showed that the infinite set $T = \{1, x, x^2, \ldots, x^n, \ldots\}$ in $R[x]$, is linearly independent. It is clear that each polynomial in $R[x]$ can be expressed as a linear combination of elements in T. Hence, T spans $R[x]$. Therefore, T is a basis of $R[x]$. By the remarks preceding Definition 10.6, $R[x]$ cannot have a finite basis, so $R[x]$ is not a finite dimensional vector space.

(10.7) theorem Let $\mathcal{V} \neq \{\mathbf{0}\}$ be a vector space spanned by a finite set $T = \{\mathbf{U}_1, \mathbf{U}_2, \ldots, \mathbf{U}_m\}$. Then T contains a subset that is a basis of $\mathcal{V}$. In particular, $\mathcal{V}$ is finite dimensional and has dimension $k \leq m$.

PROOF Since T is finite and $\mathcal{V} \neq \{\mathbf{0}\}$, there exists a maximal independent set $S = \{\mathbf{U}_{i_1}, \mathbf{U}_{i_2}, \ldots, \mathbf{U}_{ik}\}$ with $k \geq 1$, contained in T. That is, S is linearly independent, and any subset of T properly containing S is linearly dependent. If some element $\mathbf{U}_j$ of T is not in S, then $\{\mathbf{U}_j, \mathbf{U}_{i_1}, \ldots, \mathbf{U}_{ik}\}$ is a linearly dependent set. Thus, there are real numbers $r, r_1, r_2, \ldots, r_k$, not all zero, such that

$$r \cdot \mathbf{U}_j + r_1 \cdot \mathbf{U}_{i_1} + r_2 \cdot \mathbf{U}_{i_2} + \cdots + r_k \cdot \mathbf{U}_{ik} = \mathbf{0}.$$

The number r is not zero, for otherwise the linear independence of $\{\mathbf{U}_{i_1}, \mathbf{U}_{i_2}, \ldots, \mathbf{U}_{ik}\}$ is contradicted. Therefore,

$$\mathbf{U}_j = \frac{-r_1}{r} \cdot \mathbf{U}_{i_1} + \frac{-r_2}{r} \cdot \mathbf{U}_{i_2} + \cdots + \frac{-r_k}{r} \cdot \mathbf{U}_{ik}.$$

Hence, every element of T can be written as a linear combination of elements of S. Since T spans $\mathcal{V}$, every element of $\mathcal{V}$ can be written as a linear combination of elements in T. Thus, every element of $\mathcal{V}$ has an expression as a linear combination of elements in S. That is, S spans $\mathcal{V}$. Since S is linearly independent, S is a basis of $\mathcal{V}$. The last statement in the theorem follows from Definition 10.6 and the fact that S is a subset of T. ∎

EXAMPLE 5 The proof of Theorem 10.7 involved selecting a maximal independent set contained in a finite set of (nonzero) vectors in a vector space $\mathcal{V}$. We give an example of how this process can be carried out. Consider the set $T = \{(2, 1, 2), (0, 1, -1), (4, 3, 3), (-2, 0, 1)\}$ of vectors in $\mathcal{V}_3$. To choose a maximal independent set S contained in T, we begin with the vector $(2, 1, 2)$. We next consider the set $\{(2, 1, 2), (0, 1, -1)\}$. Suppose t_1 and t_2 are real numbers such that $t_1 \cdot (2, 1, 2) + t_2 \cdot (0, 1, -1) = (0, 0, 0)$. Then $2t_1 = 0$, $t_1 + t_2 = 0$, and $2t_1 - t_2 = 0$. These equations imply $t_1 = t_2 = 0$, so that the set $\{(2, 1, 2), (0, 1, -1)\}$ is linearly independent. Now consider the set $\{(2, 1, 2), (0, 1, -1), (4, 3, 3)\}$. Since $(4, 3, 3) = 2 \cdot (2, 1, 2) + 1 \cdot (0, 1, -1)$, this set is linearly dependent. Thus, we reject (4, 3, 3), and consider the set $\{(2, 1, 2), (0, 1, -1), (-2, 0, 1)\}$. If t_1, t_2, and t_3 are real numbers such that $t_1 \cdot (2, 1, 2) + t_2 \cdot (0, 1, -1) + t_3 \cdot (-2, 0, 1) = (0, 0, 0)$, then $2t_1 - 2t_3 = 0$, $t_1 + t_2 = 0$, and $2t_1 - t_2 + t_3 = 0$. Solving these equations, we find that $t_1 = t_2 = t_3 = 0$ is the only solution. Hence, the set $S = \{(2, 1, 2), (0, 1, -1), (-2, 0, 1)\}$ is linearly independent. The only subset of T that properly contains S is T itself, and T is a linearly dependent set since it contains the set $\{(2, 1, 2), (0, 1, -1), (4, 3, 3)\}$ which is linearly dependent. Thus, S is a maximal independent set contained in T. There are, of course, other maximal independent sets contained in T; for example, $\{(2, 1, 2), (4, 3, 3), (-2, 0, 1)\}$. The fact that any two maximal independent sets S contained in a set T have the same number of elements is a consequence of the proof of Theorem 10.7.

Suppose that S is a subset of a finite dimensional vector space $\mathcal{V}$, and that S contains the same number of vectors as the dimension of $\mathcal{V}$. The following result shows that to prove S is a basis of $\mathcal{V}$, it is sufficient to check only one of the two conditions of Definition 10.4.

(10.8) corollary Let $\mathcal{V}$ be a finite dimensional vector space of dimension k.
(a) Any set of k linearly independent vectors is a basis.
(b) Any set of k vectors that spans $\mathcal{V}$ is a basis.

PROOF (a) Let $\{\mathbf{U}_1, \mathbf{U}_2, \ldots, \mathbf{U}_k\}$ be linearly independent. Since $\mathcal{V}$ has dimension k, it follows from Definition 10.6 and Theorem 10.3, that $\{\mathbf{U}, \mathbf{U}_1, \mathbf{U}_2, \ldots, \mathbf{U}_k\}$ is a linearly dependent set, where $\mathbf{U}$ is any vector in $\mathcal{V}$. By the argument used in the proof of Theorem 10.7, $\mathbf{U}$ can be expressed as a linear combination of $\mathbf{U}_1, \mathbf{U}_2, \ldots, \mathbf{U}_k$. Hence, $\{\mathbf{U}_1, \mathbf{U}_2, \ldots, \mathbf{U}_k\}$ spans $\mathcal{V}$, and is therefore a basis of $\mathcal{V}$.

(b) Let $\{\mathbf{U}_1, \mathbf{U}_2, \ldots, \mathbf{U}_k\}$ span $\mathcal{V}$. By Theorem 10.7, this set contains a subset S that is a basis of $\mathcal{V}$. If S were a proper subset, then $\mathcal{V}$ would have dimension less than k. Hence, $S = \{\mathbf{U}_1, \mathbf{U}_2, \ldots, \mathbf{U}_k\}$. ▮

(10.9) corollary A subspace $\mathcal{S}$ is a proper subspace of a finite dimensional vector space $\mathcal{V}$ if and only if the dimension of $\mathcal{S}$ is less than that of $\mathcal{V}$.

The proof of Corollary 10.9 is left as an exercise (Exercise 10.15).

If $\{\mathbf{U}_1, \mathbf{U}_2, \ldots, \mathbf{U}_n\}$ is a basis of a vector space $\mathcal{V}$, then by Theorem 10.5, every vector $\mathbf{V}$ in $\mathcal{V}$ has a unique expression

$$\mathbf{V} = r_1 \cdot \mathbf{U}_1 + r_2 \cdot \mathbf{U}_2 + \cdots + r_n \cdot \mathbf{U}_n.$$

The coefficients $r_1, r_2, \ldots, r_n$ are called the *coordinates* of $\mathbf{V}$ with respect to the basis $\{\mathbf{U}_1, \mathbf{U}_2, \ldots, \mathbf{U}_n\}$. For example, the coordinates of a vector $\mathbf{V} \in \mathcal{V}_n$ with respect to the basis $\{\mathbf{E}_1^{(n)}, \mathbf{E}_2^{(n)}, \ldots, \mathbf{E}_n^{(n)}\}$ are just the components of $\mathbf{V}$.

exercises

10.1 Which of the following sets of vectors in $\mathcal{E}_3$ are linearly dependent?
(a) $\{\mathbf{I} - 2 \cdot \mathbf{J} + \mathbf{K}, 3 \cdot \mathbf{I} + \mathbf{J} - 5 \cdot \mathbf{K}\}$
(b) $\{-\mathbf{I} + 3 \cdot \mathbf{J} - 2 \cdot \mathbf{K}, 2 \cdot \mathbf{I} + \mathbf{K}, 6 \cdot \mathbf{J} - 3 \cdot \mathbf{K}\}$
(c) $\{\mathbf{I} - \mathbf{J}, \mathbf{J} - \mathbf{K}, \mathbf{K} - \mathbf{I}\}$
(d) $\{\mathbf{I}, \mathbf{I} + \mathbf{J}, \mathbf{I} + \mathbf{J} + \mathbf{K}\}$
(e) $\{4 \cdot \mathbf{I} - 3 \cdot \mathbf{J} + \mathbf{K}, \frac{1}{2} \cdot \mathbf{I} + \frac{1}{3} \cdot \mathbf{J} + \frac{1}{5} \cdot \mathbf{K}, -\mathbf{I} - \mathbf{J} + 5 \cdot \mathbf{K}, 2 \cdot \mathbf{I} - \mathbf{J}\}$
(f) $\{\mathbf{I}, \mathbf{K}, \mathbf{0}\}$

10.2 Let $\mathfrak{F}_\infty$ be the vector space of all real-valued functions defined for all real numbers x. Which of the following sets of functions in $\mathfrak{F}_\infty$ are linearly dependent?
(a) $\{f_1(x) = 2 \sin^2 x, f_2(x) = -\cos^2 x, f_3(x) = 5\}$
(b) $\{f_1(x) = e^x, f_2(x) = xe^x, f_3(x) = x^2e^x\}$
(c) $\{f_1(x) = e^{2x}, f_2(x) = \sin x\}$
(d) $\{f_1(x) = x + 3x^2, f_2(x) = x^2, f_3(x) = 1 + x^2, f_4(x) = 2x - 3x^2\}$

10.3 Which of the following sets of vectors in $\mathcal{V}_3$ are linearly independent?
(a) $\{(1, 0, 1), (-3, 2, 6), (4, 5, -2)\}$
(b) $\{(-\frac{1}{2}, \frac{2}{3}, \frac{3}{4}), (-5, 16, 6), (-\frac{1}{3}, 2, \frac{1}{4})\}$
(c) $\{(1, -4, 2), (3, -5, 1), (2, 7, 8), (-1, 1, 1)\}$
(d) $\{(1, 0, 0), (1, 1, 0)\}$
(e) $\{(1, 1, 1)\}$
(f) $\{(1, 1, 1), (0, 0, 0)\}$

10.4 Prove that if T is a set of vectors in a vector space $\mathcal{V}$ and $S \subseteq T$, where S is a linearly dependent set, then T is a linearly dependent set.

10.5 Prove that any nonempty subset of a linearly independent set of vectors in a vector space $\mathcal{V}$ is linearly independent.

10.6 Prove that every set of $n + 1$ vectors in $\mathcal{V}_n$ is linearly dependent.

10.7 Let $\mathcal{S}$ be the subspace of $R[x]$ consisting of those polynomials of degree at most three (see Example 5, §9). Prove that every set of five polynomials in $\mathcal{S}$ is linearly dependent.

10.8 Which of the sets of vectors listed in Exercise 10.1 form a basis of $\mathcal{E}_3$?

10.9 Which of the sets of vectors listed in Exercise 10.3 form a basis of $\mathcal{V}_3$?

10.10 (a) Prove that $\{(2, -\frac{1}{2}, 1), (3, 2, 1), (0, 1, 1)\}$ is a basis of $\mathcal{V}_3$. (b) Prove that $\{(1, 1, 1, 1), (0, 1, 1, 1), (0, 0, 1, 1), (0, 0, 0, 1)\}$ is a basis of $\mathcal{V}_4$. (c) Prove that $\{1\}$ is a basis of R (see Exercise 9.2). (d) Prove that $\{1, i\}$ is a basis of C (see Example 3, §9).

10.11 Let $\mathcal{S}$ be the subspace of $R[x]$ consisting of those polynomials of degree at most k (see Example 5, §9). Find a basis of $\mathcal{S}$.

10.12 Prove that the space of functions $\mathfrak{F}_\infty$ is not a finite dimensional vector space.

***10.13** Let $\mathcal{V}$ be a finite dimensional vector space of dimension n. Prove that any set of k linearly independent vectors in $\mathcal{V}$ ($k \leq n$) is contained in a basis of $\mathcal{V}$.

*10.14 Let $\mathcal{S}$ and $\mathcal{T}$ be subspaces of a finite dimensional space $\mathcal{V}$. (a) Prove the following relation between the dimensions of $\mathcal{S}$, $\mathcal{T}$, $\mathcal{S} + \mathcal{T}$, and $\mathcal{S} \cap \mathcal{T}$:

$$\dim(\mathcal{S} + \mathcal{T}) = \dim \mathcal{S} + \dim \mathcal{T} - \dim(\mathcal{S} \cap \mathcal{T}).$$

(b) Prove that $\dim(\mathcal{S} \oplus \mathcal{T}) = \dim \mathcal{S} + \dim \mathcal{T}$.

10.15 Prove Corollary 10.9.

11 euclidean vector spaces

We obtained the definition of a real vector space (Definition 9.2) by abstracting the properties of addition and scalar multiplication of vectors in $\mathcal{E}_3$. The notion of the inner product in $\mathcal{E}_3$ can also be generalized. The first step is to list the characteristic properties of inner multiplication in $\mathcal{E}_3$.

(11.1) theorem Inner multiplication in $\mathcal{E}_3$ is a mapping from ordered pairs of vectors in $\mathcal{E}_3$ to the real numbers which satisfies the following conditions for vectors **U, V, W** in $\mathcal{E}_3$ and real numbers r and s:

(a) $\mathbf{U} \circ (r \cdot \mathbf{V} + s \cdot \mathbf{W}) = r(\mathbf{U} \circ \mathbf{V}) + s(\mathbf{U} \circ \mathbf{W})$;

(b) $\mathbf{U} \circ \mathbf{V} = \mathbf{V} \circ \mathbf{U}$;

(c) $\mathbf{U} \circ \mathbf{U} > 0$ if $\mathbf{U} \neq \mathbf{0}$, and $\mathbf{0} \circ \mathbf{0} = 0$.

PROOF Condition (a) of Theorem 11.1 follows from Corollaries 7.3 and 7.4. In fact,

$$\mathbf{U} \circ (r \cdot \mathbf{V} + s \cdot \mathbf{W}) = \mathbf{U} \circ (r \cdot \mathbf{V}) + \mathbf{U} \circ (s \cdot \mathbf{W}) = r(\mathbf{U} \circ \mathbf{V}) + s(\mathbf{U} \circ \mathbf{W}).$$

Condition (b) is an immediate consequence of Definition 4.1 of the inner product, and $\mathbf{U} \circ \mathbf{U} = |\mathbf{U}|^2$ implies (c). ∎

The identity in Theorem 11.1 (a) states that the inner product is *linear*. By (b), the inner product is *symmetric,* and by (c), it is *positive definite*.

We abstract the properties of the inner product in $\mathcal{E}_3$, and make the following definition for any finite dimensional vector space $\mathcal{V}$.

(11.2) definition Let $\mathcal{V}$ be a finite dimensional vector space. Let $\mathcal{I}$ be a mapping from ordered pairs of vectors in $\mathcal{V}$ to the real numbers. That is, for each ordered pair (**U, V**) of vectors in $\mathcal{V}$, $\mathcal{I}$(**U, V**) is a unique real number. The mapping $\mathcal{I}$ satisfies the following conditions for any vectors **U, V, W** in $\mathcal{V}$ and any real numbers r and s:

(a) $\mathcal{I}(\mathbf{U}, r \cdot \mathbf{V} + s \cdot \mathbf{W}) = r\mathcal{I}(\mathbf{U}, \mathbf{V}) + s\mathcal{I}(\mathbf{U}, \mathbf{W})$;

(b) $\mathcal{I}(\mathbf{U}, \mathbf{V}) = \mathcal{I}(\mathbf{V}, \mathbf{U})$;

(c) $\mathcal{I}(\mathbf{U}, \mathbf{U}) > 0$ if $\mathbf{U} \neq \mathbf{0}$, and $\mathcal{I}(\mathbf{0}, \mathbf{0}) = 0$.

Then $\mathcal{I}$(**U, V**) is called an *inner product* of **U** and **V**, and the space $\mathcal{V}$ equipped with such an inner product is called a *Euclidean vector space*.

The linearity condition (a) and the symmetry condition (b) of Definition 11.2 can be combined to obtain a *bilinearity condition*:

$$\textbf{(11.3)}\quad \mathcal{I}(r_1 \cdot \mathbf{U}_1 + r_2 \cdot \mathbf{U}_2, s_1 \cdot \mathbf{W}_1 + s_2 \cdot \mathbf{W}_2) = r_1 s_1 \mathcal{I}(\mathbf{U}_1, \mathbf{W}_1) + r_1 s_2 \mathcal{I}(\mathbf{U}_1, \mathbf{W}_2) + r_2 s_1 \mathcal{I}(\mathbf{U}_2, \mathbf{W}_2) + r_2 s_2 \mathcal{I}(\mathbf{U}_2, \mathbf{W}_2)$$

for vectors $\mathbf{U}_1$, $\mathbf{U}_2$, $\mathbf{W}_1$, $\mathbf{W}_2$ and real numbers r_1, r_2, s_1, s_2.

The proof of 11.3 is left as an exercise for the reader (Exercise 11.2). The identity 11.3 is called a bilinearity condition because it states that an inner product $\mathcal{I}(\mathbf{U}, \mathbf{V})$ is linear in each of its two arguments. Thus, an inner product in a Euclidean vector space is bilinear, symmetric, and positive definite.

We obtain useful identities from 11.3 by making particular choices of the numbers r_1, r_2, s_1, and s_2. For example, with $r_1 = 1$, $r_2 = 1$, $s_1 = 1$, and $s_2 = 0$, we obtain

$$\mathcal{I}(\mathbf{U}_1 + \mathbf{U}_2, \mathbf{W}_1) = \mathcal{I}(\mathbf{U}_1, \mathbf{W}_1) + \mathcal{I}(\mathbf{U}_2, \mathbf{W}_1).$$

Similarly, with $r_1 = 1$, $r_2 = 0$, $s_1 = 1$, and $s_2 = 1$, we get

$$\mathcal{I}(\mathbf{U}_1, \mathbf{W}_1 + \mathbf{W}_2) = \mathcal{I}(\mathbf{U}_1, \mathbf{W}_1) + \mathcal{I}(\mathbf{U}_1, \mathbf{W}_2).$$

Finally, $r_2 = 0$ and $s_2 = 0$ yield

$$\mathcal{I}(r_1 \cdot \mathbf{U}_1, s_1 \cdot \mathbf{W}_1) = r_1 s_1 \mathcal{I}(\mathbf{U}_1, \mathbf{W}_1).$$

EXAMPLE 1 An inner product can be defined for any finite dimensional vector space $\mathcal{V}$. Let $B = \{\mathbf{U}_1, \mathbf{U}_2, \ldots, \mathbf{U}_n\}$ be a basis of $\mathcal{V}$, and let

$$\begin{aligned}\mathbf{U} &= r_1 \cdot \mathbf{U}_1 + r_2 \cdot \mathbf{U}_2 + \cdots + r_n \cdot \mathbf{U}_n\\ \mathbf{W} &= s_1 \cdot \mathbf{U}_1 + s_2 \cdot \mathbf{U}_2 + \cdots + s_n \cdot \mathbf{U}_n\end{aligned}$$

be any vectors in $\mathcal{V}$. Define

$$\mathcal{I}(\mathbf{U}, \mathbf{W}) = r_1 s_1 + r_2 s_2 + \cdots + r_n s_n.$$

Then conditions (a), (b), and (c) of Definition 11.2 can easily be verified (see Exercise 11.1). We call this inner product the B inner product of $\mathcal{V}$. In particular, if $\mathcal{V} = \mathcal{V}_n$ and $\mathbf{U} = (a_1, a_2, \ldots, a_n)$, $\mathbf{W} = (b_1, b_2, \ldots, b_n)$ are vectors in $\mathcal{V}_n$, then $\mathcal{I}(\mathbf{U}, \mathbf{W}) = a_1 b_1 + a_2 b_2 + \cdots + a_n b_n$ is the $E^{(n)}$ inner product in $\mathcal{V}_n$.

In Example 1, the value of the B inner product of two vectors in $\mathcal{V}$ depends upon the choice of the basis B. Thus, it is evident that different inner products can be defined on the same space $\mathcal{V}$. When we refer to a Euclidean vector space, we will assume that a particular mapping $\mathcal{I}$ satisfying Definition 11.2 is given and is fixed throughout the discussion.

The result 4.3 suggests that in a Euclidean vector space $\mathcal{V}$, we define the *length* of a vector $\mathbf{U}$ to be $|\mathbf{U}| = [\mathcal{I}(\mathbf{U}, \mathbf{U})]^{1/2}$. A vector $\mathbf{U}$ is a *unit vector* if $|\mathbf{U}| = 1$. We note that

$$\begin{aligned}|r \cdot \mathbf{U}| &= [\mathcal{I}(r \cdot \mathbf{U}, r \cdot \mathbf{U})]^{1/2} = [r^2 \mathcal{I}(\mathbf{U}, \mathbf{U})]^{1/2}\\ &= |r|[\mathcal{I}(\mathbf{U}, \mathbf{U})]^{1/2} = |r|\,|\mathbf{U}|.\end{aligned}$$

Thus if $\mathbf{U}$ is any nonzero vector in $\mathcal{V}$, then $(1/|\mathbf{U}|) \cdot \mathbf{U}$ is a unit vector.

Following 4.4, the vectors $\mathbf{U}_1$ and $\mathbf{U}_2$ in $\mathcal{V}$ are defined to be *orthogonal* if $\mathcal{J}(\mathbf{U}_1, \mathbf{U}_2) = 0$. Since

$$\mathcal{J}(\mathbf{0}, \mathbf{U}) = \mathcal{J}(\mathbf{0} + \mathbf{0}, \mathbf{U}) = \mathcal{J}(\mathbf{0}, \mathbf{U}) + \mathcal{J}(\mathbf{0}, \mathbf{U})$$

implies that $\mathcal{J}(\mathbf{0}, \mathbf{U}) = \mathbf{0}$, it follows that $\mathbf{0}$ and any vector $\mathbf{U}$ are orthogonal. Furthermore, $\mathcal{J}(r \cdot \mathbf{U}, s \cdot \mathbf{W}) = rs\mathcal{J}(\mathbf{U}, \mathbf{W})$ implies that if $r \neq 0$, $s \neq 0$ then $\mathbf{U}$ and $\mathbf{W}$ are orthogonal if and only if $r \cdot \mathbf{U}$ and $s \cdot \mathbf{W}$ are orthogonal.

Let $\mathcal{S}$ be a subspace of a Euclidean vector space $\mathcal{V}$. Let $\mathfrak{J}$ be the set of all vectors in $\mathcal{V}$ that are orthogonal to every vector in $\mathcal{S}$. We can show that $\mathfrak{J}$ is a subspace of $\mathcal{V}$. Let $\mathbf{U}$ and $\mathbf{V}$ be in $\mathfrak{J}$ and let $\mathbf{W}$ be any vector in $\mathcal{S}$. Then $\mathcal{J}(\mathbf{U}, \mathbf{W}) = 0$ and $\mathcal{J}(\mathbf{V}, \mathbf{W}) = 0$. Therefore,

$$\mathcal{J}(\mathbf{U} + \mathbf{V}, \mathbf{W}) = \mathcal{J}(\mathbf{U}, \mathbf{W}) + \mathcal{J}(\mathbf{V}, \mathbf{W}) = 0 + 0 = 0,$$

so that $\mathbf{U} + \mathbf{V}$ is orthogonal to any vector $\mathbf{W}$ in $\mathcal{S}$. Hence, $\mathbf{U} + \mathbf{W}$ is in $\mathfrak{J}$. Furthermore,

$$\mathcal{J}(r \cdot \mathbf{U}, \mathbf{W}) = r\mathcal{J}(\mathbf{U}, \mathbf{W}) = (r)(0) = 0.$$

Thus, if $\mathbf{U} \in \mathfrak{J}$, then $r \cdot \mathbf{U} \in \mathfrak{J}$. By Theorem 9.5, $\mathfrak{J}$ is a subspace of $\mathcal{V}$. The subspace $\mathfrak{J}$ is called the *orthogonal complement* of the subspace $\mathcal{S}$.

In Example 1, the basis $B = \{\mathbf{U}_1, \mathbf{U}_2, \ldots, \mathbf{U}_n\}$ has the property that $|\mathbf{U}_i| = 1$ for $i = 1, 2, \ldots, n$, and $\mathcal{J}(\mathbf{U}_i, \mathbf{U}_j) = 0$ for all i, j with $i \neq j$. This suggests the following definition.

(11.4) definition Let $\mathcal{V}$ be a Euclidean vector space with inner product $\mathcal{J}$. A basis $\{\mathbf{U}_1, \mathbf{U}_2, \ldots, \mathbf{U}_n\}$ of $\mathcal{V}$ such that $|\mathbf{U}_i| = 1$ for $i = 1, 2, \ldots, n$, and $\mathcal{J}(\mathbf{U}_i, \mathbf{U}_j) = 0$ for all i, j with $i \neq j$, is called an *orthonormal basis* of $\mathcal{V}$.

Thus, if $B = \{\mathbf{U}_1, \mathbf{U}_2, \ldots, \mathbf{U}_n\}$ is a basis of a vector space $\mathcal{V}$ of dimension n, and the inner product $\mathcal{J}$ is the B inner product as described in Example 1, then B is an orthonormal basis of $\mathcal{V}$ (see Exercise 11.4). Of course, with respect to a given inner product $\mathcal{J}$, some bases of $\mathcal{V}$ are orthonormal and some are not. This remark is illustrated by the following example.

EXAMPLE 2 Let $\mathcal{J}$ be the $E^{(3)}$ inner product in $\mathcal{V}_3$ as described in Example 1. Then for any two vectors (a_1, a_2, a_3) and (b_1, b_2, b_3) in $\mathcal{V}_3$,

$$\mathcal{J}[(a_1, a_2, a_3), (b_1, b_2, b_3)] = a_1b_1 + a_2b_2 + a_3b_3.$$

The basis $E^{(3)}$ is orthonormal. Indeed,

$$\mathcal{J}[(1, 0, 0), (0, 1, 0)] = 1(0) + 0(1) + 0(0) = 0,$$
$$\mathcal{J}[(1, 0, 0), (0, 0, 1)] = 1(0) + 0(0) + 0(1) = 0,$$

and

$$\mathcal{J}[(0, 1, 0), (0, 0. 1)] = 0(0) + 1(0) + 0(1) = 0.$$

Thus, the vectors in the basis $E^{(3)}$ are mutually orthogonal. Moreover,

$|(1, 0, 0)| = \mathcal{I}[(1, 0, 0), (1, 0, 0)]^{1/2} = [1(1) + 0(0) + 0(0)]^{1/2} = 1^{1/2} = 1$, so that (1, 0, 0) has length 1. Similarly, (0, 1, 0) and (0, 0, 1) are unit vectors.

Now $B = \{(1, 0, 0), (1, 1, 0), (1, 1, 1)\}$ is also a basis of $\mathcal{V}_3$. However, the basis B is not orthonormal with respect to the $E^{(3)}$ inner product $\mathcal{I}$. For example, $\mathcal{I}[(1, 0, 0), (1, 1, 0)] = 1(1) + 0(1) + 0(0) = 1 \neq 0$, so that (1, 0, 0) and (1, 1, 0) are not orthogonal. In fact, no two vectors of B are orthogonal. Also, only (1, 0, 0) is a unit vector.

Finally, $C = \{(1/\sqrt{2}, 1/\sqrt{2}, 0), (-1/\sqrt{2}, 1/\sqrt{2}, 0), (0, 0, 1)\}$ is a basis of $\mathcal{V}_3$ which is orthonormal with respect to the $E^{(3)}$ inner product. The details of showing that C is orthonormal are left to the reader (Exercise 11.5).

We have seen above that if we start with a given basis of a finite dimensional vector space $\mathcal{V}$, we can define an inner product on $\mathcal{V}$ so that the given basis is orthonormal. We now turn the problem around and assume that $\mathcal{V}$ is a Euclidean vector space with a given inner product $\mathcal{I}$, and ask whether there is some basis of $\mathcal{V}$ that is orthonormal with respect to the given inner product. In order to prove that this is indeed the case, we need the following preliminary result.

(11.5) theorem If the nonzero vectors $\mathbf{U}_1, \mathbf{U}_2, \ldots, \mathbf{U}_k$ in a Euclidean vector space $\mathcal{V}$ are mutually orthogonal, then they are linearly independent.

PROOF Assume there is a relation

$$r_1 \cdot \mathbf{U}_1 + r_2 \cdot \mathbf{U}_2 + \cdots + r_k \cdot \mathbf{U}_k = \mathbf{0}.$$

Then for each $i = 1, 2, \ldots, k$,

$$0 = \mathcal{I}(\mathbf{U}_i, \mathbf{0}) = \mathcal{I}(\mathbf{U}_i, r_1 \cdot \mathbf{U}_1 + r_2 \cdot \mathbf{U}_2 + \cdots + r_k \cdot \mathbf{U}_k)$$

$$= r_1\mathcal{I}(\mathbf{U}_i, \mathbf{U}_1) + r_2\mathcal{I}(\mathbf{U}_i, \mathbf{U}_2) + \cdots + r_k\mathcal{I}(\mathbf{U}_i, \mathbf{U}_k)$$

$$= r_i\mathcal{I}(\mathbf{U}_i, \mathbf{U}_i)$$

since $\mathcal{I}(\mathbf{U}_i, \mathbf{U}_j) = 0$ if $i \neq j$. By (c) of Definition 11.2 $\mathcal{I}(\mathbf{U}_i, \mathbf{U}_i) \neq 0$. Hence, $r_i = 0$ for $i = 1, 2, \ldots, k$. Therefore, $\{\mathbf{U}_1, \mathbf{U}_2, \ldots, \mathbf{U}_k\}$ is a linearly independent set. ∎

The following theorem not only shows that a Euclidean vector space $\mathcal{V}$ has an orthonormal basis, but the method of its proof provides an algorithm or step-by-step procedure for replacing any basis of a subspace $\mathcal{S}$ of $\mathcal{V}$ by an orthonormal basis of $\mathcal{S}$. The method is called the *Gram–Schmidt orthogonalization process*. An example of its use is given at the end of the section. Since the method of proof of Theorem 11.6 is based on a step by step process, the formal proof of the theorem naturally proceeds by mathematical induction.

(11.6) theorem If $\{\mathbf{U}_1, \mathbf{U}_2, \ldots, \mathbf{U}_k\}$ is a linearly independent set of vectors in a Euclidean vector space $\mathcal{V}$, then the subspace $\mathcal{S}$ of $\mathcal{V}$ spanned by $\{\mathbf{U}_1, \mathbf{U}_2, \ldots, \mathbf{U}_k\}$ has an orthonormal basis.

PROOF If $k = 1$, let $\mathbf{V}_1 = (1/|\mathbf{U}_1|) \cdot \mathbf{U}_1$. Then $\mathbf{V}_1$ is an orthonormal basis of the subspace spanned by $\{\mathbf{U}_1\}$. Assume that the statement of the theorem is true for any linearly independent set of $k - 1$ vectors ($k > 1$). Let $\{\mathbf{V}_1, \mathbf{V}_2, \ldots, \mathbf{V}_{k-1}\}$ be an orthonormal basis of the subspace of $\mathcal{V}$ spanned by $\{\mathbf{U}_1, \mathbf{U}_2, \ldots, \mathbf{U}_{k-1}\}$. Define

$$\mathbf{V}_k^* = \mathbf{U}_k - \mathcal{I}(\mathbf{U}_k, \mathbf{V}_1) \cdot \mathbf{V}_1 - \mathcal{I}(\mathbf{U}_k, \mathbf{V}_2) \cdot \mathbf{V}_2 - \cdots - \mathcal{I}(\mathbf{U}_k, \mathbf{V}_{k-1}) \cdot \mathbf{V}_{k-1}.$$

Then $\mathbf{V}_k^* \neq \mathbf{0}$, for otherwise $\mathbf{U}_k$ is in the subspace spanned by $\{\mathbf{U}_1, \mathbf{U}_2, \ldots, \mathbf{U}_{k-1}\}$, contradicting the linear independence of $\{\mathbf{U}_1, \mathbf{U}_2, \ldots, \mathbf{U}_{k-1}, \mathbf{U}_k\}$. Moreover,

$$\begin{aligned}\mathcal{I}(\mathbf{V}_k^*, \mathbf{V}_i) &= \mathcal{I}[\mathbf{U}_k - \mathcal{I}(\mathbf{U}_k, \mathbf{V}_1) \cdot \mathbf{V}_1 - \mathcal{I}(\mathbf{U}_k, \mathbf{V}_2) \cdot \mathbf{V}_2 \\ &\qquad - \cdots - \mathcal{I}(\mathbf{U}_k, \mathbf{V}_{k-1}) \cdot \mathbf{V}_{k-1}, \mathbf{V}_i] \\ &= \mathcal{I}(\mathbf{U}_k, \mathbf{V}_i) - \mathcal{I}(\mathbf{U}_k, \mathbf{V}_i)\mathcal{I}(\mathbf{V}_i, \mathbf{V}_i) \\ &= \mathcal{I}(\mathbf{U}_k, \mathbf{V}_i) - \mathcal{I}(\mathbf{U}_k, \mathbf{V}_i) = 0\end{aligned}$$

for $i = 1, 2, \ldots, k - 1$, since $\{\mathbf{V}_1, \mathbf{V}_2, \ldots, \mathbf{V}_{k-1}\}$ is an orthonormal basis. Therefore, $\mathbf{V}_k^*$ is orthogonal to $\mathbf{V}_i$ for $i = 1, 2, \ldots, k - 1$, and the vectors $\mathbf{V}_1, \mathbf{V}_2, \ldots, \mathbf{V}_{k-1}, \mathbf{V}_k^*$ are mutually orthogonal.

By Theorem 11.5, $\{\mathbf{V}_1, \mathbf{V}_2, \ldots, \mathbf{V}_{k-1}, \mathbf{V}_k^*\}$ is a linearly independent set contained in the subspace $\mathcal{S}$ of $\mathcal{V}$ spanned by $\{\mathbf{U}_1, \mathbf{U}_2, \ldots, \mathbf{U}_k\}$. By Corollary 10.8, $\{\mathbf{V}_1, \mathbf{V}_2, \ldots, \mathbf{V}_{k-1}, \mathbf{V}_k^*\}$ is a basis of $\mathcal{S}$. Replacing $\mathbf{V}_k^*$ by the unit vector $\mathbf{V}_k = (1/|\mathbf{V}_k^*|) \cdot \mathbf{V}_k^*$, the set $\{\mathbf{V}_1, \mathbf{V}_2, \ldots, \mathbf{V}_{k-1}, \mathbf{V}_k\}$ is an orthonormal basis of $\mathcal{S}$. This completes the induction and the proof of the theorem. ▮

The final result of this section shows that relative to an orthonormal basis, any inner product is given by the familiar formula in terms of the coordinates of the vectors. That is, if $\mathcal{I}$ is an inner product in a Euclidean vector space and $B = \{\mathbf{V}_1, \mathbf{V}_2, \ldots, \mathbf{V}_n\}$ is an orthonormal basis of $\mathcal{V}$ with respect to $\mathcal{I}$, then $\mathcal{I}$ is just the B inner product.

(11.7) theorem Each subspace $\mathcal{S} \neq \{\mathbf{0}\}$ of a Euclidean vector space $\mathcal{V}$ has an orthonormal basis. If $\{\mathbf{V}_1, \mathbf{V}_2, \ldots, \mathbf{V}_k\}$ is an orthonormal basis of $\mathcal{S}$ and

$$\begin{aligned}\mathbf{U} &= a_1 \cdot \mathbf{V}_1 + a_2 \cdot \mathbf{V}_2 + \cdots + a_k \cdot \mathbf{V}_k \\ \mathbf{W} &= b_1 \cdot \mathbf{V}_1 + b_2 \cdot \mathbf{V}_2 + \cdots + b_k \cdot \mathbf{V}_k\end{aligned}$$

then $\mathcal{I}(\mathbf{U}, \mathbf{W}) = a_1b_1 + a_2b_2 + \cdots + a_kb_k$.

PROOF If $\mathcal{V}$ has dimension n, then by Theorem 10.7, $\mathcal{S}$ has a basis of k vectors for some $k \leq n$. By Theorem 11.6, $\mathcal{S}$ has an orthonormal basis $\{\mathbf{V}_1, \mathbf{V}_2, \ldots, \mathbf{V}_k\}$. Suppose

$$\mathbf{U} = a_1 \cdot \mathbf{V}_1 + a_2 \cdot \mathbf{V}_2 + \cdots + a_k \cdot \mathbf{V}_k$$

and

$$\mathbf{W} = b_1 \cdot \mathbf{V}_1 + b_2 \cdot \mathbf{V}_2 + \cdots + b_k \cdot \mathbf{V}_k.$$

Then by 11.3,

$$\begin{aligned}\mathcal{I}(\mathbf{U}, \mathbf{W}) &= a_1b_1\mathcal{I}(\mathbf{V}_1, \mathbf{V}_1) + a_1b_2\mathcal{I}(\mathbf{V}_1, \mathbf{V}_2) + \cdots + a_1b_k\mathcal{I}(\mathbf{V}_1, \mathbf{V}_k)\\ &\quad + a_2b_1\mathcal{I}(\mathbf{V}_2, \mathbf{V}_1) + a_2b_2\mathcal{I}(\mathbf{V}_2, \mathbf{V}_2) + \cdots + a_2b_k\mathcal{I}(\mathbf{V}_2, \mathbf{V}_k)\\ &\quad + \cdots + a_kb_1\mathcal{I}(\mathbf{V}_k, \mathbf{V}_1) + a_kb_2\mathcal{I}(\mathbf{V}_k, \mathbf{V}_2)\\ &\quad + \cdots + a_kb_k\mathcal{I}(\mathbf{V}_k, \mathbf{V}_k)\\ &= a_1b_1 + a_2b_2 + \cdots + a_kb_k.\end{aligned}$$

∎

In practice, it is convenient to modify the Gram–Schmidt orthogonalization process used in the proof of Theorem 11.6 by first obtaining a basis of mutually orthogonal vectors and then replacing these vectors by unit vectors.

EXAMPLE 3 Let $\mathcal{S}$ be the subspace of $\mathcal{V}_5$ spanned by $\mathbf{U}_1 = (0, 1, 3, 0, 4)$, $\mathbf{U}_2 = (1, -2, 0, 1, 6)$, and $\mathbf{U}_3 = (0, 0, 0, 1, 1)$. These vectors are linearly independent, so that $\{\mathbf{U}_1, \mathbf{U}_2, \mathbf{U}_3\}$ is a basis of $\mathcal{S}$. The inner product in $\mathcal{V}_5$ is the $E^{(3)}$ inner product defined in Example 1. Let $\mathbf{V}_1^* = \mathbf{U}_1$. Define

$$\begin{aligned}\mathbf{V}_2^* &= \mathbf{U}_2 - \left[\frac{\mathcal{I}(\mathbf{U}_2, \mathbf{V}_1^*)}{\mathcal{I}(\mathbf{V}_1^*, \mathbf{V}_1^*)}\right] \cdot \mathbf{V}_1^*\\ &= (1, -2, 0, 1, 6) - \left(\frac{11}{13}\right) \cdot (0, 1, 3, 0, 4)\\ &= \left(1, -\frac{37}{13}, -\frac{33}{13}, 1, \frac{34}{13}\right).\end{aligned}$$

Then $\mathcal{I}(\mathbf{V}_1^*, \mathbf{V}_2^*) = 0$. Define

$$\begin{aligned}\mathbf{V}_3^* &= \mathbf{U}_3 - \left[\frac{\mathcal{I}(\mathbf{U}_3, \mathbf{V}_1^*)}{\mathcal{I}(\mathbf{V}_1^*, \mathbf{V}_1^*)}\right] \cdot \mathbf{V}_1^* - \left[\frac{\mathcal{I}(\mathbf{U}_3, \mathbf{V}_2^*)}{\mathcal{I}(\mathbf{V}_2^*, \mathbf{V}_2^*)}\right] \cdot \mathbf{V}_2^*\\ &= (0, 0, 0, 1, 1) - \left(\frac{2}{13}\right) \cdot (0, 1, 3, 0, 4)\\ &\quad - \left(\frac{47}{304}\right) \cdot \left(1, -\frac{37}{13}, -\frac{33}{13}, 1, \frac{34}{13}\right)\\ &= \left(\frac{1}{3952}\right) \cdot (-611, 1131, -273, 3341, -78).\end{aligned}$$

The vectors $\mathbf{V}_1^*, \mathbf{V}_2^*, \mathbf{V}_3^*$ are mutually orthogonal. Therefore,

$$\begin{aligned}\left(\frac{1}{|\mathbf{V}_1^*|}\right) \cdot \mathbf{V}_1^* &= \left(\frac{1}{\sqrt{26}}\right) \cdot (0, 1, 3, 0, 4)\\ \left(\frac{1}{|\mathbf{V}_2^*|}\right) \cdot \mathbf{V}_2^* &= \left(\frac{1}{4\sqrt{247}}\right) \cdot (13, -37, -33, 13, 34)\end{aligned}$$

and

$$\left(\frac{1}{|\mathbf{V}_3^*|}\right) \cdot \mathbf{V}_3^* = \left(\frac{1}{52\sqrt{4769}}\right) \cdot (-611,\ 1131,\ -273,\ 3341,\ -78)$$

is an orthonormal basis of $\mathcal{S}$.

exercises

11.1 Prove that the B inner product in $\mathcal{V}$, defined in Example 1, satisfies conditions (a), (b), and (c) of Definition 11.2.

11.2 Use (a) and (b) of Definition 11.2 to prove the bilinearity condition 11.3.

11.3 Show that the ordinary inner product in $\mathcal{E}_3$ is the B inner product with respect to the basis $B = \{\mathbf{I}, \mathbf{J}, \mathbf{K}\}$.

11.4 Let $\mathcal{I}$ be the B inner product for a given basis $B = \{\mathbf{U}_1, \mathbf{U}_2, \ldots, \mathbf{U}_n\}$ of a vector space $\mathcal{V}$ of dimension n. Show that B is an orthonormal basis with respect to $\mathcal{I}$.

11.5 Show that the basis $C = \{(1\sqrt{2}, 1/\sqrt{2}, 0), (-1/\sqrt{2}, 1/\sqrt{2}, 0), (0, 0, 1)\}$ of $\mathcal{V}_3$, given in Example 2, is an orthonormal basis of $\mathcal{V}_3$ with respect to the $E^{(3)}$ inner product in $\mathcal{V}_3$.

11.6 Which of the following sets of vectors in $\mathcal{E}_3$ are orthonormal bases with respect to the ordinary inner product in $\mathcal{E}_3$?

(a) $\{\mathbf{I}, \mathbf{J}, \mathbf{K}\}$

(b) $\{-\mathbf{I}, \mathbf{J}, -\mathbf{K}\}$

(c) $\{\mathbf{I}, \mathbf{I} - \mathbf{J}, \mathbf{I} + \mathbf{K}\}$

(d) $\{1/2 \cdot \mathbf{I} + \sqrt{3}/2 \cdot \mathbf{J}, -\sqrt{3}/2 \cdot \mathbf{I} + 1/2 \cdot \mathbf{J}, \mathbf{K}\}$

(e) $\{1/3 \cdot \mathbf{I} + \sqrt{5}/3 \cdot \mathbf{J} + \sqrt{3}/3 \cdot \mathbf{K}, \sqrt{5}/3 \cdot \mathbf{I} - 1/3 \cdot \mathbf{J}, \mathbf{K}\}$

11.7 Let $S = \{\mathbf{U}_1, \mathbf{U}_2, \ldots, \mathbf{U}_k\}$ be any set of vectors in a Euclidean vector space $\mathcal{V}$. Prove that the set of all vectors in $\mathcal{V}$ that are orthogonal to every vector in S is a subspace $\mathcal{J}$ of $\mathcal{V}$. Prove that $\mathcal{J}$ is the orthogonal complement of the subspace $\mathcal{S}$ of $\mathcal{V}$ that is spanned by S.

***11.8** Let $\mathcal{S}$ be a subspace of a Euclidean vector space $\mathcal{V}$ and let $\mathcal{J}$ be the orthogonal complement of $\mathcal{S}$. Prove that $\mathcal{V} = \mathcal{S} \oplus \mathcal{J}$.

***11.9** Prove that the following inequalities hold in a Euclidean vector space:

(a) $[\mathcal{I}(\mathbf{U}, \mathbf{V})]^2 \leq |\mathbf{U}|^2\,|\mathbf{V}|^2$ (the Schwarz inequality)

(b) $|\mathbf{U} + \mathbf{V}| \leq |\mathbf{U}| + |\mathbf{V}|$ (the triangle inequality)

11.10 Construct an orthonormal basis for the subspace of $\mathcal{V}_4$ spanned by each of the following sets of vectors:

(a) $\{(3, -5, 2, 1)\}$

(b) $\{(\frac{1}{2}, 3, 0, 2), (1, -\frac{1}{2}, 3, 1)\}$

(c) $\{(1, 0, 1, 1), (0, 2, 1, -1), (2, 1, -1, 0)\}$

(d) $\{(1, 1, 1, 1), (1, 1, 1, 0), (1, 1, 0, 0)\}$

3

systems of linear equations

12 elementary transformations

Consider the problem of deciding whether a set of vectors $\mathbf{U}_1, \mathbf{U}_2, \ldots, \mathbf{U}_n$ in $\mathcal{V}_n$ is a linearly dependent set. We know that the vectors

$$\mathbf{U}_1 = (a_{1,1}, a_{2,1}, \ldots, a_{n,1}), \quad \mathbf{U}_2 = (a_{1,2}, a_{2,2}, \ldots, a_{n,2}), \ldots, \\ \mathbf{U}_m = (a_{1,m}, a_{2,m}, \ldots, a_{n,m})$$

form a linearly dependent set if and only if there exist real numbers $x_1, x_2, \ldots, x_m$, not all zero, such that

$$x_1 \cdot (a_{1,1}, a_{2,1}, \ldots, a_{n,1}) + x_2 \cdot (a_{1,2}, a_{2,2}, \ldots, a_{n,2}) \\ + \cdots + x_m \cdot (a_{1,m}, a_{2,m}, \ldots, a_{n,m}) = (0, 0, \ldots, 0).$$

That is, $\{\mathbf{U}_1, \mathbf{U}_2, \ldots, \mathbf{U}_m\}$ is a linearly dependent set if and only if the system of linear equations

$$\begin{aligned} a_{1,1}x_1 + a_{1,2}x_2 + \cdots + a_{1,m}x_m &= 0 \\ a_{2,1}x_1 + a_{2,2}x_2 + \cdots + a_{2,m}x_m &= 0 \\ &\vdots \\ a_{n,1}x_1 + a_{n,2}x_2 + \cdots + a_{n,m}x_m &= 0 \end{aligned}$$

has a solution other than $x_1 = x_2 = \cdots = x_m = 0$.

Let $\mathbf{V} = (b_1, b_2, \ldots, b_n)$ be a given vector in $\mathcal{V}_n$. Then $\mathbf{V}$ is in the subspace of $\mathcal{V}_n$ spanned by $\{\mathbf{U}_1, \mathbf{U}_2, \ldots, \mathbf{U}_m\}$ if and only if there exist real numbers $x_1, x_2, \ldots, x_m$ such that

$$x_1 \cdot (a_{1,1}, a_{2,1}, \ldots, a_{n,1}) + x_2 \cdot (a_{1,2}, a_{2,2}, \ldots, a_{n,2}) \\ + \cdots + x_m \cdot (a_{1,m}, a_{2,m}, \ldots, a_{n,m}) = (b_1, b_2, \ldots, b_n).$$

This latter condition is equivalent to the existence of a solution of the system of linear equations

$$\begin{aligned} a_{1,1}x_1 + a_{1,2}x_2 + \cdots + a_{1,m}x_m &= b_1 \\ a_{2,1}x_1 + a_{2,2}x_2 + \cdots + a_{2,m}x_m &= b_2 \\ &\vdots \\ a_{n,1}x_1 + a_{n,2}x_2 + \cdots + a_{n,m}x_m &= b_n. \end{aligned} \tag{12–1}$$

The problem of solving such systems of n linear equations in m unknowns is not confined to the study of vector spaces. It is a problem that occurs in many branches and applications of mathematics.

The real numbers $a_{i,j}$ and b_i in equations (12–1) are called the *coefficients* of the equations. The system (12–1) is *consistent* if there exist real numbers $x_1, x_2, \ldots, x_m$ that satisfy the equations. Such a set of real numbers is called a *solution* of the system. If the system has no solutions, it is *inconsistent*. A complete analysis of a system of linear equations includes stating criteria for consistency and, when the system is consistent, describing a systematic method for finding all solutions.

(12.1) definition Let

$$\begin{aligned} a_{1,1}x_1 + a_{1,2}x_2 + \cdots + a_{1,m}x_m &= b_1 \\ a_{2,1}x_1 + a_{2,2}x_2 + \cdots + a_{2,m}x_m &= b_2 \\ &\vdots \\ a_{n,1}x_1 + a_{n,2}x_2 + \cdots + a_{n,m}x_m &= b_n \end{aligned}$$

and

$$\begin{aligned} d_{1,1}x_1 + d_{1,2}x_2 + \cdots + d_{1,m}x_m &= e_1 \\ d_{2,1}x_1 + d_{2,2}x_2 + \cdots + d_{2,m}x_m &= e_2 \\ &\vdots \\ d_{p,1}x_1 + d_{p,2}x_2 + \cdots + d_{p,m}x_m &= e_p \end{aligned}$$

be systems of n and p linear equations in m unknowns with real coefficients. The systems are *equivalent* if every solution of the first system is a solution of the second system, and vice versa.

For example, the system

$$\begin{aligned} 3x_1 - 2x_2 + x_3 &= -6 \\ 2x_1 + 5x_2 - x_3 &= 0 \\ 7x_1 - \tfrac{3}{2}x_2 + \tfrac{3}{2}x_3 &= -12 \end{aligned} \tag{12–2}$$

is equivalent to the system

$$\begin{aligned} 3x_1 - 2x_2 + x_3 &= -6 \\ 2x_1 + 5x_2 - x_3 &= 0 \end{aligned} \tag{12–3}$$

since we can show that every set of three real numbers that satisfies the first two equations of (12–2) satisfies the third equation as well.

Our method for analyzing the system of equations (12–1) entails replacing the given system by a new equivalent system whose consistency can be determined by inspection. Furthermore, when this new system is consistent, all solutions can be easily computed. The first step in this process is to describe certain elementary operations on the equations that can be defined in terms of operations on a set of vectors in a real vector space.

Let $(\mathbf{U}_1, \mathbf{U}_2, \ldots, \mathbf{U}_n)$ be an ordered n-tuple of vectors in a vector space $\mathcal{V}$. The three *elementary transformations* on the given set of vectors are described as follows:

Type I Replace $(\mathbf{U}_1, \mathbf{U}_2, \ldots, \mathbf{U}_i, \ldots, \mathbf{U}_j, \ldots, \mathbf{U}_n)$ by $(\mathbf{U}_1, \mathbf{U}_2, \ldots, \mathbf{U}_j, \ldots, \mathbf{U}_i, \ldots, \mathbf{U}_n)$ where $1 \leq i < j \leq n$. That is, two of the vectors are interchanged in the given n-tuple.

Type II Replace $(\mathbf{U}_1, \mathbf{U}_2, \ldots, \mathbf{U}_i, \ldots, \mathbf{U}_j, \ldots, \mathbf{U}_n)$ by $(\mathbf{U}_1, \mathbf{U}_2, \ldots, \mathbf{U}_i, \ldots, c \cdot \mathbf{U}_i + \mathbf{U}_j, \ldots, \mathbf{U}_n)$, where $i \neq j$ and c is a real number. Here, the vector $\mathbf{U}_j$ is replaced by the vector $c \cdot \mathbf{U}_i + \mathbf{U}_j$, with $i \neq j$.

Type III Replace $(\mathbf{U}_1, \mathbf{U}_2, \ldots, \mathbf{U}_i, \ldots, \mathbf{U}_n)$ by $(\mathbf{U}_1, \mathbf{U}_2, \ldots, c \cdot \mathbf{U}_i, \ldots, \mathbf{U}_n)$, where c is a nonzero real number. Thus, a vector $\mathbf{U}_i$ is replaced by its scalar multiple $c \cdot \mathbf{U}_i$, where c is a nonzero real number.

Note that in each case every vector of the new n-tuple is a linear combination of the vectors in the original n-tuple. Moreover, it is easy to see that each of the original vectors is also a linear combination of the vectors in the new n-tuple. This is obvious for an elementary transformation of Type I. For a Type II transformation, it is sufficient to show that $\mathbf{U}_j$ is a linear combination of the vectors $\mathbf{U}_1, \mathbf{U}_2, \ldots, \mathbf{U}_i, \ldots, c \cdot \mathbf{U}_i + \mathbf{U}_j, \ldots, \mathbf{U}_n$. This is indeed the case, since $\mathbf{U}_j = (-c) \cdot \mathbf{U}_i + 1 \cdot (c \cdot \mathbf{U}_i + \mathbf{U}_j)$. For Type III transformations, since $\mathbf{U}_i = (1/c) \cdot (c \cdot \mathbf{U}_i)$, $c \neq 0$, each vector in the original n-tuple is a linear combination of the new set of vectors obtained by a Type III elementary transformation. Thus, by Theorem 9.8, a given set of n vectors S in a vector space $\mathcal{V}$ spans the same subspace of $\mathcal{V}$ as the set T of vectors obtained from S by an elementary transformation. The following theorem is an immediate consequence of this result.

(12.2) theorem Let S be an ordered set of n vectors in a vector space $\mathcal{V}$, and let T be a set of vectors obtained from S by means of a sequence of elementary transformations. That is, there are ordered sets of vectors S_0, $S_1, \ldots, S_t$ such that S_0 is S and S_t is T, and for each k $(0 < k \leq t)$, the set S_k is obtained from the set S_{k-1} by an elementary transformation. Then the sets S and T span the same subspace of $\mathcal{V}$.

The system of linear equations (12–1) determines a subspace of the vector space $\mathcal{V}_{m+1}$ spanned by the vectors

$$\mathbf{U}_1 = (a_{1,1}, a_{1,2}, \ldots, a_{1,m}, b_1), \quad \mathbf{U}_2 = (a_{2,1}, a_{2,2}, \ldots, a_{2,m}, b_2), \ldots, \\ \mathbf{U}_n = (a_{n,1}, a_{n,2}, \ldots, a_{n,m}, b_n).$$

This subspace of $\mathcal{V}_{m+1}$ is called the *row space* of the system. For example, the row space of the system (12–2) is the subspace of $\mathcal{V}_4$ spanned by $(3, -2, 1, -6)$, $(2, 5, -1, 0)$, and $(7, -\frac{3}{2}, \frac{3}{2}, -12)$.

Suppose that a sequence of elementary transformations is performed on the vectors $(\mathbf{U}_1, \mathbf{U}_2, \ldots, \mathbf{U}_n)$ spanning the row space of (12–1), yielding a new set $(\mathbf{U}_1^*, \mathbf{U}_2^*, \ldots, \mathbf{U}_n^*)$, where $\mathbf{U}_i^* = (a_{i,1}^*, a_{i,2}^*, \ldots, a_{i,n}^*, b_i^*)$, for $i = 1, 2, \ldots, n$. Thus, a new system of n linear equations in m unknowns is determined:

$$\begin{aligned} a_{1,1}^* x_1 + a_{1,2}^* x_2 + \cdots + a_{1,m}^* x_m &= b_1^* \\ a_{2,1}^* x_1 + a_{2,2}^* x_2 + \cdots + a_{2,m}^* x_m &= b_2^* \\ &\vdots \\ a_{n,1}^* x_1 + a_{n,2}^* x_2 + \cdots + a_{n,m}^* x_m &= b_n^*. \end{aligned} \qquad (12\text{–}4)$$

The row space of this new system is spanned by the vectors $\mathbf{U}_1^*, \mathbf{U}_2^*, \ldots, \mathbf{U}_n^*$. It follows from Theorem 12.2 that the systems (12–1) and (12–4) have the same row space. The elementary transformations performed on the vectors $(\mathbf{U}_1, \mathbf{U}_2, \ldots, \mathbf{U}_n)$ of system (12–1) may be described directly in terms of the equations. A Type I elementary transformation interchanges two equations, a Type II transformation multiplies one of the equations by a constant and adds this equation to another equation, and a transformation of Type III multiplies one of the equations by a nonzero constant. Thus, the new system (12–4) is obtained from (12–1) by a sequence of elementary algebraic operations.

EXAMPLE 1 Consider the system

$$\begin{aligned} \tfrac{1}{2}x_1 - 3x_2 + 2x_3 + x_4 &= 0 \\ 0x_1 + \tfrac{1}{3}x_2 + 5x_3 - 3x_4 &= 5 \\ 4x_1 + 6x_2 + 0x_3 - 2x_4 &= -1. \end{aligned} \qquad (12\text{–}5)$$

The row space of system (12–5) is the subspace of $\mathcal{V}_5$ spanned by the vectors

$$\begin{aligned} \mathbf{U}_1 &= (\tfrac{1}{2}, -3, 2, 1, 0) \\ \mathbf{U}_2 &= (0, \tfrac{1}{3}, 5, -3, 5) \\ \mathbf{U}_3 &= (4, 6, 0, -2, -1). \end{aligned}$$

We perform the following sequence of elementary transformations on $(\mathbf{U}_1, \mathbf{U}_2, \mathbf{U}_3)$. Multiply $\mathbf{U}_1$ by 2:

$$\left.\begin{matrix} \mathbf{U}_1 \\ \mathbf{U}_2 \\ \mathbf{U}_3 \end{matrix}\right\} \rightarrow \left\{\begin{aligned} &(1, -6, 4, 2, 0) &= \mathbf{U}_1^{(1)} \\ &(0, \tfrac{1}{3}, 5, -3, 5) &= \mathbf{U}_2^{(1)} \\ &(4, 6, 0, -2, -1) &= \mathbf{U}_3^{(1)} \end{aligned}\right.$$

Multiply $\mathbf{U}_1^{(1)}$ by -4 and add to $\mathbf{U}_3^{(1)}$:

$$\left.\begin{matrix} \mathbf{U}_1^{(1)} \\ \mathbf{U}_2^{(1)} \\ \mathbf{U}_3^{(1)} \end{matrix}\right\} \rightarrow \left\{\begin{aligned} &(1, -6, 4, 2, 0) &= \mathbf{U}_1^{(2)} \\ &(0, \tfrac{1}{3}, 5, -3, 5) &= \mathbf{U}_2^{(2)} \\ &(0, 30, -16, -10, -1) &= \mathbf{U}_3^{(2)} \end{aligned}\right.$$

Multiply $\mathbf{U}_2^{(2)}$ by 3:

$$\left.\begin{matrix}\mathbf{U}_1^{(2)}\\ \mathbf{U}_2^{(2)}\\ \mathbf{U}_3^{(2)}\end{matrix}\right\} \rightarrow \begin{cases}(1, -6, 4, 2, 0) & = \mathbf{U}_1^{(3)}\\ (0, 1, 15, -9, 15) & = \mathbf{U}_2^{(3)}\\ (0, 30, -16, -10, -1) & = \mathbf{U}_3^{(3)}\end{cases}$$

Multiply $\mathbf{U}_2^{(3)}$ by -30 and add to $\mathbf{U}_3^{(3)}$:

$$\left.\begin{matrix}\mathbf{U}_1^{(3)}\\ \mathbf{U}_2^{(3)}\\ \mathbf{U}_3^{(3)}\end{matrix}\right\} \rightarrow \begin{cases}(1, -6, 4, 2, 0) & = \mathbf{U}_1^{(4)}\\ (0, 1, 15, -9, 15) & = \mathbf{U}_2^{(4)}\\ (0, 0, -466, 260, -451) & = \mathbf{U}_3^{(4)}\end{cases}$$

Multiply $\mathbf{U}_3^{(4)}$ by $-1/466$:

$$\left.\begin{matrix}\mathbf{U}_1^{(4)}\\ \mathbf{U}_2^{(4)}\\ \mathbf{U}_3^{(4)}\end{matrix}\right\} \rightarrow \begin{cases}(1, -6, 4, 2, 0) & = \mathbf{U}_1^{(5)}\\ (0, 1, 15, -9, 15) & = \mathbf{U}_2^{(5)}\\ (0, 0, 1, -130/233, 451/466) & = \mathbf{U}_3^{(5)}\end{cases}$$

The new system of equations is

$$\begin{aligned} x_1 - 6x_2 + 4x_3 + 2x_4 &= 0\\ x_2 + 15x_3 - 9x_4 &= 15\\ x_3 - {}^{130}\!/_{233}x_4 &= {}^{451}\!/_{466}. \end{aligned} \tag{12-6}$$

The row space of system (12–6) is spanned by $(1, -6, 4, 2, 0)$, $(0, 1, 15, -9, 15)$, and $(0, 0, 1, -130/233, 451/466)$. It is the same subspace of $\mathcal{V}_5$ as the row space of the original system (12–5).

In §13 we will prove that a sequence of elementary transformations replaces a system of linear equations by an equivalent system. Thus, in Example 1, the system (12–6) is equivalent to the system (12–5).

exercises

12.1 Determine which of the following systems of equations are consistent. Find all solutions of the consistent systems.

(a) $\begin{aligned} 3x_1 - 5x_2 - 2x_3 &= 0\\ x_1 + 9x_2 - x_3 &= 0\\ 2x_1 + 4x_2 - 7x_3 &= 0 \end{aligned}$

(b) $\begin{aligned} 6x_1 - 5x_2 &= 14\\ 2x_1 + x_2 &= 0\\ x_1 - 3x_2 &= 1 \end{aligned}$

(c) $\begin{aligned} x_1 - 4x_2 + 2x_3 &= 10\\ 3x_1 - x_2 + x_3 &= 5 \end{aligned}$

(d) $\begin{aligned} -2x_1 + 4x_2 + 6x_3 &= 1\\ x_1 \qquad\quad - x_3 &= 1\\ -x_1 + 4x_2 + 5x_3 &= 1 \end{aligned}$

12.2 Prove that the systems of equations (12–2) and (12–3) are equivalent.

* **12.3** Suppose that the ordered set of vectors $(\mathbf{U}_1^*, \mathbf{U}_2^*, \ldots, \mathbf{U}_n^*)$ is obtained from the ordered set $(\mathbf{U}_1, \mathbf{U}_2, \ldots, \mathbf{U}_n)$ by an elementary transformation. Prove that there is an elementary transformation of the same type that takes the set $(\mathbf{U}_1^*, \mathbf{U}_2^*, \ldots, \mathbf{U}_n^*)$ into $(\mathbf{U}_1, \mathbf{U}_2, \ldots, \mathbf{U}_n)$.

12.4 Perform a sequence of elementary transformations on the vectors spanning the row space of equations (12–6) to obtain the vectors that span the row space of equations (12–5).

13 systems of equations in echelon form

By Theorem 12.2, a system of linear equations obtained from a given system by a sequence of elementary transformations has the same row space as the given system. Then, by the following theorem, these systems are equivalent.

(13.1) theorem If two systems of linear equations in m unknowns have the same row space, the two systems are equivalent.

PROOF Let the given systems be

$$\begin{aligned} a_{1,1}x_1 + a_{1,2}x_2 + \cdots + a_{1,m}x_m &= b_1 \\ a_{2,1}x_1 + a_{2,2}x_2 + \cdots + a_{2,m}x_m &= b_2 \\ &\vdots \\ a_{n,1}x_1 + a_{n,2}x_2 + \cdots + a_{n,m}x_m &= b_n \end{aligned} \tag{13–1}$$

and

$$\begin{aligned} d_{1,1}x_1 + d_{1,2}x_2 + \cdots + d_{1,m}x_m &= e_1 \\ d_{2,1}x_1 + d_{2,2}x_2 + \cdots + d_{2,m}x_m &= e_2 \\ &\vdots \\ d_{p,1}x_1 + d_{p,2}x_2 + \cdots + d_{p,m}x_m &= e_p. \end{aligned} \tag{13–2}$$

Let $x_1 = c_1, x_2 = c_2, \ldots, x_m = c_m$ be a solution of (13–1). Since the row space of system (13–2) is the same as that of (13–1), it follows that each vector $(d_{i,1}, d_{i,2}, \ldots, d_{i,m}, e_i)$ for $i = 1, 2, \ldots, p$ is a linear combination of the vectors $(a_{j,1}, a_{j,2}, \ldots, a_{j,m}, b_j)$, $j = 1, 2, \ldots, n$. Thus, for $i = 1, 2, \ldots, p$, there exist real numbers $r_1, r_2, \ldots, r_n$ that depend on i such that

$$\begin{aligned} (d_{i,1}, d_{i,2}, \ldots, d_{i,m}, e_i) &= r_1 \cdot (a_{1,1}, a_{1,2}, \ldots, a_{1,m}, b_1) \\ &\quad + r_2 \cdot (a_{2,1}, a_{2,2}, \ldots, a_{2,m}, b_2) \\ &\quad + \cdots + r_n \cdot (a_{n,1}, a_{n,2}, \ldots, a_{n,m}, b_n) \\ &= (r_1a_{1,1} + r_2a_{2,1} + \cdots + r_na_{n,1}, r_1a_{1,2} + r_2a_{2,2} \\ &\quad + \cdots + r_na_{n,2}, \ldots, r_1a_{1,m} + r_2a_{2,m} \\ &\quad + \cdots + r_na_{n,m}, r_1b_1 + r_2b_2 + \cdots + r_nb_n). \end{aligned}$$

Therefore,

$$\begin{aligned} d_{i,1} &= r_1a_{1,1} + r_2a_{2,1} + \cdots + r_na_{n,1} \\ d_{i,2} &= r_1a_{1,2} + r_2a_{2,2} + \cdots + r_na_{n,2} \\ &\vdots \\ d_{i,m} &= r_1a_{1,m} + r_2a_{2,m} + \cdots + r_na_{n,m} \end{aligned}$$

and

$$e_i = r_1b_1 + r_2b_2 + \cdots + r_nb_n.$$

Hence,

$$\begin{aligned} d_{i,1}c_1 + d_{i,2}c_2 + \cdots + d_{i,m}c_m \\ &= (r_1a_{1,1} + r_2a_{2,1} + \cdots + r_na_{n,1})c_1 \\ &\quad + (r_1a_{1,2} + r_2a_{2,2} + \cdots + r_na_{n,2})c_2 \\ &\quad + \cdots + (r_1a_{1,m} + r_2a_{2,m} + \cdots + r_na_{n,m})c_m \\ &= r_1(a_{1,1}c_1 + a_{1,2}c_2 + \cdots + a_{1,m}c_m) \\ &\quad + r_2(a_{2,1}c_1 + a_{2,2}c_2 + \cdots + a_{2,m}c_m) \\ &\quad + \cdots + r_n(a_{n,1}c_1 + a_{n,2}c_2 + \cdots + a_{n,m}c_n). \end{aligned}$$

Since $c_1, c_2, \ldots, c_m$ is a solution of (13–1),

$$a_{j,1}c_1 + a_{j,2}c_2 + \cdots + a_{j,m}c_m = b_j$$

for $j = 1, 2, \ldots, n$. Hence,

$$d_{i,1}c_1 + d_{i,2}c_2 + \cdots + d_{i,m}c_m = r_1b_1 + r_2b_2 + \cdots + r_nb_n = e_i$$

for $i = 1, 2, \ldots, p$. Therefore, $c_1, c_2, \ldots, c_m$ is a solution of the system (13–2). Conversely, since each vector $(a_{j,1}, a_{j,2}, \ldots, a_{j,m}, b_j)$ for $j = 1, 2, \ldots, n$ is a linear combination of the vectors $(d_{i,1}, d_{i,2}, \ldots, d_{i,m}, e_i)$, $i = 1, 2, \ldots, p$, a similar argument shows that every solution of the system (13–2) is a solution of (13–1). Therefore, the systems are equivalent. ∎

(13.2) corollary Let S' be a system of linear equations obtained from a system S of linear equations by a sequence of elementary transformations. Then the systems S and S' are equivalent.

PROOF The vectors that span the row space of the system S' are obtained, by a sequence of elementary transformations from the vectors that span the row space of the system S. By Theorem 12.2, the systems S and S' have the same row space. By Theorem 13.1, the systems S and S' are equivalent. ∎

Thus, in Example 1, §12, the systems (12–6) and (12–5) have exactly the same solutions. The solutions of (12–6) are easily obtained. Every real solution is given by

$$\begin{aligned} x_4 &= r \\ x_3 &= {}^{451}/_{466} + {}^{130}/_{233}r \\ x_2 &= {}^{225}/_{466} + {}^{147}/_{233}r \\ x_1 &= {}^{227}/_{233} - {}^{104}/_{233}r \end{aligned}$$

where r is any real number. Our next step is to show that every system of linear equations can be carried into a new system with the advantages of (12–6) by a sequence of elementary transformations. The precise description of a system of n linear equations in m unknowns that has a special form clearly exhibiting the properties of the system is somewhat complicated. However, the concepts involved are quite simple, as a careful study of examples will show.

(13.3) definition A system of linear equations (12–1) is in *echelon form* if there exists an integer, k, $0 \leq k \leq n$, with the following properties:
(a) If $k < i \leq n$, then $a_{i,j} = 0$ for all j;
(b) if $k > 0$, then there exists an increasing sequence of positive integers $1 \leq m_1 < m_2 < \cdots < m_k \leq m$ such that $a_{i,j} = 0$ for $j < m_i$, and $a_{i,m_i} = 1 (1 \leq i \leq k)$.

More informally, condition (a) of Definition 13.3 states that in each equation after the kth equation the coefficients of the unknowns are all zero. Condition (b) gives the form of the first k equations with nonzero coefficients. This condition states that in the first equation, $x_{m_1} (m_1 \geq 1)$ has coefficient 1, and this is the first nonzero coefficient in the first equation; in the second equation, $x_{m_2} (m_2 > m_1)$ has coefficient 1, and this is the first nonzero coefficient in the second equation, and in general, in the ith equation $(1 \leq i \leq k)$, $x_{m_i} (m_i > m_{i-1})$ has coefficient 1, and this is the first nonzero coefficient in the ith equation.

Note that if the coefficients of $x_1, x_2, \ldots, x_m$ are zero in every equation of the system (12–1), then the system is in echelon form with $k = 0$. For example,

$$\begin{aligned} 0x_1 + 0x_2 &= -1 \\ 0x_1 + 0x_2 &= 2 \end{aligned}$$

is in echelon form. The system (12–6) is in echelon form with $k = n = 3$, $m_1 = 1$, $m_2 = 2$, $m_3 = 3$. The system

$$\begin{aligned} x_1 + 3x_2 + 0x_3 + 0x_4 &= 2 \\ 0x_1 + 0x_2 + x_3 + 6x_4 &= 1 \\ 0x_1 + 0x_2 + 0x_3 + 0x_4 &= 0 \end{aligned}$$

is in echelon form with $k = 2$, $m_1 = 1$, $m_2 = 3$. Notice that this system could be written

$$\begin{aligned} x_1 + 3x_2 + 0x_3 + 0x_4 &= 2 \\ x_3 + 6x_4 &= 1 \\ 0 &= 0. \end{aligned}$$

Except for the trivial case where $k = 0$, a typical system in echelon form can be written

$$\begin{aligned} x_{m_1} + a_{1,m_1+1}\, x_{m_1+1} + a_{1,m_1+2}\, x_{m_1+2} + \cdots + a_{1,m}x_m &= b_1 \\ x_{m_2} + a_{2,m_2+1}\, x_{m_2+1} + \cdots + a_{2,m}x_m &= b_2 \\ &\vdots \\ x_{m_k} + a_{k,m_k+1}\, x_{m_k+1} + \cdots + a_{k,m}x_m &= b_k \qquad (13\text{–}3) \\ 0 &= b_{k+1} \\ &\vdots \\ 0 &= b_n. \end{aligned}$$

Of course, if $k = n$, the equations

$$\begin{aligned} 0 &= b_{k+1} \\ &\vdots \\ 0 &= b_n \end{aligned}$$

do not appear.

Example 1 of §12 illustrated how a system of equations could be carried into a system in echelon form by a sequence of elementary transformations on the set of vectors that span the row space of the given system. It is fairly evident from this example that there is a process by which any system of linear equations can be transformed into a system in echelon form in a finite number of steps. For those who insist on formal proofs of evident propositions, one is given for the following theorem. Such a proof inevitably uses mathematical induction. Further examples of the reduction of systems of equations to echelon form follow the theorem.

(13.4) theorem A system of linear equations can be transformed into an equivalent system in echelon form by a sequence of elementary transformations.

PROOF As we observed in §12, we may describe directly, in terms of the equations, any elementary transformation on the vectors spanning the row space of a given system of equations. The proof of this theorem is by mathematical induction on the number n of equations in the given system (12–1). If $n = 1$, then the system consists of the single equation

$$a_{1,1}x_1 + a_{1,2}x_2 + \cdots + a_{1,m}x_m = b_1.$$

If $a_{i,j} = 0$ for $j = 1, 2, \ldots, m$, then this system is in echelon form with $k = 0$. Otherwise, let t be the least positive integer such that $a_{1,t} \neq 0$. Then a Type III elementary transformation, which multiplies the equation by $1/a_{1,t}$, puts the system in echelon form with $k = n = 1$ and $m_1 = t$. Therefore, suppose that $n > 1$ and that any system of $n - 1$ equations can be carried into echelon form by a sequence of elementary transformations. If $a_{i,j} = 0$ for all i and j in system (12–1), then the system is in echelon form with $k = 0$. Otherwise, let t be the least positive integer such that $a_{i,t} \neq 0$ for some i. Thus, if $j < t$, then $a_{i,j} = 0$ for $i = 1, 2, \ldots, n$.

Now if $a_{1,t} = 0$, a Type I transformation, which interchanges the first and ith equations, yields a system in which the coefficient of x_t in the first equation is $a_{i,t} \neq 0$. Moreover, for every equation, the coefficient of x_j is zero if $j < t$. The coefficient of x_t in the first equation can be changed to 1 by a Type III transformation, which multiplies the first equation by $1/a_{i,t}$. The system now has the form

$$\begin{aligned} 0x_1 + \cdots + 0x_{t-1} + 1x_t + a^*_{1,t+1}x_{t+1} + \cdots + a^*_{1,m}x_m &= b^*_1 \\ 0x_1 + \cdots + 0x_{t-1} + a^*_{2,t}x_t + a^*_{2,t+1}x_{t+1} + \cdots + a^*_{2,m}x_m &= b^*_2 \\ &\vdots \\ 0x_1 + \cdots + 0x_{t-1} + a^*_{n,t}x_t + a^*_{n,t+1}x_{t+1} + \cdots + a^*_{n,m}x_m &= b^*_n. \end{aligned}$$

(Of course, if $t = 1$, the terms with coefficient 0 do not appear.) A sequence of Type II transformations that add to the ith equation the first equation multiplied by $-a^*_{i,t}$ yields a system in which the coefficient of x_t is zero in every equation except the first. The last $n - 1$ equations then have the form

$$\begin{aligned} 0x_1 + \cdots + 0x_{t-1} + 0x_t + a^{**}_{2,t+1}x_{t+1} + \cdots + a^{**}_{2,m}x_m &= b^{**}_2 \\ &\vdots \\ 0x_1 + \cdots + 0x_{t-1} + 0x_t + a^{**}_{n,t+1}x_{t+1} + \cdots + a^{**}_{n,m}x_m &= b^{**}_n. \end{aligned}$$

By the induction hypothesis, the latter system can be carried to echelon form by a sequence of elementary transformations, which may be regarded as transformations on the complete system of n equations that do not affect the first equation. Moreover, the coefficients of $x_1, \ldots, x_t$ remain zero when the last $n - 1$ equations undergo any elementary transformation. Therefore, the final system of n equations is in echelon form. The fact that the new system of equations is equivalent to the original system is a consequence of Corollary 13.2. ∎

EXAMPLE 1 Consider the system of equations

$$\begin{aligned} \tfrac{1}{2}x_1 + \tfrac{3}{2}x_2 - \tfrac{1}{2}x_3 \qquad\qquad + x_5 &= 1 \\ 2x_1 + 6x_2 + x_3 + 6x_4 + 4x_5 &= 13 \\ -3x_1 - 9x_2 \qquad - 6x_4 - 6x_5 &= -18. \end{aligned} \tag{13–4}$$

Let us perform a sequence of elementary transformations to carry this system into echelon form. To begin, multiply the first equation by 2:

$$\begin{aligned} x_1 + 3x_2 - x_3 \qquad + 2x_5 &= 2 \\ 2x_1 + 6x_2 + x_3 + 6x_4 + 4x_5 &= 13 \\ -3x_1 - 9x_2 \qquad - 6x_4 - 6x_5 &= -18. \end{aligned}$$

Next, multiply the first equation by -2 and add it to the second equation; then multiply the first equation by 3 and add it to the third equation:

$$\begin{aligned} x_1 + 3x_2 - x_3 \qquad + 2x_5 &= 2 \\ 3x_3 + 6x_4 \qquad &= 9 \\ -3x_3 - 6x_4 \qquad &= -12. \end{aligned}$$

Finally, add the second equation to the third and then multiply the second equation by $\tfrac{1}{3}$:

$$\begin{aligned} x_1 + 3x_2 - x_3 \qquad + 2x_5 &= 2 \\ x_3 + 2x_4 \qquad &= 3 \\ 0 &= -3. \end{aligned} \tag{13–5}$$

The system (13–5) is now in echelon form with $k = 2$, $m_1 = 1$, $m_2 = 3$. Since the third equation is never satisfied, there are no values for x_1, x_2, x_3, x_4, x_5 that satisfy the equations. That is, the equations are inconsistent. Thus, the original system (13–4), which is equivalent to (13–5), has no solution.

EXAMPLE 2 Consider the system of equations

$$\begin{aligned} 3x_1 + 2x_2 - x_3 + 4x_4 &= 6 \\ -2x_1 + x_2 + 5x_3 + x_4 &= 0 \\ x_1 - 4x_2 + 2x_3 + 8x_4 &= 2 \\ 5x_1 + x_2 - 3x_3 - 2x_4 &= 1. \end{aligned} \tag{13–6}$$

We transform system (13–6) into echelon form by a sequence of elementary transformations. First, interchange the first and third equations:

$$\begin{aligned} x_1 - 4x_2 + 2x_3 + 8x_4 &= 2 \\ -2x_1 + x_2 + 5x_3 + x_4 &= 0 \\ 3x_1 + 2x_2 - x_3 + 4x_4 &= 6 \\ 5x_1 + x_2 - 3x_3 - 2x_4 &= 1. \end{aligned}$$

Multiply the first equation by 2 and add it to the second equation; then multiply the first equation by -3 and add it to the third equation; finally, multiply the first equation by -5 and add it to the fourth equation:

$$\begin{aligned} x_1 - 4x_2 + 2x_3 + 8x_4 &= 2 \\ -7x_2 + 9x_3 + 17x_4 &= 4 \\ 14x_2 - 7x_3 - 20x_4 &= 0 \\ 21x_2 - 13x_3 - 42x_4 &= -9. \end{aligned}$$

Multiply the second equation by 2 and add it to the third equation; then multiply the second equation by 3 and add it to the fourth equation:

$$\begin{aligned} x_1 - 4x_2 + 2x_3 + 8x_4 &= 2 \\ -7x_2 + 9x_3 + 17x_4 &= 4 \\ 11x_3 + 14x_4 &= 8 \\ 14x_3 + 9x_4 &= 3. \end{aligned}$$

Multiply the third equation by $-{}^{14}\!/_{11}$ and add it to the fourth equation:

$$\begin{aligned} x_1 - 4x_2 + 2x_3 + 8x_4 &= 2 \\ -7x_2 + 9x_3 + 17x_4 &= 4 \\ 11x_3 + 14x_4 &= 8 \\ -{}^{97}\!/_{11}x_4 &= -{}^{79}\!/_{11}. \end{aligned}$$

Finally multiply the second equation by $-{}^{1}\!/_{7}$; multiply the third equation by ${}^{1}\!/_{11}$; multiply the fourth equation by $-{}^{11}\!/_{97}$:

$$\begin{aligned} x_1 - 4x_2 + 2x_3 + 8x_4 &= 2 \\ x_2 - {}^{9}\!/_{7}x_3 - {}^{17}\!/_{7}x_4 &= -{}^{4}\!/_{7} \\ x_3 + {}^{14}\!/_{11}x_4 &= {}^{8}\!/_{11} \\ x_4 &= {}^{79}\!/_{97}. \end{aligned} \tag{13–7}$$

The resulting system (13–7) is in echelon form with $k = n = m = 4$, $m_1 = 1$, $m_2 = 2$, $m_3 = 3$, $m_4 = 4$. The system (13–7) has a unique solution which can be found by substituting the value of x_4 from the fourth equation into the third equation to solve for x_3, then substituting the values of x_3 and x_4 in the second equation to solve for x_2, and finally, substituting the values of x_2, x_3, and x_4 in the first equation to solve for x_1. This solution is the unique solution of the original system (13–6), since (13–6) and (13–7) are equivalent systems by Corollary 13.2.

Theorem 13.4, which completes the description of the process for solving a system of linear equations, reduces the problem of solving general systems to that of solving systems in echelon form. Moreover, the proof of Theorem 13.4 provides the method (or algorithm) for replacing a given system of equations by an equivalent system in echelon form. It is evident from the examples that systems in echelon form are easily solved.

In general, there are many echelon forms into which a system of equations may be transformed. However, by Corollary 13.2, all of these systems are equivalent.

EXAMPLE 3 Consider the system of equations

$$\begin{aligned} x_1 + x_2 + x_3 &= 1 \\ x_1 + x_2 - x_3 &= 2 \\ x_1 - x_2 - x_3 &= 3. \end{aligned}$$

This system can be carried by elementary transformations into each of the following systems, both of which are in echelon form:

$$\begin{aligned} x_1 + x_2 - x_3 &= 2 \\ x_2 \qquad &= -\tfrac{1}{2} \\ x_3 &= -\tfrac{1}{2} \end{aligned} \qquad\qquad \begin{aligned} x_1 - x_2 - x_3 &= 3 \\ x_2 + x_3 &= -1 \\ x_3 &= -\tfrac{1}{2}. \end{aligned}$$

It is evident from either echelon form that the given system has a unique solution $x_3 = -\frac{1}{2}$, $x_2 = -\frac{1}{2}$, $x_1 = 2$.

exercises

13.1 Transform each of the systems of equations of Exercise 12.1 into echelon form, and then solve the systems.

13.2 Put the following systems of equations into echelon form. State the values of k and m_i, $i = 1, 2, \ldots, k$ (see Definition 13.3). Solve the systems.

(a)
$$\begin{aligned} 2x_1 - 7x_2 + 5x_3 - 8x_4 &= 7 \\ 4x_1 - 14x_2 + 10x_3 - 5x_4 &= 2 \end{aligned}$$

(b)
$$\begin{aligned} 4x_1 \qquad\quad + 6x_3 - 2x_4 &= 4 \\ 3x_1 - 7x_2 \qquad\quad + 2x_4 &= 5 \\ x_2 - 7x_3 + x_4 &= 2 \end{aligned}$$

(c)
$$\begin{aligned} 3x_1 \qquad\quad - x_3 &= 0 \\ x_1 + x_2 - 2x_3 &= 0 \\ 2x_1 - 2x_2 + 8x_3 &= 0 \\ 2x_1 - x_2 - 4x_3 &= 0 \end{aligned}$$

(d)
$$\begin{aligned} 5x_1 + 2x_2 - 7x_3 &= 1 \\ 7x_1 - x_2 + 2x_3 &= 0 \\ 2x_1 + 5x_2 - x_3 &= 5 \end{aligned}$$

(e)
$$\begin{aligned} 2x_1 - 3x_2 + x_3 - x_4 &= 1 \\ -x_1 + 5x_2 + 2x_3 + 5x_4 &= 5 \\ 5x_1 + 9x_2 - 14x_3 + 8x_4 &= -5 \\ 3x_1 + x_2 - 4x_3 + 2x_4 &= -1 \\ x_1 - x_2 + x_3 - x_4 &= 2 \end{aligned}$$

(f)
$$\begin{aligned} x_3 - 4x_4 &= 1 \\ 2x_2 \qquad + x_4 &= \tfrac{1}{2} \\ x_1 - x_2 + x_3 \qquad &= 0 \\ 3x_1 \qquad\qquad - x_4 &= 2 \end{aligned}$$

13.3 Complete the solution of the system (13–7) in Example 2.

13.4 In Example 3, find sequences of elementary transformations that carry the given system of equations into the echelon forms given in the example. Find other echelon forms for the given system.

14 theory of linear systems; homogeneous systems

By the results of §13, the study of systems of linear equations can be replaced by the study of systems in echelon form. For systems in echelon form, we refer to the integer k of Definition 13.3 as the *associated integer* k. This associated integer k of a system in echelon form is just the number of equations in which the unknowns of the system appear.

EXAMPLE 1 The system of equations (13–5) in Example 1, §13, is in echelon form with associated integer $k = 2$. The vectors that span the row space of the system (13–5) are $\mathbf{U}_1 = (1, 3, -1, 0, 2, 2)$, $\mathbf{U}_2 = (0, 0, 1, 2, 0, 3)$, and $\mathbf{U}_3 = (0, 0, 0, 0, 0, -3)$. Suppose that c_1, c_2, and c_3 are real numbers such that

$$c_1 \cdot \mathbf{U}_1 + c_2 \cdot \mathbf{U}_2 + c_3 \cdot \mathbf{U}_3 = \mathbf{0}.$$

We have

$$\begin{aligned} c_1 \cdot (1, 3, -1, 0, 2, 2) &+ c_2 \cdot (0, 0, 1, 2, 0, 3) + c_3 \cdot (0, 0, 0, 0, 0, -3) \\ &= (c_1, 3c_1, -c_1, 0, 2c_1, 2c_1) + (0, 0, c_2, 2c_2, 0, 3c_2) \\ &\quad + (0, 0, 0, 0, 0, -3c_3) \\ &= (c_1, 3c_1, -c_1 + c_2, 2c_2, 2c_1, 2c_1 + 3c_2 - 3c_3) \\ &= (0, 0, 0, 0, 0, 0). \end{aligned}$$

Therefore,

$$\begin{aligned} c_1 &= 0 \\ 3c_1 &= 0 \\ -c_1 + c_2 &= 0 \\ 2c_2 &= 0 \\ 2c_1 &= 0 \\ 2c_1 + 3c_2 - 3c_3 &= 0. \end{aligned}$$

These equations imply $c_1 = c_2 = c_3 = 0$. Hence, the vectors $\mathbf{U}_1$, $\mathbf{U}_2$, and $\mathbf{U}_3$ are linearly independent, so that the dimension of the row space of the system is $k + 1 = 3$. As we observed earlier, the system (13–5) is inconsistent.

EXAMPLE 2 The system of equations (13–7) in Example 2, §13, is in echelon form with associated integer $k = 4$. The vectors that span the row space of the system (13–7) are $\mathbf{U}_1 = (1, -4, 2, 8, 2)$, $\mathbf{U}_2 = (0, 1, -9/7, -17/7, -4/7)$, $\mathbf{U}_3 = (0, 0, 1, 14/11, 8/11)$, and $\mathbf{U}_4 =$

$(0, 0, 0, 1, {}^{79}\!/_{97})$. If c_1, c_2, c_3, and c_4 are real numbers such that

$$c_1 \cdot \mathbf{U}_1 + c_2 \cdot \mathbf{U}_2 + c_3 \cdot \mathbf{U}_3 + c_4 \cdot \mathbf{U}_4 = \mathbf{0},$$

we obtain the following equations:

$$\begin{aligned} c_1 &= 0 \\ -4c_1 + \quad c_2 &= 0 \\ 2c_1 - {}^{9}\!/_{7}c_2 + \quad c_3 &= 0 \\ 8c_1 - {}^{17}\!/_{7}c_2 + {}^{14}\!/_{11}c_3 + \quad c_4 &= 0 \\ 2c_1 - {}^{4}\!/_{7}c_2 + {}^{8}\!/_{11}c_3 + {}^{79}\!/_{97}c_4 &= 0. \end{aligned}$$

These equations imply $c_1 = c_2 = c_3 = c_4 = 0$. Therefore, $\{\mathbf{U}_1, \mathbf{U}_2, \mathbf{U}_3, \mathbf{U}_4\}$ is a linearly independent set and it follows that the dimension of the row space of the system is $k = 4$. Since (13–7) is a system in $m = 4$ unknowns, we have $k = m$. As we discovered in the example, the system (13–7) is consistent, and has a unique solution.

EXAMPLE 3 The system

$$\begin{aligned} x_1 - 2x_2 + x_3 &= 4 \\ x_3 &= 2 \end{aligned}$$

is in echelon form with associated integer $k = 2$. The vectors that span the row space of the system are $\mathbf{U}_1 = (1, -2, 1, 4)$ and $\mathbf{U}_2 = (0, 0, 1, 2)$; if $c_1 \cdot \mathbf{U}_1 + c_2 \cdot \mathbf{U}_2 = \mathbf{0}$, then

$$\begin{aligned} c_1 &= 0 \\ -2c_1 &= 0 \\ c_1 + \quad c_2 &= 0 \\ 4c_1 + 2c_2 &= 0. \end{aligned}$$

Hence, $c_1 = c_2 = 0$, and the vectors $\mathbf{U}_1$ and $\mathbf{U}_2$ are linearly independent. Thus, the dimension of the row space of the system is $k = 2$. In this system, the number of unknowns is $m = 3$, so that here $k < m$. If r is any real number, then $x_1 = 2 + 2r$, $x_2 = r$, $x_3 = 2$ is a solution of the system. Hence, the system is consistent, but it does not have a unique solution.

The above examples illustrate the various statements in the following theorem. Moreover, these examples also suggest the method for proving the theorem.

(14.1) theorem Let S be a system of n equations in m unknowns, and let S be in echelon form with associated integer k. The dimension of the row space of S is either k or $k + 1$. The system is consistent if and only if this dimension is k. If the system is consistent, then its solution is unique if and only if $k = m$.

PROOF The vectors that span the row space of a system S in echelon form are [see equations (13–3)]

$$\begin{aligned}
\mathbf{U}_1 &= (0, \ldots, 0, 1, a_{1,m_1+1}, \ldots, a_{1,m}, b_1)\\
\mathbf{U}_2 &= (0, \ldots, 0, 1, a_{2,m_2+1}, \ldots, a_{2,m}, b_2)\\
&\vdots\\
\mathbf{U}_k &= (0, \ldots, 0, 1, a_{k,m_k+1}, \ldots, a_{k,m}, b_k)\\
\mathbf{U}_{k+1} &= (0, \ldots, 0, b_{k+1})\\
&\vdots\\
\mathbf{U}_n &= (0, \ldots, 0, b_n).
\end{aligned}$$

It is convenient to consider the following three mutually exclusive cases: (1) $k = n$; (2) $k < n$ and $b_{k+1} = \cdots = b_n = 0$; (3) $k < n$ and some $b_r \neq 0$ where $k + 1 \leq r \leq n$. In case (1), the row space is spanned by $\mathbf{U}_1, \mathbf{U}_2, \ldots, \mathbf{U}_k$. In case (2), the last $n - k$ vectors are zero, so that again the row space is spanned by $\mathbf{U}_1, \mathbf{U}_2, \ldots, \mathbf{U}_k$. Thus, we may consider cases (1) and (2) together. Clearly, if $k = 0$, the row space is $\{\mathbf{0}\}$ and its dimension is $k = 0$. Therefore, suppose that $k > 0$, and that

$$c_1 \cdot \mathbf{U}_1 + c_2 \cdot \mathbf{U}_2 + \cdots + c_k \cdot \mathbf{U}_k = \mathbf{0}$$

for real numbers $c_1, c_2, \ldots, c_k$. Since $m_1 < m_2 < \cdots < m_k$, it follows that

$$\begin{aligned}
c_1 &= 0\\
c_1 a_{1,m_2} + c_2 &= 0\\
c_1 a_{1,m_3} + c_2 a_{2,m_3} + c_3 &= 0 \qquad (14\text{–}1)\\
&\vdots\\
c_1 a_{1,m_k} + c_2 a_{2,m_k} + \cdots + c_{k-1} a_{k-1,m_k} + c_k &= 0.
\end{aligned}$$

Equations (14–1) imply that $c_1 = c_2 = \cdots = c_k = 0$. Thus, the vectors $\mathbf{U}_1, \mathbf{U}_2, \ldots, \mathbf{U}_k$ are linearly independent and form a basis of the row space. Therefore, in cases (1) and (2), the dimension of the row space is k.

In case (3), some $b_r \neq 0$, where $k + 1 \leq r \leq n$. If

$$c_1 \cdot \mathbf{U}_1 + c_2 \cdot \mathbf{U}_2 + \cdots + c_k \cdot \mathbf{U}_k + c_{k+1} \cdot \mathbf{U}_r = \mathbf{0},$$

then, in addition to equations (14–1), we have the equation

$$c_1 b_1 + c_2 b_2 + \cdots + c_k b_k + c_{k+1} b_r = 0.$$

Therefore, $c_1 = c_2 = \cdots = c_k = 0$ and $c_{k+1} b_r = 0$. Since $b_r \neq 0$, this implies $c_{k+1} = 0$. Hence, the set $\{\mathbf{U}_1, \mathbf{U}_2, \ldots, \mathbf{U}_k, \mathbf{U}_r\}$ is linearly independent. (If $k = 0$, this set contains the single vector $\mathbf{U}_r$.) Since each $\mathbf{U}_s$ with $k + 1 \leq s \leq n$ is a scalar multiple of $\mathbf{U}_r[\mathbf{U}_s = (b_s/b_r) \cdot \mathbf{U}_r]$, the vectors $\mathbf{U}_1, \mathbf{U}_2, \ldots, \mathbf{U}_k, \mathbf{U}_r$ span the row space. Therefore, in case (3), the row space has a basis of $k + 1$ vectors. That is, the dimension of the row space is $k + 1$.

If the dimension of the row space of the system of equations S is k, then case (1) or (2) must hold. Therefore, the given system has the form

$$\begin{aligned}
x_{m_1} + a_{1,m_1+1} x_{m_1+1} + \cdots + a_{1,m} x_m &= b_1\\
x_{m_2} + a_{2,m_2+1} x_{m_2+1} + \cdots + a_{2,m} x_m &= b_2 \qquad (14\text{–}2)\\
&\vdots\\
x_{m_k} + a_{k,m_k+1} x_{m_k+1} + \cdots + a_{k,m} x_m &= b_k.
\end{aligned}$$

By choosing arbitrary real values for the unknowns other than $x_{m_1}, x_{m_2}, \ldots, x_{m_k}$, and by solving the equations in turn for $x_{m_k}, x_{m_{k-1}}, \ldots, x_{m_1}$, we obtain a solution. Therefore, the system is consistent. However, if the dimension of the row space is $k + 1$, then the situation must be that of case (3). Thus, the system contains an equation $0 = b_r$, where $b_r \neq 0$. Since this is impossible, the equations are inconsistent.

If the system is consistent, then it is clear from equations (14–2) that the solution is unique if and only if no arbitrary choice for the values of some of the unknowns is possible; that is, if and only if the only unknowns are $x_{m_1}, x_{m_2}, \ldots, x_{m_k}$. The latter condition holds if and only if $x_{m_1} = x_1$, $x_{m_2} = x_2, \ldots, x_{m_k} = x_m$; that is, if and only if $k = m$. ∎

The system of equations (12–1) is called *homogeneous* if $b_1 = b_2 = \cdots = b_n = 0$. Suppose that a homogeneous system is carried into echelon form by a sequence of elementary transformations. Let $b_1^*, b_2^*, \ldots, b_n^*$ be the right-hand members of the equations in this echelon form. By observing the effect of any elementary transformation on a homogeneous system of equations, it is clear that $b_1^* = b_2^* = \cdots = b_n^* = 0$. Therefore, either case (1) or (2) of Theorem 14.1 applies, and the system is consistent. It is immediately obvious that $x_1 = 0, x_2 = 0, \ldots, x_m = 0$ is a solution of a homogeneous system. However, we have previously encountered problems where we wish to know whether a homogeneous system has a solution other than this trivial one. Theorem 14.2 provides the answer to this question.

(14.2) theorem A homogeneous system of n equations in m unknowns has a nontrivial solution if and only if the dimension k of its row space is less than m.

PROOF By Theorem 12.2, a system of equations has the same row space as any echelon form for the system. Since a homogeneous system of equations is consistent, by Theorem 14.1, the dimension of its row space is the associated integer k of an echelon form for the system. By Definition 13.3, for any system of equations in echelon form, either $k = 0$ or $k \leq m_k \leq m$. Thus, it is always the case that $k \leq m$. By Theorem 14.1, a homogeneous system has a unique solution (which must be $x_1 = 0$, $x_2 = 0, \ldots, x_m = 0$) if and only if $k = m$. Therefore, a homogeneous system has a solution other than $x_1 = 0, x_2 = 0, \ldots, x_m = 0$ if and only if $k < m$. ∎

EXAMPLE 4 Consider the homogeneous system of $n = 3$ linear equations in $m = 3$ unknowns:

$$\begin{aligned} 2x_1 - 3x_2 - x_3 &= 0 \\ x_1 + 4x_2 + 8x_3 &= 0 \\ 3x_1 + x_2 + 6x_3 &= 0. \end{aligned}$$

An echelon form for this system is

$$\begin{aligned} x_1 + 4x_2 + \quad 8x_3 &= 0 \\ x_2 + {}^{17}\!/_{11}x_3 &= 0 \\ x_3 &= 0. \end{aligned}$$

This echelon form has associated integer $k = 3$, which is the dimension of the row space of the original system. Since $k = m = 3$, the system has only the trivial solution $x_1 = x_2 = x_3 = 0$. Of course, it is immediately evident from the echelon form that the system admits only the trivial solution.

EXAMPLE 5 The following homogeneous system consists of $n = 5$ equations in $m = 4$ unknowns:

$$\begin{aligned} x_1 - 3x_2 + 4x_3 - \quad x_4 &= 0 \\ 3x_1 + \quad x_2 + 2x_3 + 4x_4 &= 0 \\ 2x_1 - 4x_2 + 6x_3 + \quad x_4 &= 0 \\ 2x_1 + 2x_2 \qquad\quad + 2x_4 &= 0 \\ 4x_1 - 4x_2 + 8x_3 \qquad\quad &= 0. \end{aligned}$$

An echelon form for this system is

$$\begin{aligned} x_1 - 3x_2 + 4x_3 - \quad x_4 &= 0 \\ x_2 - \quad x_3 + {}^3\!/_2 x_4 &= 0 \\ x_4 &= 0. \end{aligned}$$

This echelon form has associated integer $k = 3$. Thus, the dimension of the row space of the original system is $k = 3$. Since $m = 4$, we have $k < m$, so that the system has a nontrivial solution. In fact, if r is any real number, then $x_1 = -r$, $x_2 = r$, $x_3 = r$, $x_4 = 0$ is a solution of the given system. For example, $x_1 = -1$, $x_2 = 1$, $x_3 = 1$, $x_4 = 0$ is a nontrivial solution.

(14.3) corollary A homogeneous system of n equations in m unknowns has a nontrivial solution if $n < m$.

PROOF The dimension of the row space of the system is the associated integer k of an echelon form for the system. By Definition 13.3, $k \leq n < m$. By Theorem 14.2, the system has a nontrivial solution. ▮

EXAMPLE 6 The homogeneous system

$$\begin{aligned} 6x_1 - 2x_2 - 5x_3 + \quad x_4 &= 0 \\ 3x_1 + \quad x_2 + \quad x_3 - 2x_4 &= 0 \\ x_1 + 3x_2 - \quad x_3 + 4x_4 &= 0 \end{aligned}$$

has fewer equations than unknowns ($n = 3$, $m = 4$). Therefore, by Corollary 14.3, the system has a nontrivial solution. An echelon form for the system is

$$\begin{aligned} x_1 + 3x_2 - \quad x_3 + \ 4x_4 &= 0 \\ x_2 - \tfrac{1}{2}x_3 + \tfrac{7}{4}x_4 &= 0 \\ x_3 - \tfrac{4}{3}x_4 &= 0. \end{aligned}$$

If r is any real number, $x_1 = 7r$, $x_2 = -13r$, $x_3 = 16r$, $x_4 = 12r$ is a solution of the system. Thus, one nontrivial solution is $x_1 = 7$, $x_2 = -13$, $x_3 = 16$, $x_4 = 12$.

At this point, let us digress from our study of systems of equations to explain a notation which is convenient for the rest of this discussion. It is a standard notation which is useful in many mathematical expositions. The expression

$$\sum_{i=1}^{n} a_i$$

stands for the sum $a_1 + a_2 + \cdots + a_n$. For example, $\Sigma_{i=1}^4 a_i = a_1 + a_2 + a_3 + a_4$. We read $\Sigma_{i=1}^n a_i$ as "the sum of the a_i from $i = 1$ to $i = n$." The symbol Σ is called the *summation sign*. The letter i in the expression is called the *index of summation*. A different choice for the index of summation does not change the sum; thus,

$$\sum_{i=1}^{n} a_i, \ \sum_{j=1}^{n} a_j, \ \sum_{t=1}^{n} a_t$$

are the same, since they are all abbreviations of

$$a_1 + a_2 + \cdots + a_n.$$

A linear equation can be conveniently expressed using the summation notation. For example, the ith equation in a system of n linear equations in m unknowns,

$$a_{i,1}x_1 + a_{i,2}x_2 + \cdots + a_{i,m}x_m = b_i,$$

can be written as

$$\sum_{j=1}^{m} a_{i,j}x_j = b_i.$$

Other examples of the use of the summation notation are

$$\sum_{i=1}^{5} 2^i = 2^1 + 2^2 + 2^3 + 2^4 + 2^5 = 62,$$

$$\sum_{r=1}^{n} ra_r = a_1 + 2a_2 + 3a_3 + \cdots + na_n,$$

and

$$\sum_{j=1}^{m} a_{i,j}(c_j + d_j) = a_{i,1}(c_1 + d_1) + a_{i,2}(c_2 + d_2) + \cdots + a_{i,m}(c_m + d_m).$$

Returning to our discussion, we observe that the row space of a homogeneous system of linear equations

$$\begin{aligned} a_{1,1}x_1 + a_{1,2}x_2 + \cdots + a_{1,m}x_m &= 0 \\ a_{2,1}x_1 + a_{2,2}x_2 + \cdots + a_{2,m}x_m &= 0 \\ &\vdots \\ a_{n,1}x_1 + a_{n,2}x_2 + \cdots + a_{n,m}x_m &= 0 \end{aligned} \tag{14–3}$$

is the subspace of $\mathcal{V}_{m+1}$ spanned by the set of vectors $T = \{(a_{1,1}, a_{1,2}, \ldots, a_{1,m}, 0), (a_{2,1}, a_{2,2}, \ldots, a_{2,m}, 0), \ldots, (a_{n,1}, a_{n,2}, \ldots, a_{n,m}, 0)\}$. Corresponding to each vector $(a_{i,1}, a_{i,2}, \ldots, a_{i,m}, 0)$ in T, there is a unique vector $(a_{i,1}, a_{i,2}, \ldots, a_{i,m})$ in $\mathcal{V}_m$. Since the $(m + 1)$th component of each vector in T equals zero, a subset of T is a linearly independent set in $\mathcal{V}_{m+1}$ if and only if the corresponding subset of $T' = \{(a_{1,1}, a_{1,2}, \ldots, a_{1,m}), (a_{2,1}, a_{2,2}, \ldots, a_{2,m}), \ldots, (a_{n,1}, a_{n,2}, \ldots, a_{n,m})\}$ is a linearly independent set in $\mathcal{V}_m$. Therefore, the dimension k of the row space of equations (14–3) is the same as the dimension of the subspace $\mathcal{S}$ of $\mathcal{V}_m$ spanned by T'.

A solution $x_1 = c_1, x_2 = c_2, \ldots, x_m = c_m$ of (14–3) may be regarded as a vector $(c_1, c_2, \ldots, c_m)$ in $\mathcal{V}_m$. We shall refer to $(c_1, c_2, \ldots, c_m)$ as a *solution vector* of (14–3). If $(c_1, c_2, \ldots, c_m)$ and $(d_1, d_2, \ldots, d_m)$ are solution vectors of (14–3), then

$$\sum_{j=1}^{m} a_{i,j}(c_j + d_j) = \sum_{j=1}^{m} a_{i,j}c_j + \sum_{j=1}^{m} a_{i,j}d_j = 0 + 0 = 0$$

for $i = 1, 2, \ldots, n$. Hence, the sum

$$(c_1, c_2, \ldots, c_m) + (d_1, d_2, \ldots, d_m) = (c_1 + d_1, c_2 + d_2, \ldots, c_m + d_m)$$

of two solution vectors is again a solution vector. Similarly, if t is a real number and $(c_1, c_2, \ldots, c_m)$ is a solution vector of (14–3), then

$$t \cdot (c_1, c_2, \ldots, c_m) = (tc_1, tc_2, \ldots, tc_m)$$

is a solution vector, since

$$\sum_{j=1}^{m} a_{i,j}(tc_j) = t \sum_{j=1}^{m} a_{i,j}c_j = t(0) = 0.$$

Therefore, the set of all solution vectors of (14–3) is a subspace $\mathfrak{I}$ of $\mathcal{V}_m$, called the *solution space* of (14–3).

EXAMPLE 7 Since the homogeneous system of equations given in Example 4 above has only the trivial solution $x_1 = x_2 = x_3 = 0$, the solution space of the system is the zero subspace $\{(0, 0, 0)\}$ of $\mathcal{V}_3$. In Example 5, every solution vector of the given system has the form $(-r, r, r, 0)$, where r is a real number. Since $(-r, r, r, 0) = r \cdot (-1, 1, 1, 0)$, the solution space of this system is the subspace $\mathfrak{J}$ of $\mathcal{V}_4$ spanned by the vector $(-1, 1, 1, 0)$. In Example 6, $(7r, -13r, 16r, 12r)$, where r is any real number, is a solution vector. Therefore, the solution space of the system given in Example 6 is the subspace $\mathfrak{J}$ of $\mathcal{V}_4$ spanned by the vector $(7, -13, 16, 12)$.

We now regard $\mathcal{V}_m$ as a Euclidean vector space with the $E^{(m)}$ inner product (see Example 1, §11). Then a vector $(c_1, c_2, \ldots, c_m)$ is in the solution space of (14–3) if and only if it is orthogonal to each vector in the set T' that spans the subspace $\mathcal{S}$ of $\mathcal{V}_m$ mentioned above. Moreover, $(c_1, c_2, \ldots, c_m)$ is orthogonal to each vector in T' if and only if it is orthogonal to each vector in $\mathcal{S}$ (see Exercise 11.7). Therefore, the solution space $\mathfrak{J}$ of (14–3) is the orthogonal complement of $\mathcal{S}$ in $\mathcal{V}_m$, and $\mathcal{V}_m = \mathcal{S} \oplus \mathfrak{J}$. Since the dimension of $\mathcal{S}$ is k, it now follows from Exercise 10.14 that the dimension of $\mathfrak{J}$ is $m - k$. Thus, we have proved the following useful result.

(14.4) theorem If k is the dimension of the row space of a homogeneous system of n linear equations in m unknowns, then the dimension of the solution space of the system is $m - k$.

By Corollary 10.8 and Theorem 14.4, any set of $m - k$ linearly independent vectors in the solution space of the equations (14–3) is a basis of this vector space. Thus, any solution vector of (14–3) is a linear combination of $m - k$ linearly independent solution vectors of (14–3). Theorem 14.4 tells us, therefore, when we can be certain of having found all solutions of the system.

EXAMPLE 8 The homogeneous system

$$\begin{aligned} x_1 + 2x_2 + 3x_3 + 4x_4 &= 0 \\ 4x_1 + 3x_2 + 2x_3 + x_4 &= 0 \end{aligned}$$

is a system of $n = 2$ equations in $m = 4$ unknowns. An echelon form for the system is

$$\begin{aligned} x_1 + 2x_2 + 3x_3 + 4x_4 &= 0 \\ x_2 + 2x_3 + 3x_4 &= 0. \end{aligned}$$

Thus, $k = 2$ is the dimension of the row space of the given system. The dimension of the solution space $\mathfrak{J}$ is $m - k = 4 - 2 = 2$. Therefore, every solution vector is a linear combination of two linearly independent solution vectors. By inspection of the echelon form of the system, $(1, -2, 1, 0)$ and $(2, -3, 0, 1)$ are solution vectors. It is easy to check that these vectors are linearly independent in $\mathcal{V}_4$. Thus, *every* solution of the system has the form

$$r \cdot (1, -2, 1, 0) + s \cdot (2, -3, 0, 1) = (r + 2s, -2r - 3s, r, s)$$

where r and s are any real numbers.

As mentioned in the above discussion, the row space of a homogeneous system of n equations in m unknowns, may be regarded as a subspace of $\mathcal{V}_m$ (instead of $\mathcal{V}_{m+1}$). Thus, the row space $\mathcal{S}$ of the given system is the subspace of $\mathcal{V}_4$ spanned by the vectors (1, 2, 3, 4) and (0, 1, 2, 3). Every vector in $\mathcal{S}$ has the form

$$t \cdot (1, 2, 3, 4) + u \cdot (0, 1, 2, 3) = (t, 2t + u, 3t + 2u, 4t + 3u)$$

where t and u are real numbers. Computing the $E^{(4)}$ inner product,

$$\begin{aligned}(r + 2s, -2r - 3s, r, s) \circ (t, 2t + u, 3t + 2u, 4t + 3u)\\ &= (r + 2s)t + (-2r - 3s)(2t + u) + r(3t + 2u) + s(4t + 3u)\\ &= rt + 2st - 4rt - 2ru - 6st - 3su + 3rt + 2ru + 4st + 3su\\ &= 0\end{aligned}$$

we find that every vector in $\mathfrak{J}$ is orthogonal to every vector in $\mathcal{S}$. That is, $\mathfrak{J}$ is the orthogonal complement of $\mathcal{S}$, and $\mathcal{V}_4 = \mathcal{S} \oplus \mathfrak{J}$. Since $\{(1, 2, 3, 4), (0, 1, 2, 3)\}$ is a basis of $\mathcal{S}$ and $\{(1, -2, 1, 0), (2, -3, 0, 1)\}$ is a basis of $\mathfrak{J}$, it follows that $\{(1, 2, 3, 4), (0, 1, 2, 3), (1, -2, 1, 0), (2, -3, 0, -1)\}$ is a basis of $\mathcal{V}_4$ (see Exercise 14.8).

Any solution $x_1 = r_1, x_2 = r_2, \ldots, x_m = r_m$ of the general system of linear equations (12–1) can also be regarded as a vector $(r_1, r_2, \ldots, r_m)$ in $\mathcal{V}_m$. Replacing each b_i, $i = 1, 2, \ldots, n$, by 0 in the equations (12–1), we obtain the *associated homogeneous system* (14–3). The following theorem describes the structure of the solutions of the system (12–1).

(14.5) theorem Suppose that the system of linear equations (12–1) is consistent, and let $(r_1, r_2, \ldots, r_m)$ be a fixed solution vector of (12–1). Then $(s_1, s_2, \ldots, s_m)$ is a solution vector of (12–1) if and only if

$$\begin{aligned}(s_1, s_2, \ldots, s_m) &= (r_1, r_2, \ldots, r_m) + (c_1, c_2, \ldots, c_m)\\ &= (r_1 + c_1, r_2 + c_2, \ldots, r_m + c_m)\end{aligned}$$

where $(c_1, c_2, \ldots, c_m)$ is a solution vector of the associated homogeneous system (14–3).

PROOF Let $(c_1, c_2, \ldots, c_m)$ be a solution vector of the homogeneous system (14–3). Then, since $(r_1, r_2, \ldots, r_m)$ is a solution vector of (12–1), we have

$$\sum_{j=1}^{m} a_{i,j}(r_j + c_j) = \sum_{j=1}^{m} a_{i,j}r_j + \sum_{j=1}^{m} a_{i,j}c_j = b_i + 0 = b_i$$

for $i = 1, 2, \ldots, n$. Hence, $(r_1 + c_1, r_2 + c_2, \ldots, r_m + c_m)$ is a solution vector of (12–1). Conversely, suppose that $(s_1, s_2, \ldots, s_m)$ is any solution vector of (12–1). Then

$$\sum_{j=1}^{m} a_{i,j}(s_j - r_j) = \sum_{j=1}^{m} a_{i,j}s_j - \sum_{j=1}^{m} a_{i,j}r_j = b_i - b_i = 0$$

for $i = 1, 2, \ldots, n$. Let $s_j - r_j = c_j$ for $j = 1, 2, \ldots, m$. Therefore,

$$(s_1, s_2, \ldots, s_m) - (r_1, r_2, \ldots, r_m) = (c_1, c_2, \ldots, c_m)$$

is a solution vector of the homogeneous system (14–3). That is,

$$(s_1, s_2, \ldots, s_m) = (r_1, r_2, \ldots, r_m) + (c_1, c_2, \ldots, c_m).$$ ∎

Combining the results of Theorems 14.4 and 14.5, we have the following: Any solution vector of a system of n linear equations in m unknowns is the sum of a fixed solution vector of this system and a linear combination of $m - k$ linearly independent solution vectors of the associated homogeneous system, where k is the dimension of the row space of the homogeneous system.

EXAMPLE 9 The system of equations

$$\begin{aligned} x_1 + 2x_2 + 3x_3 + 4x_4 &= 2 \\ 4x_1 + 3x_2 + 2x_3 + x_4 &= 5 \end{aligned}$$

has the homogeneous system given in Example 8 as its associated homogeneous system. An echelon form for the given system is

$$\begin{aligned} x_1 + 2x_2 + 3x_3 + 4x_4 &= 2 \\ x_2 + 2x_3 + 3x_4 &= 3/5. \end{aligned}$$

If we let $x_3 = x_4 = 0$, a solution vector of the system is $(4/5, 3/5, 0, 0)$. In Example 8, we found that $\{(1, -2, 1, 0), (2, -3, 0, 1)\}$ is a linearly independent set of solutions of the associated homogeneous system. Thus, every solution of the given system has the form

$$\begin{aligned} &(4/5, 3/5, 0, 0) + r \cdot (1, -2, 1, 0) + s \cdot (2, -3, 0, 1) \\ &\qquad = (4/5 + r + 2s, 3/5 - 2r - 3s, r, s) \end{aligned}$$

where r and s are real numbers.

exercises

14.1 Which of the following systems of equations are consistent? Find all solutions of the consistent systems.

(a)
$$\begin{aligned} 3x_1 - 5x_2 + x_3 &= 1 \\ x_1 - 7x_2 + 2x_3 &= 0 \\ 3x_1 + 11x_2 - 4x_3 &= 5 \end{aligned}$$

(b)
$$\begin{aligned} x_1 - 5x_2 + 4x_3 + x_4 &= 7 \\ 2x_1 + x_2 - 3x_3 - x_4 &= 1 \\ 3x_1 - 4x_2 + x_3 - 5x_4 &= 0 \\ x_1 + x_2 + x_3 + x_4 &= 1 \end{aligned}$$

(c)
$$\begin{aligned} 3x_1 - 5x_2 &= 1/2 \\ 1/3 x_1 + x_2 &= 2/3 \\ 2x_1 - x_2 &= 1 \end{aligned}$$

(d)
$$\begin{aligned} x_1 + 3x_2 - x_3 &= 11 \\ 2x_1 - x_2 - x_3 &= 0 \\ x_1 + 2x_2 + 6x_3 &= 7 \\ x_1 + x_2 + x_3 &= 9/2 \end{aligned}$$

14.2 For each of the systems of equations in Exercise 14.1, consider the associated homogeneous systems. Which of these associated homogeneous systems have nontrivial solutions?

14.3 Find nontrivial solutions of the following systems of homogeneous equations whenever they exist. Also find the solution space of each system.

(a)
$$\begin{aligned} x_1 - \tfrac{7}{2}x_2 - \tfrac{1}{2}x_3 &= 0 \\ 3x_1 - 7x_2 + 2x_3 &= 0 \end{aligned}$$

(b)
$$\begin{aligned} x_1 + x_2 - 2x_3 &= 0 \\ 4x_1 - 3x_2 + 4x_3 &= 0 \\ 2x_1 - x_2 + x_3 &= 0 \\ x_1 - \tfrac{1}{2}x_2 - 2x_3 &= 0 \end{aligned}$$

(c)
$$\begin{aligned} - 2x_2 + 2x_3 &= 0 \\ 4x_1 - x_2 + 5x_3 &= 0 \\ 2x_1 + x_2 + x_3 &= 0 \end{aligned}$$

(d)
$$\begin{aligned} x_1 - 4x_2 - 6x_3 &= 0 \\ 2x_1 + 5x_2 - x_3 &= 0 \\ 5x_1 - 2x_2 + x_3 &= 0 \end{aligned}$$

14.4 In Exercise 14.3 find bases for the row space $\mathcal{S}$ and solution space $\mathfrak{J}$ of each system. If $\mathfrak{J} = \{(0, 0, 0)\}$, then $\mathfrak{J}$ does not have a basis.

14.5 Which of the following sets of vectors in $\mathcal{V}_4$ are linearly dependent?

(a) $\{(-2, 1, 3, 6), (0, 1, 4, -5), (2, -3, 1, 1)\}$

(b) $\{(1, 1, 1, 2). (2, -1, 5, 0), (-1, -1, 2, 1), (1, -\tfrac{1}{2}, 4, \tfrac{3}{2})\}$

(c) $\{(0, -2, 2, 0), (4, -1, 5, 0), (2, 1, 1, 0)\}$

14.6 Prove that a system of n linear equations in m unknowns (12–1) is consistent if and only if the dimension of its row space is the same as the dimension of the row space of the associated homogeneous system.

14.7 Let $\mathcal{S}$ be the subspace of $\mathcal{V}_5$ spanned by the vectors $(-2, 3, 0, -1, 0)$, $(-4, 0, 1, 2, 7)$, $(8, 6, -3, -8, -21)$. Regard $\mathcal{V}_5$ as a Euclidean vector space with the $E^{(5)}$ inner product. Find a basis of the orthogonal complement $\mathfrak{J}$ of $\mathcal{S}$ in $\mathcal{V}_5$.

14.8 Let $\mathcal{V}$ be a vector space that is the direct sum of two of its subspaces, $\mathcal{V} = \mathcal{S} \oplus \mathfrak{J}$. Prove that if B is a basis of $\mathcal{S}$ and C is a basis of $\mathfrak{J}$, then $B \cup C$ is a basis of $\mathcal{V}$.

4

linear transformations and matrices

15 definition of a linear transformation

In our study of real vector spaces (in particular, the vector space $\mathcal{E}_3$ of geometric vectors) we have encountered certain operations that satisfy a linearity condition. For example, if $\mathcal{V}$ is a vector space, and $a \cdot \mathbf{U} + b \cdot \mathbf{W}$ is a linear combination of the vectors $\mathbf{U}$ and $\mathbf{W}$ in $\mathcal{V}$, then it follows from Definition 9.2, (g) and (i), that

$$k \cdot (a \cdot \mathbf{U} + b \cdot \mathbf{W}) = a \cdot (k \cdot \mathbf{U}) + b \cdot (k \cdot \mathbf{W}) \tag{15–1}$$

for a given real number k. If we regard scalar multiplication by a fixed real number k as a mapping of the vector space $\mathcal{V}$ which sends each vector $\mathbf{V} \in \mathcal{V}$ into the vector $k \cdot \mathbf{V} \in \mathcal{V}$, and denote this mapping by D_k, then equation (15–1) can be written

$$D_k(a \cdot \mathbf{U} + b \cdot \mathbf{W}) = a \cdot D_k(\mathbf{U}) + b \cdot D_k(\mathbf{W}). \tag{15–2}$$

For $\mathbf{V} \in \mathcal{V}$, the vector $D_k(\mathbf{V}) = k \cdot \mathbf{V}$ is called the *image* of $\mathbf{V}$ by the mapping D_k. Equation (15–2) states that the image of a linear combination of vectors in $\mathcal{V}$ is the same linear combination of the images of the vectors. We describe this property (15–2) of D_k by saying that D_k is a *linear transformation* of $\mathcal{V}$ into $\mathcal{V}$.

Consider the operation of cross multiplication of vectors in $\mathcal{E}_3$. If $\mathbf{V}_0$ is a fixed vector in $\mathcal{E}_3$, then it follows from Corollary 5.4 that

$$\mathbf{V}_0 \times (a \cdot \mathbf{U} + b \cdot \mathbf{W}) = a \cdot (\mathbf{V}_0 \times U) + b \cdot (\mathbf{V}_0 \times \mathbf{W}) \tag{15–3}$$

for any vectors $\mathbf{U}, \mathbf{W} \in \mathcal{E}_3$ and real numbers a, b. Cross multiplication by a fixed vector $\mathbf{V}_0$ can be interpreted as a mapping of the vector space $\mathcal{E}_3$ which sends each vector $\mathbf{V}$ into the vector $\mathbf{V}_0 \times \mathbf{V}$. Then equation (15–3) states that this mapping is a linear transformation of $\mathcal{E}_3$ into $\mathcal{E}_3$.

Another example of a mapping that has the linearity property is inner multiplication by a fixed vector $\mathbf{V}_0$ in a Euclidean vector space $\mathcal{V}$. If $\mathbf{V} \in \mathcal{V}$, then $\mathcal{I}(\mathbf{V}_0, \mathbf{V})$ is a real number. The set R of real numbers is a real vector space with respect to ordinary addition and multiplication (see Exercise 9.2). Thus, inner multiplication by a fixed vector $\mathbf{V}_0$ is a mapping of the vector space $\mathcal{V}$ into the vector space R which sends each vector $\mathbf{V} \in \mathcal{V}$ into the real number $\mathcal{I}(\mathbf{V}_0, \mathbf{V})$. By (a) of Definition 11.12,

$$\mathcal{I}(\mathbf{V}_0, a \cdot \mathbf{U} + b \cdot \mathbf{W}) = a\mathcal{I}(\mathbf{V}_0, \mathbf{U}) + b\mathcal{I}(\mathbf{V}_0, \mathbf{W}) \qquad (15\text{–}4)$$

for all vectors $\mathbf{U}, \mathbf{W} \in \mathcal{V}$ and all real numbers a, b. Equation (15–4) is the statement that this mapping is a linear transformation of $\mathcal{V}$ into R.

The above examples suggest that mappings of vector spaces that satisfy the linearity condition exhibited by equations (15–2), (15–3), and (15–4) are important in the study of real vector spaces. This is indeed the case. The concept of a linear transformation is the central idea in linear algebra.

(15.1) definition A mapping L of a vector space $\mathcal{V}$ into a vector space $\mathcal{W}$, such that

$$L(a \cdot \mathbf{U} + b \cdot \mathbf{W}) = a \cdot L(\mathbf{U}) + b \cdot L(\mathbf{W})$$

for all $\mathbf{U}, \mathbf{W} \in \mathcal{V}$ and all real numbers a, b, is called a *linear transformation* of $\mathcal{V}$ into $\mathcal{W}$.

A mapping L of $\mathcal{V}$ into $\mathcal{W}$ is a single-valued function of $\mathcal{V}$ into $\mathcal{W}$. That is, for each $\mathbf{V} \in \mathcal{V}$, $L(\mathbf{V})$ is a unique vector in $\mathcal{W}$. The vector $L(\mathbf{V})$ is called the *image* of $\mathbf{V}$ by L. Although the symbols $\cdot$ and $+$ used for scalar multiplication and addition are the same on both sides of the equation in Definition 15.1, on the left side they stand for these operations in $\mathcal{V}$, while on the right side they stand for the operations in $\mathcal{W}$. Strictly speaking, different symbols should be used in the two members of the equation, but there is no confusion if the meaning of the notation is understood.

In Definition 15.1, the vector space $\mathcal{W}$ may be taken to be the space $\mathcal{V}$. In this case, we say that L is a linear transformation of $\mathcal{V}$ into itself. A linear transformation of $\mathcal{V}$ into itself is often called a *linear operator*. This was the situation in the first examples discussed in the beginning of this section.

Familiar geometric operations provide further examples of linear transformations.

EXAMPLE 1 A rotation of a plane through an angle φ about the origin O of a rectangular Cartesian coordinate system carries each point P_i into a point P_i' such that P_i and P_i' are on the same circle with center at O, and the arc P_iP_i' subtends the angle φ at O (see Figure 46).

Let P be a point with coordinates (x, y) that is rotated through an angle

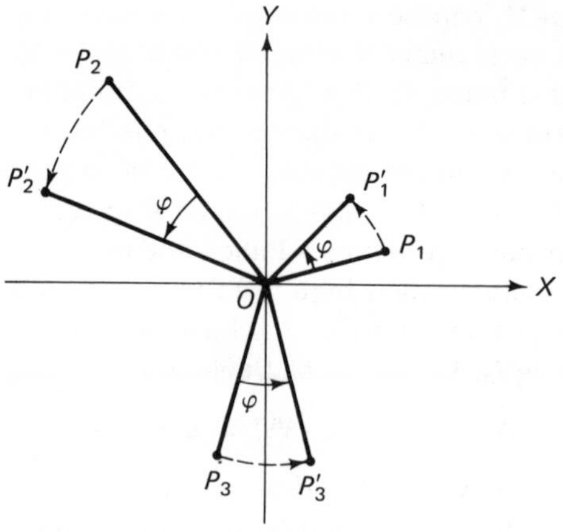

FIGURE 46

φ into a point P' with coordinates (x', y'). Let r be the distance of P from the origin, and let θ be the angle from the X axis to OP (see Figure 47). Since

$$\frac{x}{r} = \cos\theta, \qquad \frac{y}{r} = \sin\theta,$$

$$\frac{x'}{r} = \cos(\varphi + \theta) = \cos\varphi\cos\theta - \sin\varphi\sin\theta,$$

and

$$\frac{y'}{r} = \sin(\varphi + \theta) = \sin\varphi\cos\theta + \cos\varphi\sin\theta,$$

it follows that

$$\begin{aligned} x' &= x\cos\varphi - y\sin\varphi \\ y' &= x\sin\varphi + y\cos\varphi. \end{aligned} \tag{15–5}$$

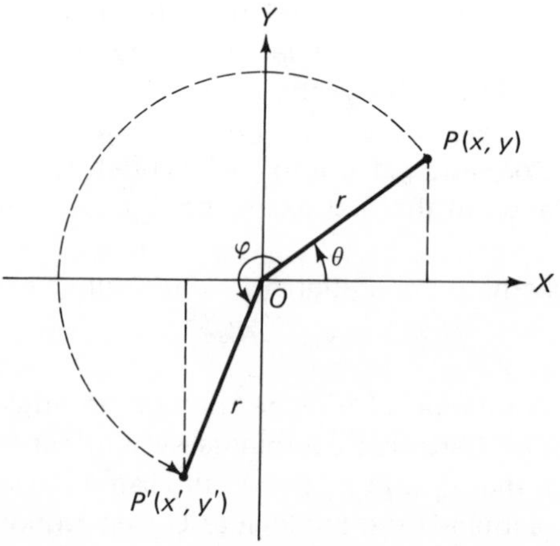

FIGURE 47

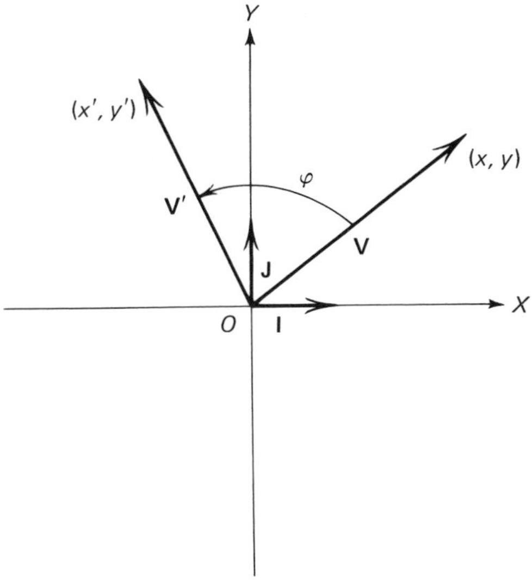

FIGURE 48

We recall that in §9, we denoted the set of all geometric vectors in the XY plane by $\mathcal{E}_2$ (see Example 7, §9). Each vector $\mathbf{V}$ in $\mathcal{E}_2$ has coordinates (x, y) with respect to the basis $\{\mathbf{I}, \mathbf{J}\}$. That is, $\mathbf{V} = x \cdot \mathbf{I} + y \cdot \mathbf{J}$. Corresponding to a rotation of the XY plane, there is a mapping of the vector space $\mathcal{E}_2$ which sends a vector $\mathbf{V}$ with coordinates (x, y) into a vector $\mathbf{V}'$ with coordinates (x', y') where x' and y' are given by equations (15–5) (see Figure 48).

Let T_φ denote the mapping of $\mathcal{E}_2$ into $\mathcal{E}_2$ corresponding to a rotation of the plane through an angle φ. Then every vector $\mathbf{V}$ in $\mathcal{E}_2$ is carried into a unique vector $T_\varphi(\mathbf{V})$, the image of $\mathbf{V}$ by T_φ. If $\mathbf{V} = x \cdot \mathbf{I} + y \cdot \mathbf{J}$, then

$$T_\varphi(\mathbf{V}) = T_\varphi(x \cdot \mathbf{I} + y \cdot \mathbf{J}) = (x \cos \varphi - y \sin \varphi) \cdot \mathbf{I} + (x \sin \varphi + y \cos \varphi) \cdot \mathbf{J}.$$

Let $\mathbf{W} = s \cdot \mathbf{I} + t \cdot \mathbf{J}$ be a vector in $\mathcal{E}_2$. Then $\mathbf{V} + \mathbf{W} = (x + s) \cdot \mathbf{I} + (y + t) \cdot \mathbf{J}$, and

$$\begin{aligned} T_\varphi(\mathbf{V} + \mathbf{W}) &= T_\varphi[(x + s) \cdot \mathbf{I} + (y + t) \cdot \mathbf{J}] \\ &= [(x + s) \cos \varphi - (y + t) \sin \varphi] \cdot \mathbf{I} \\ &\quad + [(x + s) \sin \varphi + (y + t) \cos \varphi] \cdot \mathbf{J} \\ &= [(x \cos \varphi - y \sin \varphi) + (s \cos \varphi - t \sin \varphi)] \cdot \mathbf{I} \\ &\quad + [(x \sin \varphi + y \cos \varphi) + (s \sin \varphi + t \cos \varphi)] \cdot \mathbf{J} \\ &= [(x \cos \varphi - y \sin \varphi) \cdot \mathbf{I} + (x \sin \varphi + y \cos \varphi) \cdot \mathbf{J}] \\ &\quad + [(s \cos \varphi - t \sin \varphi) \cdot \mathbf{I} + (s \sin \varphi + t \cos \varphi) \cdot \mathbf{J}] \\ &= T_\varphi(\mathbf{V}) + T_\varphi(\mathbf{W}). \end{aligned}$$

If r is a real number, then $r \cdot \mathbf{V} = (rx) \cdot \mathbf{I} + (ry) \cdot \mathbf{J}$, and

$$\begin{aligned} T_\varphi(r \cdot \mathbf{V}) &= T_\varphi[(rx) \cdot \mathbf{I} + (ry) \cdot \mathbf{J}] \\ &= (rx \cos \varphi - ry \sin \varphi) \cdot \mathbf{I} + (rx \sin \varphi + ry \cos \varphi) \cdot \mathbf{J} \\ &= r \cdot [(x \cos \varphi - y \sin \varphi) \cdot \mathbf{I} + (x \sin \varphi + y \cos \varphi) \cdot \mathbf{J}] = r \cdot T_\varphi(\mathbf{V}). \end{aligned}$$

The two conditions $T_\varphi(\mathbf{V} + \mathbf{W}) = T_\varphi(\mathbf{V}) + T_\varphi(\mathbf{W})$ and $T_\varphi(r \cdot \mathbf{V}) = r \cdot T_\varphi(\mathbf{V})$ imply

$$T_\varphi(a \cdot \mathbf{V} + b \cdot \mathbf{W}) = a \cdot T_\varphi(\mathbf{V}) + b \cdot T_\varphi(\mathbf{W}) \tag{15–6}$$

for all real numbers a, b. This is exactly the requirement of Definition 15.1. Therefore, T_φ is a linear transformation of $\mathcal{E}_2$ into $\mathcal{E}_2$.

EXAMPLE 2 Let $\mathbf{V}_0 = \overrightarrow{OP_0}$ be a fixed nonzero vector in $\mathcal{E}_3$, and let $\mathbf{V} = \overrightarrow{OP}$ be any vector in $\mathcal{E}_3$. Let θ be the angle between $\mathbf{V}_0$ and $\mathbf{V}$. The projection of $\mathbf{V}$ onto $\mathbf{V}_0$ is the vector $\mathbf{V}'$ with length $\big|\,|\mathbf{V}| \cos\theta\big|$ along the line determined by $\mathbf{V}_0$, with the direction of $\mathbf{V}_0$ if θ is acute, and with the direction of $-\mathbf{V}_0$ if θ is obtuse (see Figure 49). The reader can verify (see Exercise 15.13) that the vector $\mathbf{V}'$ is given by the formula

$$\mathbf{V}' = \left(\frac{\mathbf{V} \circ \mathbf{V}_0}{\mathbf{V}_0 \circ \mathbf{V}_0}\right) \cdot \mathbf{V}_0.$$

The mapping $P_{\mathbf{V}_0}$ of $\mathcal{E}_3$ defined by $P_{\mathbf{V}_0}(\mathbf{V}) = [(\mathbf{V} \circ \mathbf{V}_0)/(\mathbf{V}_0 \circ \mathbf{V}_0)] \cdot \mathbf{V}_0$ is called the *orthogonal projection* of $\mathcal{E}_3$ into the subspace $\mathcal{S}$ of $\mathcal{E}_3$ spanned by the vector $\mathbf{V}_0$. Using the identities satisfied by inner and scalar multiplication in $\mathcal{E}_3$, we obtain

$$\begin{aligned} P_{\mathbf{V}_0}(\mathbf{U} + \mathbf{W}) &= \left(\frac{(\mathbf{U} + \mathbf{W}) \circ \mathbf{V}_0}{\mathbf{V}_0 \circ \mathbf{V}_0}\right) \cdot \mathbf{V}_0 = \left(\frac{\mathbf{U} \circ \mathbf{V}_0}{\mathbf{V}_0 \circ \mathbf{V}_0} + \frac{\mathbf{W} \circ \mathbf{V}_0}{\mathbf{V}_0 \circ \mathbf{V}_0}\right) \cdot \mathbf{V}_0 \\ &= \left(\frac{\mathbf{U} \circ \mathbf{V}_0}{\mathbf{V}_0 \circ \mathbf{V}_0}\right) \cdot \mathbf{V}_0 + \left(\frac{\mathbf{W} \circ \mathbf{V}_0}{\mathbf{V}_0 \circ \mathbf{V}_0}\right) \cdot \mathbf{V}_0 = P_{\mathbf{V}_0}(\mathbf{U}) + P_{\mathbf{V}_0}(\mathbf{W}) \end{aligned}$$

for vectors $\mathbf{U}, \mathbf{W} \in \mathcal{E}_3$. Furthermore,

$$\begin{aligned} P_{\mathbf{V}_0}(r \cdot \mathbf{U}) &= \left(\frac{(r \cdot \mathbf{U}) \circ \mathbf{V}_0}{\mathbf{V}_0 \circ \mathbf{V}_0}\right) \cdot \mathbf{V}_0 = \left[r\left(\frac{\mathbf{U} \circ \mathbf{V}_0}{\mathbf{V}_0 \circ \mathbf{V}_0}\right)\right] \cdot \mathbf{V}_0 \\ &= r \cdot \left[\left(\frac{\mathbf{U} \circ \mathbf{V}_0}{\mathbf{V}_0 \circ \mathbf{V}_0}\right) \cdot \mathbf{V}_0\right] = r \cdot P_{\mathbf{V}_0}(\mathbf{U}) \end{aligned}$$

for $\mathbf{U} \in \mathcal{E}_3$ and any real number r. As in Example 1 above, the two conditions $P_{\mathbf{V}_0}(\mathbf{U} + \mathbf{W}) = P_{\mathbf{V}_0}(\mathbf{U}) + P_{\mathbf{V}_0}(\mathbf{W})$ and $P_{\mathbf{V}_0}(r \cdot \mathbf{U}) = r \cdot P_{\mathbf{V}_0}(\mathbf{U})$ imply the linearity condition

$$P_{\mathbf{V}_0}(a \cdot \mathbf{U} + b \cdot \mathbf{W}) = a \cdot P_{\mathbf{V}_0}(\mathbf{U}) + b \cdot P_{\mathbf{V}_0}(\mathbf{W}).$$

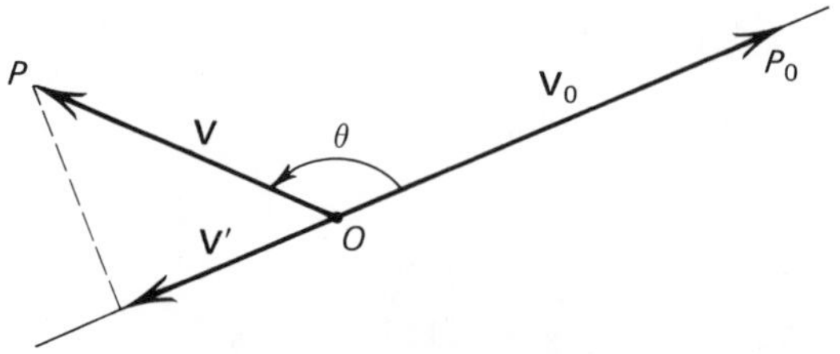

FIGURE 49

Hence, $P_{\mathbf{V}_0}$ is a linear transformation of $\mathcal{E}_3$ into the subspace $\mathcal{S}$ of $\mathcal{E}_3$ spanned by the vector $\mathbf{V}_0$.

EXAMPLE 3 Let $\mathcal{S}$ be a subspace of $\mathcal{E}_3$ spanned by the nonzero orthogonal vectors **U** and **W**. Then the mapping $P_{\mathcal{S}}$ of $\mathcal{E}_3$ defined by

$$P_{\mathcal{S}}(\mathbf{V}) = \left(\frac{\mathbf{V} \circ \mathbf{U}}{\mathbf{U} \circ \mathbf{U}}\right) \cdot \mathbf{U} + \left(\frac{\mathbf{V} \circ \mathbf{W}}{\mathbf{W} \circ \mathbf{W}}\right) \cdot \mathbf{W}$$

for $\mathbf{V} \in \mathcal{E}_3$ is an orthogonal projection of $\mathcal{E}_3$ into $\mathcal{S}$. It is left as an exercise for the reader (Exercise 15.16) to check that $P_{\mathcal{S}}$ is a linear transformation of $\mathcal{E}_3$ into $\mathcal{S}$. In particular, if **U** and **W** are the unit coordinate vectors **I** and **J**, then $\mathcal{S} = \mathcal{E}_2$, and the mapping is a linear transformation of $\mathcal{E}_3$ into $\mathcal{E}_2$. In Exercise 15.17, the reader is asked to show that if **V** has coordinates (x, y, z) with respect to $\{\mathbf{I}, \mathbf{J}, \mathbf{K}\}$ then $P_{\mathcal{E}_2}(\mathbf{V})$ has coordinates $(x, y, 0)$. Let ℓ be a line perpendicular to the XY plane at a point $(x, y, 0)$. Then every geometric vector $\overrightarrow{OP}$ with terminal point P on ℓ is mapped by $P_{\mathcal{E}_2}$ into the vector in the XY plane with terminal point $(x, y, 0)$. Thus, $P_{\mathcal{E}_2}$ maps infinitely many different vectors in $\mathcal{E}_3$ into the same vector in $\mathcal{E}_2$.

exercises

15.1 Prove that equation (15–1) follows from the identities 9.2(g) and 9.2(i) in the definition of a vector space $\mathcal{V}$.

15.2 Let D_k be the linear transformation of a vector space $\mathcal{V}$ defined by $D_k(\mathbf{V}) = k \cdot \mathbf{V}$ for $\mathbf{V} \in \mathcal{V}$.
(a) Let $\mathcal{V} = \mathcal{V}_4$. Compute the following vectors: $D_5[(2, 1, 0, -1)]$; $D_{1/2}[(-4, 6, -3, 2)]$; $D_{\sqrt{2}}[(2\sqrt{2}, 0, -\sqrt{2}, \sqrt{3})]$.
(b) Let $\mathcal{V} = \mathcal{E}_3$. Compute the following vectors: $D_4(\mathbf{I} + \mathbf{J} - \mathbf{K})$; $D_7(2 \cdot \mathbf{I} - 3 \cdot \mathbf{J} + 5 \cdot \mathbf{K})$; $D_0(6 \cdot \mathbf{I} + 10 \cdot \mathbf{J} + 3 \cdot \mathbf{K})$.
(c) Let $\mathcal{V} = R[x]$. Compute the following vectors: $D_{1/3}(3x^2 + 6x - 9)$; $D_2(7x^3 + 2x^2 + 1)$; $D_1(4x^{33} + 2x^{11} - x)$.
(d) Let $\mathcal{V} = \mathcal{V}_3$. Show that equation (15–2) is satisfied when $k = \frac{1}{2}$, $a = 4$, $b = 6$, $\mathbf{U} = (1, 2, 7)$, and $\mathbf{V} = (-3, 0, 1)$.

15.3 Show that if $k \neq 0$, every vector **V** in a vector space $\mathcal{V}$ can be expressed $\mathbf{V} = D_k(\mathbf{U})$ for some $\mathbf{U} \in \mathcal{V}$. Also show that $D_k(\mathbf{U}) = \mathbf{0}$ implies $\mathbf{U} = \mathbf{0}$.

15.4 For real numbers k and l, D_k, D_l, D_{k+l}, and D_{kl} are linear transformations of a vector space $\mathcal{V}$ into itself. Show that $D_k(\mathbf{U}) + D_l(\mathbf{U}) = D_{k+l}(\mathbf{U})$, and $D_k[D_l(\mathbf{U})] = D_{kl}(\mathbf{U})$ for every $\mathbf{U} \in \mathcal{V}$.

15.5 Let $\mathbf{V}_0 = a \cdot \mathbf{I} + b \cdot \mathbf{J} + c \cdot \mathbf{K}$ be a given vector in $\mathcal{E}_3$. Let L be the linear transformation defined by $L(\mathbf{V}) = \mathbf{V}_0 \times \mathbf{V}$ for all $\mathbf{V} \in \mathcal{E}_3$. (a) If $\mathbf{V} = x \cdot \mathbf{I} + y \cdot \mathbf{J} + z \cdot \mathbf{K}$ and $L(\mathbf{V}) = x' \cdot \mathbf{I} + y' \cdot \mathbf{J} + z' \cdot \mathbf{K}$, compute the coordinates (x', y', z') of $L(\mathbf{V})$ in terms of the coordinates (x, y, z) of **V**. (b) Compute $L(\mathbf{V})$ for the following vectors: $\mathbf{V} = 2 \cdot \mathbf{I} - 3 \cdot \mathbf{J} + \mathbf{K}$; $\mathbf{V} = \mathbf{I} + \mathbf{J} - \mathbf{K}$; $\mathbf{V} = 3 \cdot \mathbf{J} - 5 \cdot \mathbf{K}$; $\mathbf{V} = -\mathbf{I} + 2 \cdot \mathbf{J}$.

15.6 Let $\mathbf{V}_0$ be a fixed nonzero vector in $\mathcal{E}_3$. (a) Prove or disprove the following statement: Every $\mathbf{V} \in \mathcal{E}_3$ can be expressed $\mathbf{V} = \mathbf{V}_0 \times \mathbf{U}$ for some $\mathbf{U} \in \mathcal{E}_3$.

(b) Suppose that $\mathbf{V} \circ \mathbf{V}_0 = 0$. Show that there is a vector $\mathbf{U} \in \mathcal{E}_3$ such that $\mathbf{U} \circ \mathbf{V}_0 = 0$ and $\mathbf{V} = \mathbf{V}_0 \times \mathbf{U}$. (c) Show that for every $\mathbf{U} \in \mathcal{E}_3$, $\mathbf{V}_0 \times \mathbf{U}$ is in the orthogonal complement of the subspace of $\mathcal{E}_3$ spanned by $\mathbf{V}_0$.

15.7 Let $\mathbf{V}_0$ be a fixed nonzero vector in $\mathcal{E}_3$. Show that every real number r can be expressed $r = \mathbf{V}_0 \circ \mathbf{U}$ for some $\mathbf{U} \in \mathcal{E}_3$.

15.8 Let T_φ be the linear transformation of $\mathcal{E}_2$ into itself described in Example 1. Compute the following vectors for $\varphi = 45°$: $T_\varphi(-\mathbf{I} + \mathbf{J})$, $T_\varphi(2 \cdot \mathbf{I} + 4 \cdot \mathbf{J})$, $T_\varphi(3 \cdot \mathbf{I} - 5 \cdot \mathbf{J})$. Compute these same vectors for $\varphi = 90°$, $\varphi = 150°$, and $\varphi = 270°$.

15.9 Show that equation (15–6) is satisfied when $\varphi = 30°$, $a = -3$, $b = 2$, $\mathbf{V} = 2 \cdot \mathbf{I} + \frac{1}{2} \cdot \mathbf{J}$, and $\mathbf{W} = -3 \cdot \mathbf{I} + 5 \cdot \mathbf{J}$.

15.10 Show that equation (15–6) follows from the identities $T_\varphi(\mathbf{V} + \mathbf{W}) = T_\varphi(\mathbf{V}) + T_\varphi(\mathbf{W})$ and $T_\varphi(r \cdot \mathbf{V}) = r \cdot T_\varphi(\mathbf{V})$.

15.11 Show that every vector $\mathbf{V} \in \mathcal{E}_2$ can be expressed $\mathbf{V} = T_\varphi(\mathbf{U})$ for some $\mathbf{U} \in \mathcal{E}_2$. Show that $T_\varphi(\mathbf{V}) = \mathbf{0}$ if and only if $\mathbf{V} = \mathbf{0}$.

15.12 If φ and ξ are given angles, then T_φ, T_ξ, and $T_{\varphi+\xi}$ are linear transformations of $\mathcal{E}_2$ into itself. Show that $T_\varphi[T_\xi(\mathbf{U})] = T_{\varphi+\xi}(\mathbf{U})$ for every $\mathbf{U} \in \mathcal{E}_2$.

15.13 In Example 2, prove that the orthogonal projection $\mathbf{V}'$ of a vector $\mathbf{V} \in \mathcal{E}_3$ into the subspace of $\mathcal{E}_3$ spanned by a fixed vector $\mathbf{V}_0$ is given by the formula

$$\mathbf{V}' = [(\mathbf{V} \circ \mathbf{V}_0)/(\mathbf{V}_0 \circ \mathbf{V}_0)] \cdot \mathbf{V}_0.$$

15.14 Find the orthogonal projections of the following vectors into the subspace of $\mathcal{E}_3$ spanned by the vector $\mathbf{V}_0 = \mathbf{I} + \mathbf{J} + \mathbf{K}$:
$2 \cdot \mathbf{I} - 3 \cdot \mathbf{J} + 5 \cdot \mathbf{K}$; $\mathbf{I} + 2 \cdot \mathbf{J}$; $5 \cdot \mathbf{I} + 2 \cdot \mathbf{K}$; $3 \cdot \mathbf{I} + 3 \cdot \mathbf{J} + 3 \cdot \mathbf{K}$; $-5 \cdot \mathbf{I} + 3 \cdot \mathbf{J} + 2 \cdot \mathbf{K}$.

15.15 Show that $P_{\mathbf{V}_0}(\mathbf{V}) = \mathbf{0}$ if and only if $\mathbf{V}$ is orthogonal to $\mathbf{V}_0$.

15.16 In Example 3, prove that $P_{\mathcal{S}}$ is a linear transformation of $\mathcal{E}_3$ into $\mathcal{S}$.

15.17 In Example 3, show that $P_{\mathcal{E}_2}(x \cdot \mathbf{I} + y \cdot \mathbf{J} + z \cdot \mathbf{K}) = x \cdot \mathbf{I} + y \cdot \mathbf{J}$ for every vector $\mathbf{V} = x \cdot \mathbf{I} + y \cdot \mathbf{J} + z \cdot \mathbf{K}$ in $\mathcal{E}_3$.

15.18 Find the orthogonal projections in $\mathcal{E}_2$ of the following vectors in $\mathcal{E}_3$: $5 \cdot \mathbf{I} - 3 \cdot \mathbf{J} + \mathbf{K}$; $\mathbf{I} + \mathbf{J} - \mathbf{K}$; $6 \cdot \mathbf{K}$; $7 \cdot \mathbf{I} + 2 \cdot \mathbf{J}$; $-\mathbf{I} + 3 \cdot \mathbf{J} - 5 \cdot \mathbf{K}$.

15.19 Let $\mathcal{S}$ be the subspace of $\mathcal{E}_3$ spanned by the nonzero orthogonal vectors $\mathbf{U}$ and $\mathbf{W}$. Prove that $P_{\mathcal{S}}(\mathbf{V}) = \mathbf{0}$ if and only if $\mathbf{V}$ is orthogonal to $\mathbf{U}$ and $\mathbf{V}$ is orthogonal to $\mathbf{W}$.

15.20 Show that $\mathbf{U} = 2 \cdot \mathbf{I} + \mathbf{J} - 5 \cdot \mathbf{K}$ and $\mathbf{W} = \mathbf{I} + 3 \cdot \mathbf{J} + \mathbf{K}$ are orthogonal vectors in $\mathcal{E}_3$. Find the orthogonal projections of the following vectors in the subspace $\mathcal{S}$ of $\mathcal{E}_3$ spanned by $\mathbf{U}$ and $\mathbf{W}$: $-6 \cdot \mathbf{I} - 3 \cdot \mathbf{J} + 15 \cdot \mathbf{K}$; $2 \cdot \mathbf{I} + 4 \cdot \mathbf{J} + 6 \cdot \mathbf{K}$; $\mathbf{I} + \mathbf{J}$; $\mathbf{J}$; $(48/5) \cdot \mathbf{I} + (-21/5) \cdot \mathbf{J} + 3 \cdot \mathbf{K}$.

15.21 In Example 3, show that $\mathbf{V} - P_{\mathcal{S}}(\mathbf{V})$ is orthogonal to both $\mathbf{U}$ and $\mathbf{W}$.

15.22 In Example 3, let $\mathbf{U}$ and $\mathbf{W}$ have coordinates (a, b, c) and (d, e, f) respectively, with respect to $\{\mathbf{I}, \mathbf{J}, \mathbf{K}\}$. Derive the equations that give the coordinates (x', y', z') of $P_{\mathcal{S}}(\mathbf{V})$ in terms of the coordinates (x, y, z) of $\mathbf{V}$.

15.23 Let L be a mapping of $\mathcal{E}_3$ into $\mathcal{E}_3$ that sends a vector $\mathbf{V} = x \cdot \mathbf{I} + y \cdot \mathbf{J} + z \cdot \mathbf{K}$ into a vector $L(\mathbf{V}) = x' \cdot \mathbf{I} + y' \cdot \mathbf{J} + z' \cdot \mathbf{K}$, where

$$x' = a_{1,1}x + a_{1,2}y + a_{1,3}z$$
$$y' = a_{2,1}x + a_{2,2}y + a_{2,3}z$$
$$z' = a_{3,1}x + a_{3,2}y + a_{3,3}z$$

for real numbers $a_{i,j}$, $i = 1, 2, 3$, $j = 1, 2, 3$. Prove that L is a linear transformation of $\mathcal{E}_3$ into $\mathcal{E}_3$.

16 properties of linear transformations

The condition $L(a \cdot \mathbf{U} + b \cdot \mathbf{W}) = a \cdot L(\mathbf{U}) + b \cdot L(\mathbf{W})$ in Definition 15.1 of a linear transformation L of a vector space $\mathcal{V}$ into a vector space $\mathcal{W}$ is easily seen to be equivalent to the two conditions

$$L(\mathbf{U} + \mathbf{W}) = L(\mathbf{U}) + L(\mathbf{W}) \tag{16-1}$$

and

$$L(a \cdot \mathbf{U}) = a \cdot L(\mathbf{U}) \tag{16-2}$$

for all **U, W** in $\mathcal{V}$ and every real number a. Thus, we say that a linear transformation L "preserves" vector addition and scalar multiplication.

By (16–1),

$$L(\mathbf{0}) = L(\mathbf{0} + \mathbf{0}) = L(\mathbf{0}) + L(\mathbf{0}).$$

Subtracting the vector $L(\mathbf{0})$ from both sides of this equation, we find that $L(\mathbf{0})$ is the zero vector of $\mathcal{W}$. That is, $L(\mathbf{0}) = \mathbf{0}$ (where the same symbol $\mathbf{0}$ is used to denote both the zero vector in $\mathcal{V}$ and the zero vector in $\mathcal{W}$). By (16–2),

$$L(-\mathbf{U}) = L((-1) \cdot \mathbf{U}) = (-1) \cdot L(\mathbf{U}) = -L(\mathbf{U}).$$

Thus, the image of the negative of a vector is the negative of the image of the vector.

If $\mathcal{S}$ is a subspace of $\mathcal{V}$, then $L(\mathcal{S})$ is the set of all vectors in $\mathcal{W}$ that are images of vectors in $\mathcal{S}$. That is,

$$L(\mathcal{S}) = \{L(\mathbf{U}) | \mathbf{U} \in \mathcal{S}\}.$$

If $\mathfrak{J}$ is a subspace of $\mathcal{W}$, then $L^{-1}(\mathfrak{J})$ is the set of all vectors in $\mathcal{V}$ whose images by L are in $\mathfrak{J}$. Thus,

$$L^{-1}(\mathfrak{J}) = \{\mathbf{U} | L(\mathbf{U}) \in \mathfrak{J}\}.$$

The set of vectors $L(\mathcal{S})$ is called the *image* of $\mathcal{S}$, and $L^{-1}(\mathfrak{J})$ is called the *complete inverse image* of $\mathfrak{J}$.

(16.1) theorem Let L be a linear transformation of $\mathcal{V}$ into $\mathcal{W}$, and let $\mathcal{S}$ be a subspace of $\mathcal{V}$ and $\mathfrak{J}$ be a subspace of $\mathcal{W}$. Then $L(\mathcal{S})$ is a subspace of $\mathcal{W}$ and $L^{-1}(\mathfrak{J})$ is a subspace of $\mathcal{V}$.

PROOF Since $\mathbf{0} \in \mathcal{S}$, $L(\mathcal{S})$ is nonempty. Suppose that $\mathbf{U}'$ and $\mathbf{W}'$ are in $L(\mathcal{S})$. Then $\mathbf{U}' = L(\mathbf{U})$ and $\mathbf{W}' = L(\mathbf{W})$ for some $\mathbf{U}, \mathbf{W} \in \mathcal{S}$. Since $\mathcal{S}$ is a subspace of $\mathcal{V}$, it follows that $\mathbf{U} + \mathbf{W} \in \mathcal{S}$ and $a \cdot \mathbf{U} \in \mathcal{S}$ for any real number a. Therefore, by (16–1) and (16–2),

$$\mathbf{U}' + \mathbf{W}' = L(\mathbf{U}) + L(\mathbf{W}) = L(\mathbf{U} + \mathbf{W}) \in L(\mathcal{S})$$

and

$$a \cdot \mathbf{U}' = a \cdot L(\mathbf{U}) = L(a \cdot \mathbf{U}) \in L(\mathcal{S}).$$

Hence, by Theorem 9.5, $L(\mathcal{S})$ is a subspace of $\mathcal{W}$.

Since $\mathbf{0} \in \mathcal{T}$, and $L(\mathbf{0}) = \mathbf{0}$, it follows that $\mathbf{0} \in L^{-1}(\mathcal{T})$. Thus, $L^{-1}(\mathcal{T})$ is nonempty. Suppose that $\mathbf{U}$ and $\mathbf{W}$ are in $L^{-1}(\mathcal{T})$. Then $L(\mathbf{U})$ and $L(\mathbf{W})$ are in $\mathcal{T}$. Therefore, $L(\mathbf{U}) + L(\mathbf{W}) \in \mathcal{T}$, and $a \cdot L(\mathbf{U}) \in \mathcal{T}$ for any real number a, since $\mathcal{T}$ is a subspace of $\mathcal{W}$. Again by (16–1) and (16–2),

$$L(\mathbf{U} + \mathbf{W}) = L(\mathbf{U}) + L(\mathbf{W}) \in \mathcal{T}$$

and

$$L(a \cdot \mathbf{U}) = a \cdot L(\mathbf{U}) \in \mathcal{T}.$$

Hence, $\mathbf{U} + \mathbf{W} \in L^{-1}(\mathcal{T})$ and $a \cdot \mathbf{U} \in L^{-1}(\mathcal{T})$. By Theorem 9.5, $L^{-1}(\mathcal{T})$ is a subspace of $\mathcal{V}$. ∎

If we take $\mathcal{S} = \mathcal{V}$ in Theorem 16.1, then $L(\mathcal{V})$ is a subspace of $\mathcal{W}$ which is called the *range space* of L. Let $\mathcal{T} = \{\mathbf{0}\}$ be the zero subspace of $\mathcal{W}$. Then, by Theorem 16.1, $L^{-1}(\{\mathbf{0}\})$ is a subspace of $\mathcal{V}$, which is called the *null space* or *kernel* of L. In other words, the elements of $L^{-1}(\{\mathbf{0}\})$ are just those vectors sent into $\mathbf{0}$ by the linear transformation L.

EXAMPLE 1 Let D_k be the linear transformation of a vector space $\mathcal{V}$ into itself defined by $D_k(\mathbf{V}) = k \cdot \mathbf{V}$ for $k \neq 0$. We will show that $D_k(\mathcal{S}) = \mathcal{S}$ for every subspace $\mathcal{S}$ of $\mathcal{V}$. In particular, $D_k(\mathcal{V}) = \mathcal{V}$, so that the range space of D_k is $\mathcal{V}$. In addition, we will show that $D_k^{-1}(\mathcal{T}) = \mathcal{T}$ for every subspace $\mathcal{T}$ of $\mathcal{V}$. For $\mathcal{T} = \{\mathbf{0}\}$, we obtain $D_k^{-1}(\{\mathbf{0}\}) = \{\mathbf{0}\}$. That is, the null space of D_k is $\{\mathbf{0}\}$.

Let $\mathcal{S}$ be a subspace of $\mathcal{V}$ and let $\mathbf{V} \in \mathcal{S}$. Then $\mathbf{U} = (1/k) \cdot \mathbf{V} \in \mathcal{S}$, and $D_k(\mathbf{U}) = k \cdot [(1/k) \cdot \mathbf{V}] = \mathbf{V}$, with $\mathbf{U} \in \mathcal{S}$. That is, $\mathbf{V} \in D_k(\mathcal{S})$. We have shown that if $\mathbf{V} \in \mathcal{S}$, then $\mathbf{V} \in D_k(\mathcal{S})$. Therefore, $\mathcal{S} \subseteq D_k(\mathcal{S})$. On the other hand, suppose that $\mathbf{V} \in D_k(\mathcal{S})$. Then $\mathbf{V} = D_k(\mathbf{U}) = k \cdot \mathbf{U}$, with $\mathbf{U} \in \mathcal{S}$. Therefore, $\mathbf{V} = k \cdot \mathbf{U} \in \mathcal{S}$. Thus, if $\mathbf{V} \in D_k(\mathcal{S})$, then $\mathbf{V} \in \mathcal{S}$. Hence, $D_k(\mathcal{S}) \subseteq \mathcal{S}$. Since $\mathcal{S} \subseteq D_k(\mathcal{S})$ and $D_k(\mathcal{S}) \subseteq \mathcal{S}$, we have $D_k(\mathcal{S}) = \mathcal{S}$.

Let $\mathcal{T}$ be a subspace of $\mathcal{V}$. If $\mathbf{V} \in \mathcal{T}$, then $D_k(\mathbf{V}) = k \cdot \mathbf{V} \in \mathcal{T}$. Hence, $\mathbf{V} \in D_k^{-1}(\mathcal{T})$. Therefore, $\mathcal{T} \subseteq D_k^{-1}(\mathcal{T})$. If $\mathbf{V} \in D_k^{-1}(\mathcal{T})$, then $D_k(\mathbf{V}) = k \cdot \mathbf{V} \in \mathcal{T}$. Therefore, $\mathbf{V} = (1/k) \cdot (k \cdot \mathbf{V}) \in \mathcal{T}$. Hence, $D_k^{-1}(\mathcal{T}) \subseteq \mathcal{T}$. Since $\mathcal{T} \subseteq D_k^{-1}(\mathcal{T})$ and $D_k^{-1}(\mathcal{T}) \subseteq \mathcal{T}$, it follows that $D_k^{-1}(\mathcal{T}) = \mathcal{T}$.

EXAMPLE 2 Let L be the linear transformation of $\mathcal{E}_3$ into $\mathcal{E}_3$ defined by $L(\mathbf{V}) = \mathbf{V}_0 \times \mathbf{V}$, for a fixed nonzero vector $\mathbf{V}_0 \in \mathcal{E}_3$ (see §15). Since $L(\mathbf{V})$ is orthogonal to $\mathbf{V}_0$, it follows that every vector in the range space $L(\mathcal{E}_3)$ of L is in the plane π through O perpendicular to $\mathbf{V}_0$. The vectors in the plane π form a subspace of $\mathcal{E}_3$ that we have previously denoted by $\mathcal{E}_\pi$ (see

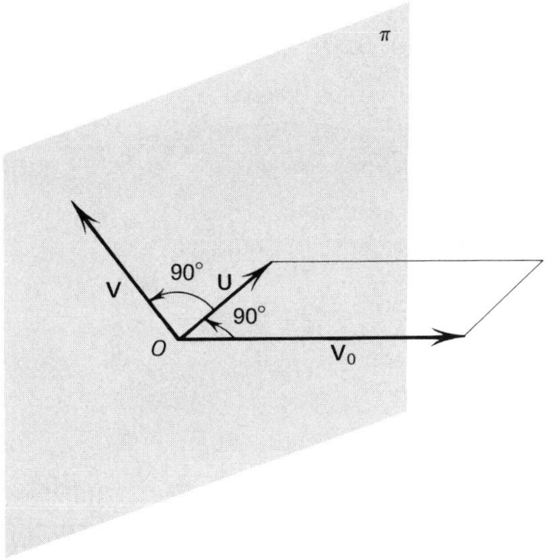

FIGURE 50

Example 1, §9). Thus, $L(\mathcal{E}_3) \subseteq \mathcal{E}_\pi$. Suppose, on the other hand, that **V** is any vector in $\mathcal{E}_\pi$. There is a vector **U** in $\mathcal{E}_\pi$ such that $|\mathbf{U}| = |\mathbf{V}|/|\mathbf{V}_0|$, **U** is orthogonal to **V**, and the vectors $\mathbf{V}_0$, **U**, and **V** form a right-handed system (see Figure 50). Then **V** is orthogonal to both **U** and $\mathbf{V}_0$. Therefore, **V** has the direction of $\mathbf{V}_0 \times \mathbf{U}$. Since $|\mathbf{V}_0 \times \mathbf{U}| = |\mathbf{V}_0|\,|\mathbf{U}| \sin 90° = |\mathbf{V}_0|\,|\mathbf{V}|/|\mathbf{V}_0| = |\mathbf{V}|$, the length of **V** is equal to the length of $\mathbf{V}_0 \times \mathbf{U}$. Hence, $\mathbf{V} = \mathbf{V}_0 \times \mathbf{U} = L(\mathbf{U}) \in L(\mathcal{E}_3)$. We have shown that if $\mathbf{V} \in \mathcal{E}_\pi$, then $\mathbf{V} \in L(\mathcal{E}_3)$. That is, $\mathcal{E}_\pi \subseteq L(\mathcal{E}_3)$. Combining this with the above result $L(\mathcal{E}_3) \subseteq \mathcal{E}_\pi$, we have that $L(\mathcal{E}_3)$, the range space of L, is $\mathcal{E}_\pi$.

Since $L(\mathbf{V}) = \mathbf{V}_0 \times \mathbf{V} = \mathbf{0}$ if and only if **V** and $\mathbf{V}_0$ are collinear (see Exercise 5.5), it follows that $L^{-1}(\{\mathbf{0}\})$, the null space of L, is the subspace of $\mathcal{E}_3$ spanned by $\mathbf{V}_0$.

Finally, we note that $\mathcal{E}_\pi$ is the orthogonal complement of the subspace of $\mathcal{E}_3$ spanned by $\mathbf{V}_0$ (see Exercise 11.8). Thus, $L(\mathcal{E}_3)$ is the orthogonal complement of $L^{-1}(\{\mathbf{0}\})$, so that $\mathcal{E}_3 = L(\mathcal{E}_3) \oplus L^{-1}(\{\mathbf{0}\})$.

EXAMPLE 3 Let L be the mapping of $\mathcal{V}_3$ into $\mathcal{V}_5$ defined by

$$L[(a_1, a_2, a_3)] = (a_1 - a_2, 0, a_1 - a_3, a_2, 0).$$

The reader can verify that L is a linear transformation. Let $\mathfrak{J}$ be the subspace of $\mathcal{V}_5$ with basis (1, 0, 0, 0, 0), (0, 0, 1, 0, 0), and (0, 0, 0, 1, 0). Clearly, $L(\mathcal{V}_3) \subseteq \mathfrak{J}$. Let

$$c_1 \cdot (1, 0, 0, 0, 0) + c_2 \cdot (0, 0, 1, 0, 0) + c_3 \cdot (0, 0, 0, 1, 0) = (c_1, 0, c_2, c_3, 0)$$

be any element of $\mathfrak{J}$. Then

$$L[(c_1 + c_3, c_3, c_1 + c_3 - c_2)] = (c_1, 0, c_2, c_3, 0).$$

Therefore, $L(\mathcal{V}_3) = \mathcal{J}$, and $\mathcal{J}$ is the range space of L. Suppose that $L[(a_1, a_2, a_3)] = (0, 0, 0, 0, 0)$. Then

$$\begin{aligned} a_1 - a_2 \quad &= 0 \\ a_1 \quad - a_3 &= 0 \\ a_2 \quad &= 0. \end{aligned}$$

This homogeneous system of equations has only the trivial solution $a_1 = a_2 = a_3 = 0$. Hence, the null space of L is $\{(0, 0, 0)\} = \{\mathbf{0}\}$.

EXAMPLE 4 Let $\mathcal{V} = R[x]$ (see Example 1, §9). For

$$f(x) = a_n x^n + a_{n-1}x^{n-1} + \cdots + a_1 x + a_0 \in R[x],$$

let

$$D(f(x)) = na_n x^{n-1} + (n-1)a_{n-1}x^{n-2} + \cdots + 2a_2 x + a_1.$$

That is, $D(f(x))$ is the ordinary derivative of $f(x)$. It follows from the properties of the derivative that D is a linear transformation of $R[x]$ into $R[x]$. Since

$$D\left[\left(\frac{a_n}{n+1}\right)x^{n+1} + \left(\frac{a_{n-1}}{n}\right)x^n + \cdots + \left(\frac{a_1}{2}\right)x^2 + a_0 x\right] = f(x)$$

for every $f(x) \in R[x]$, it follows that the range space of D is $R[x]$. Furthermore, $D(g(x)) = 0$ if and only if $g(x)$ is a constant polynomial. Therefore, the null space of D is the set of all constant polynomials in $R[x]$.

(16.2) definition A linear transformation L of a vector space $\mathcal{V}$ into a vector space $\mathcal{W}$ is *nonsingular* if $L^{-1}(\{\mathbf{0}\}) = \{\mathbf{0}\}$; L is *singular* if $L^{-1}(\{\mathbf{0}\}) \neq \{\mathbf{0}\}$.

The linear transformation D_k, $k \neq 0$, discussed in Example 1, and the linear transformation L of $\mathcal{V}_3$ into $\mathcal{V}_5$ discussed in Example 3 are both nonsingular. The linear transformations described in Examples 2 and 4 are singular.

We recall that a *one-to-one correspondence of a set S onto a set T* is a mapping (or function) f of S into T with the properties (i) if $t \in T$, there is an element $s \in S$ such that $f(s) = t$, and (ii) if $s_1, s_2 \in S$ and $f(s_1) = f(s_2)$, then $s_1 = s_2$. The first condition (i) states that f is an *onto* mapping; that is, every element of T is the image of some element in S by f. The second condition (ii) says that f is *one-to-one;* that is, distinct elements of S have images by f which are distinct elements of T. The following theorem gives a useful characterization of a nonsingular linear transformation in terms of the concept of a one-to-one correspondence.

(16.3) theorem A linear transformation L of a vector space $\mathcal{V}$ into a vector space $\mathcal{W}$ is nonsingular if and only if L is a one-to-one linear transformation of $\mathcal{V}$ onto $L(\mathcal{V})$.

PROOF By a one-to-one linear transformation of $\mathcal{V}$ onto $L(\mathcal{V})$, we mean that L is a linear transformation that, at the same time, is a one-to-one correspondence of $\mathcal{V}$ onto $L(\mathcal{V})$. By the definition of $L(\mathcal{V})$, L is a mapping of $\mathcal{V}$ onto $L(\mathcal{V})$. Suppose first that L is nonsingular. Assume that $L(\mathbf{U}) = L(\mathbf{W})$ for $\mathbf{U}, \mathbf{W} \in \mathcal{V}$. Then it follows from Definition 15.1 that

$$L(\mathbf{U} - \mathbf{W}) = L(\mathbf{U}) - L(\mathbf{W}) = \mathbf{0}.$$

Hence, $\mathbf{U} - \mathbf{W} \in L^{-1}(\{\mathbf{0}\}) = \{\mathbf{0}\}$. That is, $\mathbf{U} - \mathbf{W} = \mathbf{0}$, so that $\mathbf{U} = \mathbf{W}$. This proves that different vectors in $\mathcal{V}$ are mapped onto different vectors in $L(\mathcal{V})$. Therefore, L is a one-to-one correspondence of $\mathcal{V}$ onto $L(\mathcal{V})$. Conversely, if L is one-to-one, then $\mathbf{0} \in \mathcal{V}$ is the only vector mapped onto $\mathbf{0} \in L(\mathcal{V})$. Hence, $L^{-1}(\{\mathbf{0}\}) = \{\mathbf{0}\}$, and L is nonsingular. ▌

Another characterization of a nonsingular linear transformation can be given if $\mathcal{V}$ is finite dimensional.

(16.4) theorem Let L be a linear transformation of a finite dimensional vector space $\mathcal{V}$ into a vector space $\mathcal{W}$. Then L is nonsingular if and only if the dimension of $L(\mathcal{V})$ is equal to the dimension of $\mathcal{V}$.

PROOF Let $S = \{\mathbf{U}_1, \mathbf{U}_2, \ldots, \mathbf{U}_n\}$ be a basis of $\mathcal{V}$, and consider the set $T = \{L(\mathbf{U}_1), L(\mathbf{U}_2), \ldots, L(\mathbf{U}_n)\} \subseteq L(\mathcal{V})$. Each vector $\mathbf{V} \in \mathcal{V}$ has an expression

$$\mathbf{V} = r_1 \cdot \mathbf{U}_1 + r_2 \cdot \mathbf{U}_2 + \cdots + r_n \cdot \mathbf{U}_n.$$

Therefore,

$$L(\mathbf{V}) = r_1 \cdot L(\mathbf{U}_1) + r_2 \cdot L(\mathbf{U}_2) + \cdots + r_n \cdot L(\mathbf{U}_n).$$

Hence, the set T spans $L(\mathcal{V})$.

Now suppose that L is nonsingular. If

$$\begin{aligned}\mathbf{0} &= r_1 \cdot L(\mathbf{U}_1) + r_2 \cdot L(\mathbf{U}_2) + \cdots + r_n \cdot L(\mathbf{U}_n)\\ &= L(r_1 \cdot \mathbf{U}_1 + r_2 \cdot \mathbf{U}_2 + \cdots + r_n \cdot \mathbf{U}_n),\end{aligned}$$

then since $L^{-1}(\{\mathbf{0}\}) = \{\mathbf{0}\}$,

$$r_1 \cdot \mathbf{U}_1 + r_2 \cdot \mathbf{U}_2 + \cdots + r_n \cdot \mathbf{U}_n = \mathbf{0}.$$

This implies that $r_1 = r_2 = \cdots = r_n = 0$, since S is a basis of $\mathcal{V}$. Therefore, T is a linearly independent set in $L(\mathcal{V})$ that spans $L(\mathcal{V})$. That is, T is a basis of $L(\mathcal{V})$. Hence, $\dim L(\mathcal{V}) = n = \dim \mathcal{V}$.

Conversely, assume that $\dim L(\mathcal{V}) = \dim \mathcal{V} = n$. Since T spans $L(\mathcal{V})$, it follows from Corollary 10.8 that T is a basis of $L(\mathcal{V})$. Therefore, if

$$\begin{aligned}\mathbf{0} = L(\mathbf{V}) &= L(r_1 \cdot \mathbf{U}_1 + r_2 \cdot \mathbf{U}_2 + \cdots + r_n \cdot \mathbf{U}_n)\\ &= r_1 \cdot L(\mathbf{U}_1) + r_2 \cdot L(\mathbf{U}_2) + \cdots + r_n \cdot L(\mathbf{U}_n),\end{aligned}$$

we have $r_1 = r_2 = \cdots = r_n = 0$; hence, $\mathbf{V} = \mathbf{0}$. We have proved that $L^{-1}(\{\mathbf{0}\}) = \{\mathbf{0}\}$. That is, L is nonsingular. ▌

(16.5) corollary Let L be a linear transformation of a finite dimensional vector space $\mathcal{V}$ into itself. Then L is nonsingular if and only if $L(\mathcal{V}) = \mathcal{V}$.

PROOF If L is nonsingular, then, by Theorem 16.4, the dimension of $L(\mathcal{V})$ is equal to the dimension of $\mathcal{V}$. Therefore, $L(\mathcal{V}) = \mathcal{V}$ by Corollary 10.9. Conversely, if $L(\mathcal{V}) = \mathcal{V}$, then $\dim L(\mathcal{V}) = \dim \mathcal{V}$. Hence, by Theorem 16.4, L is nonsingular. ∎

(16.6) definition If L is a nonsingular linear transformation of $\mathcal{V}$ into $\mathcal{W}$ such that $L(\mathcal{V}) = \mathcal{W}$, then L is called an *isomorphism* of $\mathcal{V}$ onto $\mathcal{W}$. We say that $\mathcal{V}$ is *isomorphic* to $\mathcal{W}$.

In other words, a linear transformation L of $\mathcal{V}$ and $\mathcal{W}$ is an isomorphism if it is both one-to-one and onto.

If $\mathcal{V}$ is isomorphic to $\mathcal{W}$, then $\mathcal{V}$ and $\mathcal{W}$ are essentially the same as algebraic systems. For not only is there a one-to-one correspondence between $\mathcal{V}$ and $\mathcal{W}$ (that is, $\mathcal{V}$ and $\mathcal{W}$ are equivalent as sets), but also the conditions (16–1) and (16–2) imply that the operations of addition and scalar multiplication are "preserved" by the correspondence. Any property of $\mathcal{V}$ that can be stated as an identity in terms of the operations of addition and scalar multiplication is also a property of $\mathcal{W}$, and vice versa. The following theorem shows the importance of Example 4, §9.

(16.7) theorem A finite dimensional vector space $\mathcal{V}$ of dimension n is isomorphic to $\mathcal{V}_n$.

PROOF Let $\{\mathbf{U}_1, \mathbf{U}_2, \ldots, \mathbf{U}_n\}$ be a basis of $\mathcal{V}$. Then every vector $\mathbf{V} \in \mathcal{V}$ has a unique expression

$$\mathbf{V} = r_1 \cdot \mathbf{U}_1 + r_2 \cdot \mathbf{U}_2 + \cdots + r_n \cdot \mathbf{U}_n.$$

Define a mapping L of $\mathcal{V}$ into $\mathcal{V}_n$ by

$$L(\mathbf{V}) = L(r_1 \cdot \mathbf{U}_1 + r_2 \cdot \mathbf{U}_2 + \cdots + r_n \cdot \mathbf{U}_n) = (r_1, r_2, \ldots, r_n).$$

The proof that L is an isomorphism of $\mathcal{V}$ onto $\mathcal{V}_n$ is left as an exercise (Exercise 16.11). ∎

It follows from the isomorphism of Theorem 16.7 that addition and scalar multiplication of vectors in a finite dimensional vector space $\mathcal{V}$ can be performed in terms of the coordinates of the vectors with respect to some basis of $\mathcal{V}$. Let $\{\mathbf{U}_1, \mathbf{U}_2, \ldots, \mathbf{U}_n\}$ be a basis of $\mathcal{V}$ and let

$$\mathbf{V} = r_1 \cdot \mathbf{U}_1 + r_2 \cdot \mathbf{U}_2 + \cdots + r_n \cdot \mathbf{U}_n$$
$$\mathbf{W} = s_1 \cdot \mathbf{U}_1 + s_2 \cdot \mathbf{U}_2 + \cdots + s_n \cdot \mathbf{U}_n.$$

By the isomorphism of Theorem 16.7, $\mathbf{V}$ corresponds to $(r_1, r_2, \ldots, r_n) \in \mathcal{V}_n$ and $\mathbf{W}$ corresponds to $(s_1, s_2, \ldots, s_n) \in \mathcal{V}_n$. Hence, $\mathbf{V} + \mathbf{W}$ corresponds to

$$(r_1, r_2, \ldots, r_n) + (s_1, s_2, \ldots, s_n) = (r_1 + s_1, r_2 + s_2, \ldots, r_n + s_n)$$

where $r_1 + s_1, r_2 + s_2, \ldots, r_n + s_n$ are the coordinates of $\mathbf{V} + \mathbf{W}$.

Similarly, for any real number r, $r \cdot \mathbf{V}$ corresponds to

$$r \cdot (r_1, r_2, \ldots, r_n) = (rr_1, rr_2, \ldots, rr_n)$$

where $rr_1, rr_2, \ldots, rr_n$ are the coordinates of $r \cdot \mathbf{V}$.

exercises

16.1 Show that condition $L(a \cdot \mathbf{U} + b \cdot \mathbf{W}) = a \cdot L(\mathbf{U}) + b \cdot L(\mathbf{W})$ in Definition 15.1 is equivalent to the two conditions (16–1) and (16–2).

16.2 Let L be a linear transformation of $\mathcal{V}$ into $\mathcal{W}$, and let $\mathcal{S}$ be a subspace of $\mathcal{V}$. Show that $L^{-1}[L(\mathcal{S})] \supseteq \mathcal{S}$. Show that $L^{-1}[L(\mathcal{V})] = L^{-1}(\mathcal{W})$.

16.3 Verify that the mapping L of $\mathcal{V}_3$ into $\mathcal{V}_5$ given in Example 3 is a linear transformation.

16.4 The mapping L of $\mathcal{E}_2$ into $\mathcal{E}_2$ that sends a vector $\mathbf{V}$ with coordinates (x, y) into a vector $L(\mathbf{V})$ with coordinates (x', y'), where

$$x' = ax + by$$
$$y' = cx + dy$$

for real numbers a, b, c, d, is a linear transformation. Prove that L is nonsingular if and only if $ad - bc \neq 0$.

16.5 Let L be the linear transformation of a Euclidean vector space $\mathcal{V}$ into R defined by $L(\mathbf{V}) = \mathcal{I}(\mathbf{V}_0, \mathbf{V})$ for a fixed nonzero vector $\mathbf{V}_0 \in \mathcal{V}$ (see §15). Prove that $L(\mathcal{V}) = R$ and that $L^{-1}(\{\mathbf{0}\})$ is the orthogonal complement of the subspace of $\mathcal{V}$ spanned by $\mathbf{V}_0$. Show that there is a one-dimensional subspace $\mathcal{S}$ of $\mathcal{V}$ such that $L(\mathcal{S}) = L(\mathcal{V})$.

16.6 Let T_φ be the linear transformation of $\mathcal{E}_2$ into itself described in Example 1, §15. Prove that T_φ is a nonsingular linear transformation of $\mathcal{E}_2$ onto $\mathcal{E}_2$.

16.7 Let $P_{\mathbf{V}_0}$ be the orthogonal projection of $\mathcal{E}_3$ into the subspace $\mathcal{S}$ of $\mathcal{E}_3$ spanned by the nonzero vector $\mathbf{V}_0 \in \mathcal{E}_3$ (see Example 2, §15). Find the range space and null space of $P_{\mathbf{V}_0}$.

16.8 Let $P_\mathcal{S}$ be the orthogonal projection of $\mathcal{E}_3$ into the subspace of $\mathcal{S}$ of $\mathcal{E}_3$ spanned by the nonzero orthogonal vectors $\mathbf{U}$ and $\mathbf{W}$ (see Example 3, §15). Find the range space and null space of $P_\mathcal{S}$. Show that $P_\mathcal{S}(\mathcal{S}) = P_\mathcal{S}(\mathcal{E}_3)$.

16.9 Show that if $\mathcal{V}$ and $\mathcal{W}$ are finite dimensional vector spaces such that $\dim \mathcal{W} < \dim \mathcal{V}$, then every linear transformation of $\mathcal{V}$ into $\mathcal{W}$ is singular.

16.10 Let L be an isomorphism of the vector space $\mathcal{V}$ onto the vector space $\mathcal{W}$. Let M be the mapping of $\mathcal{W}$ into $\mathcal{V}$ defined by $M(\mathbf{U}) = \mathbf{V}$ if $L(\mathbf{V}) = \mathbf{U}$. Prove that M is an isomorphism of $\mathcal{W}$ onto $\mathcal{V}$.

16.11 Complete the proof of Theorem 16.7.

16.12 Prove that $\mathcal{V}_n$ is not isomorphic to $\mathcal{V}_m$ if $n \neq m$.

16.13 Let L be a linear transformation of a finite dimensional vector space $\mathcal{V}$ into a vector space $\mathcal{W}$. (a) Prove that $\dim L(\mathcal{V}) \leq \dim \mathcal{V}$. (b) Prove that $\dim L^{-1}(\{\mathbf{0}\}) + \dim L(\mathcal{V}) = \dim \mathcal{V}$ (which is a generalization of Theorem 16.4).

16.14 Let L be a linear transformation of the vector space $\mathcal{V}$ into the vector space $\mathcal{W}$. Let $\mathcal{S}$ be a subspace of $\mathcal{V}$ spanned by the vectors $\mathbf{U}_1, \mathbf{U}_2, \ldots, \mathbf{U}_k$. Show that $L(\mathcal{S})$ is spanned by $L(\mathbf{U}_1), L(\mathbf{U}_2), \ldots, L(\mathbf{U}_k)$. Show that if L is nonsingular, then $L(\mathcal{S})$ is isomorphic to $\mathcal{S}$, and that in particular, $\dim L(\mathcal{S}) = \dim \mathcal{S}$.

16.15 Let L be an isomorphism of $\mathcal{E}_3$ into itself. Let $\mathcal{S}$ be the subspace of all vectors on a line ℓ through O. Show that $L(\mathcal{S})$ is a subspace of all vectors on some line ℓ' through O. Recall that $\mathcal{E}_\pi$ is the subspace of all vectors in a plane π through O. Show that $L(\mathcal{E}_\pi) = \mathcal{E}_{\pi'}$, for some plane π' through O.

17 the algebra of linear transformations

Let $\mathfrak{M}$ be the set of all linear transformations of a vector space $\mathcal{V}$ into a vector space $\mathcal{W}$. Then operations of *addition* and *scalar multiplication* by a real number can be defined in $\mathfrak{M}$ in a natural way. If L and M are in $\mathfrak{M}$, then the mappings $L + M$ and $r \cdot L$ defined by the rules

$$(L + M)(\mathbf{U}) = L(\mathbf{U}) + M(\mathbf{U}) \tag{17–1}$$
$$(r \cdot L)(\mathbf{U}) = r \cdot L(\mathbf{U}) \tag{17–2}$$

for $\mathbf{U} \in \mathcal{V}$, are readily seen to be linear transformations of $\mathcal{V}$ into $\mathcal{W}$. The mapping $\mathbf{O}$ of $\mathcal{V}$ into $\mathcal{W}$ defined by $\mathbf{O}(\mathbf{U}) = \mathbf{0}$ for all $\mathbf{U} \in \mathcal{V}$ is called the *zero mapping*. The zero mapping is a linear transformation, and

$$(L + \mathbf{O})(\mathbf{U}) = L(\mathbf{U}) + \mathbf{O}(\mathbf{U}) = L(\mathbf{U}) + \mathbf{0} = L(\mathbf{U})$$

for all $L \in \mathfrak{M}$. Thus, $L + \mathbf{O} = L$. Similarly, $\mathbf{O} + L = L$. For $L \in \mathfrak{M}$, the mapping $-L$ is defined by $(-L)(\mathbf{U}) = -L(\mathbf{U})$ for $\mathbf{U} \in \mathcal{V}$. Then $-L \in \mathfrak{M}$ and

$$L + (-L) = (-L) + L = \mathbf{O}.$$

Using these definitions, it is a routine job to prove the following theorem by verifying that the conditions (a) through (j) of Definition 9.2 are satisfied.

(17.1) theorem Let $\mathfrak{M}$ be the set of all linear transformations of a vector space $\mathcal{V}$ into a vector space $\mathcal{W}$. Then $\mathfrak{M}$ is a vector space with respect to the operations of addition and scalar multiplication defined by (17–1) and (17–2).

Let $\mathcal{V}$, $\mathcal{W}$, and $\mathcal{P}$ be vector spaces, and suppose that L is a linear transformation of $\mathcal{V}$ into $\mathcal{W}$ and that M is a linear transformation of $\mathcal{W}$ into $\mathcal{P}$. For each $\mathbf{U} \in \mathcal{V}$, $L(\mathbf{U})$ is a unique vector in $\mathcal{W}$; hence, $M[L(\mathbf{U})]$ is a unique vector in $\mathcal{P}$. Thus, if we define

$$(ML)(\mathbf{U}) = M[L(\mathbf{U})] \tag{17–3}$$

for $\mathbf{U} \in \mathcal{V}$, ML is a mapping of $\mathcal{V}$ into $\mathcal{P}$. Moreover, ML satisfies the linearity condition of Definition 15.1. Indeed,

$$\begin{aligned}(ML)(a \cdot \mathbf{U} + b \cdot \mathbf{W}) &= M[L(a \cdot \mathbf{U} + b \cdot \mathbf{W})] = M[a \cdot L(\mathbf{U}) + b \cdot L(\mathbf{W})] \\ &= a \cdot M[L(\mathbf{U})] + b \cdot M[L(\mathbf{W})] \\ &= a \cdot (ML)(\mathbf{U}) + b \cdot (ML)(\mathbf{W})\end{aligned}$$

using the facts that L is a linear transformation of $\mathcal{V}$ and M is a linear transformation of $\mathcal{W}$. Thus, ML is a linear transformation of $\mathcal{V}$ into $\mathcal{P}$. If $\mathcal{V} = \mathcal{W} = \mathcal{P}$, so that L and M are linear transformations of $\mathcal{V}$ into itself, then ML is a linear transformation of $\mathcal{V}$ into itself. The rule (17–3) is the usual composition of mappings and is called *multiplication* of linear transformations. The linear transformation ML is called the *product* of L and M.

EXAMPLE 1 The mappings L and M of $\mathcal{V}_5$ into itself defined by

$$L[(a_1, a_2, a_3, a_4, a_5)] = (a_1 + 2a_2, 0, a_4 - a_5, 0, a_3)$$

and

$$M[(a_1, a_2, a_3, a_4, a_5)] = (a_1, a_1 + a_2, a_1 + a_2 + a_3, 5a_4, -a_5)$$

are linear transformations. Then

$$\begin{aligned}(L + M)[(a_1, a_2, a_3, a_4, a_5)] &= (a_1 + 2a_2, 0, a_4 - a_5, 0, a_3) \\ &\quad + (a_1, a_1 + a_2, a_1 + a_2 + a_3, 5a_4, -a_5) \\ &= (2a_1 + 2a_2, a_1 + a_2, a_1 + a_2 + a_3 + a_4 - a_5, 5a_4, a_3 - a_5) \\ (10 \cdot L)[(a_1, a_2, a_3, a_4, a_5)] &= 10 \cdot (a_1 + 2a_2, 0, a_4 - a_5, 0, a_3) \\ &= (10a_1 + 20a_2, 0, 10a_4 - 10a_5, 0, 10a_3) \\ (ML)[(a_1, a_2, a_3, a_4, a_5)] &= M[(a_1 + 2a_2, 0, a_4 - a_5, 0, a_3)] \\ &= (a_1 + 2a_2, a_1 + 2a_2, a_1 + 2a_2 + a_4 - a_5, 0, -a_3) \\ (LM)[(a_1, a_2, a_3, a_4, a_5)] &= L[(a_1, a_1 + a_2, a_1 + a_2 + a_3, 5a_4, -a_5)] \\ &= (3a_1 + 2a_2, 0, 5a_4 + a_5, 0, a_1 + a_2 + a_3).\end{aligned}$$

EXAMPLE 2 Let L be the linear transformation of $\mathcal{V}_3$ into $\mathcal{V}_5$ given in Example 3, §16. The mapping M of $\mathcal{V}_5$ into $\mathcal{V}_4$ defined by

$$M[(b_1, b_2, b_3, b_4, b_5)] = (b_1, b_2, b_3, b_4)$$

is a linear transformation. Then,

$$\begin{aligned}(ML)[(a_1, a_2, a_3)] &= M[(a_1 - a_2, 0, a_1 - a_3, a_2, 0)] \\ &= (a_1 - a_2, 0, a_1 - a_3, a_2).\end{aligned}$$

EXAMPLE 3 Let $P_{\mathcal{S}}$ be the orthogonal projection of $\mathcal{E}_3$ into a subspace $\mathcal{S}$ of $\mathcal{E}_3$ spanned by the nonzero orthogonal vectors $\mathbf{U}$ and $\mathbf{W}$ (see Example 3, §15). For $\mathbf{V} \in \mathcal{E}_3$,

$$\begin{aligned}P_{\mathcal{S}}^2(\mathbf{V}) = P_{\mathcal{S}}[P_{\mathcal{S}}(\mathbf{V})] &= P_{\mathcal{S}}\left[\left(\frac{\mathbf{V} \circ \mathbf{U}}{\mathbf{U} \circ \mathbf{U}}\right) \cdot \mathbf{U} + \left(\frac{\mathbf{V} \circ \mathbf{W}}{\mathbf{W} \circ \mathbf{W}}\right) \cdot \mathbf{W}\right] \\ &= \left(\frac{\mathbf{V} \circ \mathbf{U}}{\mathbf{U} \circ \mathbf{U}}\right) \cdot P_{\mathcal{S}}(\mathbf{U}) + \left(\frac{\mathbf{V} \circ \mathbf{W}}{\mathbf{W} \circ \mathbf{W}}\right) \cdot P_{\mathcal{S}}(\mathbf{W})\end{aligned}$$

$$= \left(\frac{\mathbf{V} \circ \mathbf{U}}{\mathbf{U} \circ \mathbf{U}}\right) \cdot \left[\left(\frac{\mathbf{U} \circ \mathbf{U}}{\mathbf{U} \circ \mathbf{U}}\right) \cdot \mathbf{U} + \left(\frac{\mathbf{U} \circ \mathbf{W}}{\mathbf{W} \circ \mathbf{W}}\right) \cdot \mathbf{W}\right]$$
$$+ \left(\frac{\mathbf{V} \circ \mathbf{W}}{\mathbf{W} \circ \mathbf{W}}\right) \cdot \left[\left(\frac{\mathbf{W} \circ \mathbf{U}}{\mathbf{U} \circ \mathbf{U}}\right) \cdot \mathbf{U} + \left(\frac{\mathbf{W} \circ \mathbf{W}}{\mathbf{W} \circ \mathbf{W}}\right) \cdot \mathbf{W}\right]$$
$$= \left(\frac{\mathbf{V} \circ \mathbf{U}}{\mathbf{U} \circ \mathbf{U}}\right) \cdot \mathbf{U} + \left(\frac{\mathbf{V} \circ \mathbf{W}}{\mathbf{W} \circ \mathbf{W}}\right) \cdot \mathbf{W} = P_{\mathcal{S}}(\mathbf{V})$$

since $\mathbf{U} \circ \mathbf{W} = \mathbf{W} \circ \mathbf{U} = 0$. Therefore, $P_{\mathcal{S}}^2 = P_{\mathcal{S}}$.

Suppose that in addition to the vector spaces $\mathcal{V}$, $\mathcal{W}$, $\mathcal{P}$ and the linear transformations L of $\mathcal{V}$ into $\mathcal{W}$ and M of $\mathcal{W}$ into $\mathcal{P}$ described above, there is a space $\mathcal{Q}$ and a linear transformation N of $\mathcal{P}$ into $\mathcal{Q}$. Then ML maps $\mathcal{V}$ into $\mathcal{P}$ and N maps $\mathcal{P}$ into $\mathcal{Q}$, so that $N(ML)$ is a linear transformation of $\mathcal{V}$ into $\mathcal{Q}$. Similarly, NM maps $\mathcal{W}$ into $\mathcal{Q}$, so that $(NM)L$ is also a linear transformation of $\mathcal{V}$ into $\mathcal{Q}$. It is an immediate consequence of (17–3) that $[N(ML)](\mathbf{U}) = [(NM)L](\mathbf{U})$ for all vectors $\mathbf{U} \in \mathcal{V}$. That is, the multiplication of linear transformations is associative:

$$N(ML) = (NM)L. \tag{17–4}$$

EXAMPLE 4 Let $\mathbf{V}_0$ be a fixed nonzero vector in $\mathcal{E}_3$. Suppose that the coordinate system is chosen so that the direction of $\mathbf{V}_0$ is the positive direction of the Z axis. Then following the discussion of the cross product in §5, we have that for any vector $\mathbf{V} \in \mathcal{E}_3$, $\mathbf{V}_0 \times \mathbf{V}$ is obtained from $\mathbf{V}$ by projecting $\mathbf{V}$ onto the XY plane, rotating this projection in the XY plane through 90° about $\mathbf{V}_0$, and finally multiplying by the real number $|\mathbf{V}_0|$ (see Figure 51). Projecting $\mathbf{V}$ onto the XY plane is the orthogonal projection $P_{\mathcal{E}_2}$ of $\mathcal{E}_3$ into $\mathcal{E}_2$, rotating through 90° about $\mathbf{V}_0$ is the rotation T_{90° of $\mathcal{E}_2$ into $\mathcal{E}_2$, and multiplying by $|\mathbf{V}_0|$ is the mapping $D_{|\mathbf{V}_0|}$ of $\mathcal{E}_2$ into $\mathcal{E}_2$. Thus, if L is the mapping of $\mathcal{E}_3$ into itself defined by $L(\mathbf{V}) = \mathbf{V}_0 \times \mathbf{V}$ for $\mathbf{V} \in \mathcal{E}_3$, then

$$L = D_{|\mathbf{V}_0|}(T_{90^\circ}P_{\mathcal{E}_2}).$$

We have seen in earlier discussions that $P_{\mathcal{E}_2}$, T_{90°, and $D_{|\mathbf{V}_0|}$ are linear transformations. Hence, it follows that L is a linear transformation of $\mathcal{E}_3$ into $\mathcal{E}_2$. Of course, we have already observed in §15, that L is a linear transformation of $\mathcal{E}_3$ into itself, and in Example 2, §16, that the range space $L(\mathcal{E}_3)$ of L is $\mathcal{E}_2$.

The mapping that sends each vector of a space $\mathcal{V}$ into itself is a linear transformation of $\mathcal{V}$ into $\mathcal{V}$ called the *identity transformation* of $\mathcal{V}$. This transformation will be denoted by $I_{\mathcal{V}}$ (or simply by I if a single vector space $\mathcal{V}$ is under discussion). Thus, $I_{\mathcal{V}}(\mathbf{U}) = \mathbf{U}$ for all $\mathbf{U} \in \mathcal{V}$. If L is any linear transformation of $\mathcal{V}$ into $\mathcal{W}$, then $L(\mathbf{U}) = I_{\mathcal{W}}[L(\mathbf{U})] = L[I_{\mathcal{V}}(\mathbf{U})]$ for all $\mathbf{U} \in \mathcal{V}$. Therefore,

$$L = I_{\mathcal{W}}L = LI_{\mathcal{V}}. \tag{17–5}$$

Suppose that L is an isomorphism of $\mathcal{V}$ onto $\mathcal{W}$. For $\mathbf{U} \in \mathcal{W}$, let $M(\mathbf{U})$ be the unique vector $\mathbf{V} \in \mathcal{V}$ such that $L(\mathbf{V}) = \mathbf{U}$. Then M is an isomorphism

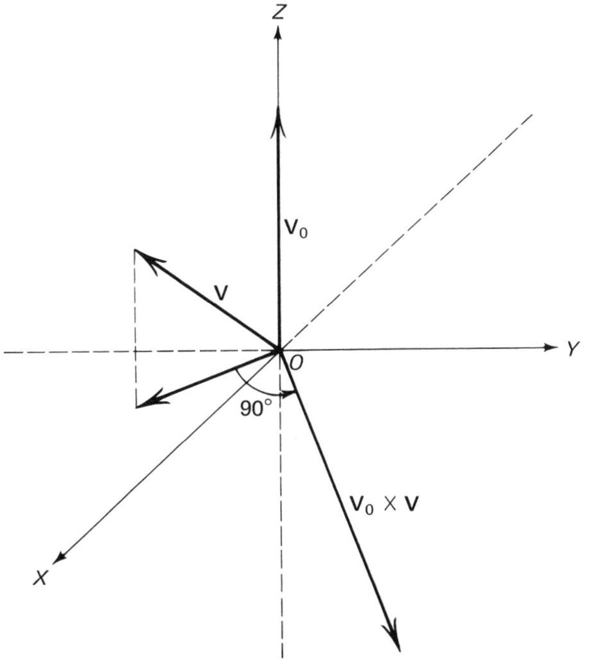

FIGURE 51

of $\mathcal{W}$ onto $\mathcal{V}$ (see Exercise 16.10). The linear transformation M is called the *inverse* of L, and we write $M = L^*$. (We use the notation L^* for the inverse of an isomorphism L of $\mathcal{V}$ of $\mathcal{W}$ instead of the more conventional notation L^{-1} to avoid confusion with the notation for the complete inverse image of a subspace of $\mathcal{W}$.) It is a consequence of the definition of L^* that $L^*L = I_{\mathcal{V}}$ and $LL^* = I_{\mathcal{W}}$. The converse of this result is given by the following theorem.

(17.2) theorem Let L be a linear transformation of $\mathcal{V}$ into $\mathcal{W}$, and let M be a linear transformation of $\mathcal{W}$ into $\mathcal{V}$ such that $ML = I_{\mathcal{V}}$ and $LM = I_{\mathcal{W}}$. Then L is an isomorphism of $\mathcal{V}$ onto $\mathcal{W}$ and $M = L^*$.

PROOF Let $\mathbf{V} \in L^{-1}(\{\mathbf{0}\})$. Then $L(\mathbf{V}) = \mathbf{0} \in \mathcal{W}$, and $ML(\mathbf{V}) = M(\mathbf{0}) = \mathbf{0} \in \mathcal{V}$. However, since $ML = I_{\mathcal{V}}$, it follows that $ML(\mathbf{V}) = \mathbf{V}$. Hence, $\mathbf{V} = \mathbf{0}$, and $L^{-1}(\{\mathbf{0}\}) = \{\mathbf{0}\}$. Thus, L is nonsingular. Let $\mathbf{U} \in \mathcal{W}$. Since $LM(\mathbf{U}) = I_{\mathcal{W}}(\mathbf{U}) = \mathbf{U}$, it follows that $M(\mathbf{U})$ is a vector in $\mathcal{V}$ such that $L[M(\mathbf{U})] = \mathbf{U}$. Therefore, $L(\mathcal{V}) = \mathcal{W}$. By Definition 16.6, L is an isomorphism of $\mathcal{V}$ onto $\mathcal{W}$. Since L is an isomorphism, the inverse transformation L^* exists. Thus, we have

$$L^* = L^*I_{\mathcal{W}} = L^*(LM) = (L^*L)M = I_{\mathcal{V}}M = M. \qquad \blacksquare$$

If L is a nonsingular transformation of a finite dimensional vector space $\mathcal{V}$ into itself, then, by Corollary 16.5, $L(\mathcal{V}) = \mathcal{V}$. Hence, L is an isomorphism of $\mathcal{V}$ onto $\mathcal{V}$ (Definition 16.6), and the inverse transformation L^* exists and is an isomorphism. Therefore, L^* is nonsingular and $L^*(\mathcal{V}) = \mathcal{V}$. Let I be the identity transformation of $\mathcal{V}$. Then $L^*L = LL^* = I$. Conversely, by Theorem 17.2, if L and M are linear transformations of $\mathcal{V}$ into $\mathcal{V}$ such that $ML = LM = I$, then L and M are nonsingular and $M = L^*$.

Let $\mathfrak{M}$ be the set of all linear transformations of a vector space $\mathcal{V}$ (not necessarily finite dimensional) into itself. In this case all three operations—addition (17–1), scalar multiplication (17–2), and multiplication (17–3)—are defined in $\mathfrak{M}$. The identity transformation I is in $\mathfrak{M}$. Let K, L, $M \in \mathfrak{M}$, and let r be a real number. Then in addition to the associativity of multiplication in $\mathfrak{M}$ (17–4) and those properties of addition and scalar multiplication that are a consequence of the fact that $\mathfrak{M}$ is a vector space (Theorem 17.1), the following identities are satisfied:

$$K(L + M) = KL + KM \tag{17–6}$$

$$(K + L)M = KM + LM \tag{17–7}$$

$$r \cdot (LM) = (r \cdot L)M = L(r \cdot M) \tag{17–8}$$

$$L = IL = LI. \tag{17–9}$$

exercises

17.1 Let L and M be linear transformations of a vector space $\mathcal{V}$ into a vector space $\mathcal{W}$. Prove that $L + M$ and $r \cdot L$, as defined by (17–1) and (17–2), are linear transformations of $\mathcal{V}$ into $\mathcal{W}$.

17.2 Let L be a linear transformation of a vector space $\mathcal{V}$ into a vector space $\mathcal{W}$. Let D_r be the linear transformation of $\mathcal{W}$ into itself defined by $D_r(\mathbf{U}) = r \cdot \mathbf{U}$ for $\mathbf{U} \in \mathcal{W}$. Prove that $r \cdot L = D_r L$.

17.3 Let L be the linear transformation of $\mathcal{E}_3$ into itself defined by $L(\mathbf{V}) = \mathbf{K} \times \mathbf{V}$ for $\mathbf{V} \in \mathcal{E}_3$, let $P_{\mathbf{I}+\mathbf{J}}$ be the orthogonal projection of $\mathcal{E}_3$ into the subspace spanned by $\mathbf{I} + \mathbf{J}$, and let D_3 be the linear transformation of $\mathcal{E}_3$ into itself defined by $D_3(\mathbf{V}) = 3 \cdot \mathbf{V}$ for $\mathbf{V} \in \mathcal{E}_3$. Compute the following vectors in $\mathcal{E}_3$: $(L + P_{\mathbf{I}+\mathbf{J}})(2 \cdot \mathbf{I} - 3 \cdot \mathbf{J} + \mathbf{K})$; $(P_{\mathbf{I}+\mathbf{J}} - D_3)(\mathbf{I} - \mathbf{K})$; $(-4 \cdot L)(3 \cdot \mathbf{J} - \mathbf{K})$; $(P_{\mathbf{I}+\mathbf{J}}L)(\mathbf{I} + \mathbf{J} + \mathbf{K})$; $(2 \cdot L - 3 \cdot P_{\mathbf{I}+\mathbf{J}} + D_3)(-4 \cdot \mathbf{I})$;
$[L(D_3 P_{\mathbf{I}+\mathbf{J}})](x \cdot \mathbf{I} + y \cdot \mathbf{J} + z \cdot \mathbf{K})$;
$(P_{\mathbf{I}+\mathbf{J}}L - LP_{\mathbf{I}+\mathbf{J}})(3 \cdot \mathbf{I} + 2 \cdot \mathbf{K})$;
$(D_3 L - LD_3)(x \cdot \mathbf{I} + y \cdot \mathbf{J} + z \cdot \mathbf{K})$.

17.4 Let T_{45° be the rotation through 45° of $\mathcal{E}_2$ into $\mathcal{E}_2$, $P_{\mathcal{E}_2}$ be the orthogonal projection of $\mathcal{E}_3$ into $\mathcal{E}_2$, and L be the linear transformation of $\mathcal{E}_3$ into R defined by $L(\mathbf{V}) = (\mathbf{I} + \mathbf{J} + \mathbf{K}) \circ \mathbf{V}$ for $\mathbf{V} \in \mathcal{E}_3$. Compute the following vectors: $[L(T_{45^\circ} P_{\mathcal{E}_2})](3 \cdot \mathbf{I} - 2 \cdot \mathbf{J} + 5 \cdot \mathbf{K})$; $(-2 \cdot T_{45^\circ})(-\mathbf{I} + 3 \cdot \mathbf{J})$; $(T_{45^\circ} - 3 \cdot P_{\mathcal{E}_2})(\mathbf{I} + \mathbf{J})$; $(LP_{\mathcal{E}_2})(\mathbf{I} - \mathbf{J} + \mathbf{K})$.

17.5 What are the inverses of the following nonsingular linear transformations of $\mathcal{E}_2$ into $\mathcal{E}_2$? (a) T_φ; (b) D_k, $k \neq 0$; (c) L defined by $L(x \cdot \mathbf{I} + y \cdot \mathbf{J}) = x' \cdot \mathbf{I} + y' \cdot \mathbf{J}$, where $x' = ax + by$, $y' = cx + dy$ and $ad - bc \neq 0$.

17.6 Let $\mathfrak{M}$ be the set of all linear transformations of a vector space $\mathcal{V}$ into a vector space $\mathcal{W}$. Prove that the zero mapping $\mathbf{O}$ of $\mathcal{V}$ into $\mathcal{W}$ is in $\mathfrak{M}$. For $L \in \mathfrak{M}$, prove that $-L \in \mathfrak{M}$. Prove that $L + (-L) = (-L) + L = \mathbf{O}$.

17.7 Prove Theorem 17.1.

17.8 Let L be a linear transformation of $\mathcal{V}$ into $\mathcal{W}$, and let M be a linear transformation of $\mathcal{W}$ into $\mathcal{P}$. Prove that the range space of ML is contained in the range space of M. Under what condition are these range spaces the same subspace of $\mathcal{P}$? Prove that the null space of L is contained in the null space of ML. Under what condition are these null spaces the same subspace of $\mathcal{V}$?

17.9 In Example 2, show that the range space of ML is a proper subspace of the range space of M, that is, $ML(\mathcal{V}_3) \subset M(\mathcal{V}_5)$. Show that the null space of L is the same subspace of $\mathcal{V}_3$ as the null space of ML, that is $L^{-1}(\{\mathbf{0}\}) = (ML)^{-1}(\{\mathbf{0}\})$. Is ML a nonsingular transformation of $\mathcal{V}_3$ into $\mathcal{V}_4$? Is M a nonsingular transformation of $\mathcal{V}_5$ into $\mathcal{V}_4$? Is ML an isomorphism of $\mathcal{V}_3$ onto $\mathcal{V}_4$?

17.10 Let L be a nonsingular linear transformation of $\mathcal{V}$ into $\mathcal{W}$, and let M be a nonsingular linear transformation of $\mathcal{W}$ into $\mathcal{P}$. Prove that ML is a nonsingular linear transformation of $\mathcal{V}$ into $\mathcal{P}$.

17.11 Let L be an isomorphism of $\mathcal{V}$ onto $\mathcal{W}$, and let L^* be the inverse of L. Prove that $L^*L = I_{\mathcal{V}}$ and $LL^* = I_{\mathcal{W}}$.

17.12 Let $\mathfrak{M}$ be the set of all linear transformations of a vector space $\mathcal{V}$ into itself. Let $K, L, M \in \mathfrak{M}$, and let I be the identity transformation of $\mathcal{V}$. Prove the identities (17–6), (17–7), (17–8), and (17–9).

18 the matrix of a linear transformation

In the remaining sections of this book, we will be concerned with linear transformations of finite dimensional vector spaces. Let $\mathcal{V}$ and $\mathcal{W}$ be finite dimensional vector spaces of dimensions n and m, respectively. Let $\{\mathbf{U}_1, \mathbf{U}_2, \ldots, \mathbf{U}_n\}$ be a basis of $\mathcal{V}$, and let $\{\mathbf{W}_1, \mathbf{W}_2, \ldots, \mathbf{W}_m\}$ be a basis of $\mathcal{W}$. Suppose that L is a linear transformation of $\mathcal{V}$ into $\mathcal{W}$. Then if $\mathbf{V}$ is any vector in $\mathcal{V}$,

$$\mathbf{V} = r_1 \cdot \mathbf{U}_1 + r_2 \cdot \mathbf{U}_2 + \cdots + r_n \cdot \mathbf{U}_n.$$

Therefore, by the linearity condition of Definition 15.1,

$$L(\mathbf{V}) = r_1 \cdot L(\mathbf{U}_1) + r_2 \cdot L(\mathbf{U}_2) + \cdots + r_n \cdot L(\mathbf{U}_n).$$

Thus, the linear transformation L is determined completely by its effect on the basis vectors of $\mathcal{V}$. Since $L(\mathbf{U}_i) \in \mathcal{W}$, we have

$$L(\mathbf{U}_i) = a_{1,i} \cdot \mathbf{W}_1 + a_{2,i} \cdot \mathbf{W}_2 + \cdots + a_{m,i} \cdot \mathbf{W}_m$$

for $i = 1, 2, \ldots, n$. The numbers $a_{1,i}, a_{2,i}, \ldots, a_{m,i}$ are the coordinates of $L(\mathbf{U}_i)$ with respect to the basis $\{\mathbf{W}_1, \mathbf{W}_2, \ldots, \mathbf{W}_m\}$. Therefore, these numbers are uniquely determined by L. We have

$$\begin{aligned} L(\mathbf{V}) &= r_1 \cdot (a_{1,1} \cdot \mathbf{W}_1 + a_{2,1} \cdot \mathbf{W}_2 + \cdots + a_{m,1} \cdot \mathbf{W}_m) \\ &\quad + r_2 \cdot (a_{1,2} \cdot \mathbf{W}_1 + a_{2,2} \cdot \mathbf{W}_2 + \cdots + a_{m,2} \cdot \mathbf{W}_m) \\ &\quad + \cdots + r_n \cdot (a_{1,n} \cdot \mathbf{W}_1 + a_{2,n} \cdot \mathbf{W}_2 + \cdots + a_{m,n} \cdot \mathbf{W}_m) \\ &= (r_1 a_{1,1} + r_2 a_{1,2} + \cdots + r_n a_{1,n}) \cdot \mathbf{W}_1 \\ &\quad + (r_1 a_{2,1} + r_2 a_{2,2} + \cdots + r_n a_{2,n}) \cdot \mathbf{W}_2 \\ &\quad + \cdots + (r_1 a_{m,1} + r_2 a_{m,2} + \cdots + r_n a_{m,n}) \cdot \mathbf{W}_m. \end{aligned}$$

Let the coordinates of $L(\mathbf{V})$ with respect to the basis $\{\mathbf{W}_1, \mathbf{W}_2, \ldots, \mathbf{W}_m\}$ of $\mathcal{W}$ be denoted by $s_1, s_2, \ldots, s_m$. Then the linear transformation L is given by the system of linear equations

$$\begin{aligned} a_{1,1}r_1 + a_{1,2}r_2 + \cdots + a_{1,n}r_n &= s_1 \\ a_{2,1}r_1 + a_{2,2}r_2 + \cdots + a_{2,n}r_n &= s_2 \\ &\vdots \\ a_{m,1}r_1 + a_{m,2}r_2 + \cdots + a_{m,n}r_n &= s_m. \end{aligned} \tag{18–1}$$

With respect to the given bases in $\mathcal{V}$ and $\mathcal{W}$, the equations (18–1) give the coordinates of $L(\mathbf{V})$ in terms of the coordinates of $\mathbf{V}$. Thus, L is completely described by the rectangular array of real numbers

$$\begin{bmatrix} a_{1,1} & a_{1,2} & \cdots & a_{1,n} \\ a_{2,1} & a_{2,2} & \cdots & a_{2,n} \\ \vdots & \vdots & & \vdots \\ a_{m,1} & a_{m,2} & \cdots & a_{m,n} \end{bmatrix} \tag{18–2}$$

which is called the *matrix* of L with respect to the given bases.

EXAMPLE 1 Let L be the linear transformation of $\mathcal{V}_3$ into $\mathcal{V}_5$ described in Example 3 of §16. Select the basis $E^{(3)} = \{\mathbf{E}_1^{(3)}, \mathbf{E}_2^{(3)}, \mathbf{E}_3^{(3)}\}$ for $\mathcal{V}_3$ and the basis $E^{(5)} = \{\mathbf{E}_1^{(5)}, \mathbf{E}_2^{(5)}, \mathbf{E}_3^{(5)}, \mathbf{E}_4^{(5)}, \mathbf{E}_5^{(5)}\}$ for $\mathcal{V}_5$. Then

$$\begin{aligned} L(\mathbf{E}_1^{(3)}) &= L[(1, 0, 0)] = (1, 0, 1, 0, 0) \\ &= 1 \cdot \mathbf{E}_1^{(5)} + 0 \cdot \mathbf{E}_2^{(5)} + 1 \cdot \mathbf{E}_3^{(5)} + 0 \cdot \mathbf{E}_4^{(5)} + 0 \cdot \mathbf{E}_5^{(5)} \\ L(\mathbf{E}_2^{(3)}) &= L[(0, 1, 0)] = (-1, 0, 0, 1, 0) \\ &= (-1) \cdot \mathbf{E}_1^{(5)} + 0 \cdot \mathbf{E}_2^{(5)} + 0 \cdot \mathbf{E}_3^{(5)} + 1 \cdot \mathbf{E}_4^{(5)} + 0 \cdot \mathbf{E}_5^{(5)} \\ L(\mathbf{E}_3^{(3)}) &= L[(0, 0, 1)] = (0, 0, -1, 0, 0) \\ &= 0 \cdot \mathbf{E}_1^{(5)} + 0 \cdot \mathbf{E}_2^{(5)} + (-1) \cdot \mathbf{E}_3^{(5)} + 0 \cdot \mathbf{E}_4^{(5)} + 0 \cdot \mathbf{E}_5^{(5)}. \end{aligned}$$

Therefore, the matrix of L, with respect to the bases $E^{(3)}$ and $E^{(5)}$, is

$$\begin{bmatrix} 1 & -1 & 0 \\ 0 & 0 & 0 \\ 1 & 0 & -1 \\ 0 & 1 & 0 \\ 0 & 0 & 0 \end{bmatrix}.$$

Hence, if $\mathbf{V} = (r_1, r_2, r_3)$ is any vector in $\mathcal{V}_3$, then $L(\mathbf{V}) = (s_1, s_2, s_3, s_4, s_5)$, where

$$\begin{aligned} 1r_1 - 1r_2 + 0r_3 &= s_1 \\ 0r_1 + 0r_2 + 0r_3 &= s_2 \\ 1r_1 + 0r_2 - 1r_3 &= s_3 \\ 0r_1 + 1r_2 + 0r_3 &= s_4 \\ 0r_1 + 0r_2 + 0r_3 &= s_5. \end{aligned}$$

Suppose that we select different bases in $\mathcal{V}_3$ and $\mathcal{V}_5$, say $B = \{(1, 0, 0), (1, 1, 0), (1, 1, 1)\}$ as a basis for $\mathcal{V}_3$ and $C = \{(1, 0, 0, 0, 0), (1, 1, 0, 0, 0), (1, 1, 1, 0, 0), (1, 1, 1, 1, 0), (1, 1, 1, 1, 1)\}$ as a basis for $\mathcal{V}_5$. Then

$$\begin{aligned} L[(1, 0, 0)] &= (1, 0, 1, 0, 0) \\ &= 1 \cdot (1, 0, 0, 0, 0) + (-1) \cdot (1, 1, 0, 0, 0) \\ &\quad + 1 \cdot (1, 1, 1, 0, 0) \\ L[(1, 1, 0)] &= (0, 0, 1, 1, 0) \\ &= (-1) \cdot (1, 1, 0, 0, 0) + 1 \cdot (1, 1, 1, 1, 0) \\ L[(1, 1, 1)] &= (0, 0, 0, 1, 0) \\ &= (-1) \cdot (1, 1, 1, 0, 0) + 1 \cdot (1, 1, 1, 1, 0). \end{aligned}$$

Therefore, the matrix of L with respect to the bases B and C is

$$\begin{bmatrix} 1 & 0 & 0 \\ -1 & -1 & 0 \\ 1 & 0 & -1 \\ 0 & 1 & 1 \\ 0 & 0 & 0 \end{bmatrix}.$$

EXAMPLE 2 Let P be the orthogonal projection of $\mathcal{E}_3$ into the subspace $\mathcal{S}$ of $\mathcal{E}_3$ spanned by the nonzero orthogonal vectors **U** and **W**. Select the basis $\{\mathbf{I}, \mathbf{J}, \mathbf{K}\}$ for $\mathcal{E}_3$ and the basis $\{\mathbf{U}, \mathbf{W}\}$ for $\mathcal{S}$. The set $\{\mathbf{U}, \mathbf{W}\}$ is linearly independent because **U** and **W** are orthogonal. Since $\{\mathbf{U}, \mathbf{W}\}$ spans $\mathcal{S}$, it follows that $\{\mathbf{U}, \mathbf{W}\}$ is a basis of $\mathcal{S}$. We have

$$\begin{aligned} P_{\mathcal{S}}(\mathbf{I}) &= \left(\frac{\mathbf{I} \circ \mathbf{U}}{\mathbf{U} \circ \mathbf{U}}\right) \cdot \mathbf{U} + \left(\frac{\mathbf{I} \circ \mathbf{W}}{\mathbf{W} \circ \mathbf{W}}\right) \cdot \mathbf{W} \\ P_{\mathcal{S}}(\mathbf{J}) &= \left(\frac{\mathbf{J} \circ \mathbf{U}}{\mathbf{U} \circ \mathbf{U}}\right) \cdot \mathbf{U} + \left(\frac{\mathbf{J} \circ \mathbf{W}}{\mathbf{W} \circ \mathbf{W}}\right) \cdot \mathbf{W} \\ P_{\mathcal{S}}(\mathbf{K}) &= \left(\frac{\mathbf{K} \circ \mathbf{U}}{\mathbf{U} \circ \mathbf{U}}\right) \cdot \mathbf{U} + \left(\frac{\mathbf{K} \circ \mathbf{W}}{\mathbf{W} \circ \mathbf{W}}\right) \cdot \mathbf{W}. \end{aligned}$$

Therefore, the matrix of $P_{\mathcal{S}}$ with respect to the bases $\{\mathbf{I}, \mathbf{J}, \mathbf{K}\}$ and $\{\mathbf{U}, \mathbf{W}\}$ is

$$\begin{bmatrix} \dfrac{\mathbf{I} \circ \mathbf{U}}{\mathbf{U} \circ \mathbf{U}} & \dfrac{\mathbf{J} \circ \mathbf{U}}{\mathbf{U} \circ \mathbf{U}} & \dfrac{\mathbf{K} \circ \mathbf{U}}{\mathbf{U} \circ \mathbf{U}} \\ \dfrac{\mathbf{I} \circ \mathbf{W}}{\mathbf{W} \circ \mathbf{W}} & \dfrac{\mathbf{J} \circ \mathbf{W}}{\mathbf{W} \circ \mathbf{W}} & \dfrac{\mathbf{K} \circ \mathbf{W}}{\mathbf{W} \circ \mathbf{W}} \end{bmatrix}.$$

If $\mathbf{U} = a \cdot \mathbf{I} + b \cdot \mathbf{J} + c \cdot \mathbf{K}$ and $\mathbf{W} = d \cdot \mathbf{I} + e \cdot \mathbf{J} + f \cdot \mathbf{K}$, then the above matrix can be written

$$\begin{bmatrix} \dfrac{a}{a^2+b^2+c^2} & \dfrac{b}{a^2+b^2+c^2} & \dfrac{c}{a^2+b^2+c^2} \\ \dfrac{d}{d^2+e^2+f^2} & \dfrac{e}{d^2+e^2+f^2} & \dfrac{f}{d^2+e^2+f^2} \end{bmatrix}.$$

If $\mathbf{V} = x \cdot \mathbf{I} + y \cdot \mathbf{J} + z \cdot \mathbf{K}$ is any vector in $\mathcal{E}_3$, then $P_{\mathcal{S}}(\mathbf{V}) = r \cdot \mathbf{U} + s \cdot \mathbf{W}$ in $\mathcal{S}$, where

$$\frac{ax}{a^2+b^2+c^2} + \frac{by}{a^2+b^2+c^2} + \frac{cz}{a^2+b^2+c^2} = r$$
$$\frac{dx}{d^2+e^2+f^2} + \frac{ey}{d^2+e^2+f^2} + \frac{fz}{d^2+e^2+f^2} = s.$$

In particular, if $\mathbf{U} = \mathbf{I}$, $\mathbf{W} = \mathbf{J}$, so that $\mathcal{S} = \mathcal{E}_2$, then the matrix of $P_{\mathcal{E}_2}$ with respect to the bases $\{\mathbf{I}, \mathbf{J}, \mathbf{K}\}$ and $\{\mathbf{I}, \mathbf{J}\}$ is

$$\begin{bmatrix} 1 & 0 & 0 \\ 0 & 1 & 0 \end{bmatrix}.$$

We have seen that the concept of a rectangular matrix arises naturally in the study of a linear transformation of a vector space. Many computational problems involving linear transformations are easily solved by operating with their associated matrices. However, matrices have other useful applications, so that it is worthwhile to study them as mathematical objects in their own right.

(18.1) definition An m by n *real matrix* A is a rectangular array

$$A = \begin{bmatrix} a_{1,1} & a_{1,2} & \cdots & a_{1,n} \\ a_{2,1} & a_{2,2} & \cdots & a_{2,n} \\ \vdots & \vdots & & \vdots \\ a_{m,1} & a_{m,2} & \cdots & a_{m,n} \end{bmatrix}$$

of m rows and n columns, where the elements $a_{i,j}$ are real numbers.

The position of each element in the array is indicated by its subscripts; that is, the element $a_{i,j}$ is in the ith row and jth column. The number m of rows and the number n of columns are called the *dimensions* of the matrix. When the dimensions of a matrix are known, we will abbreviate our notation and write $A = [a_{i,j}]$ for an m by n matrix.

Two matrices are *equal* if they have the same dimensions and the same element in each position; that is,

$$A = \begin{bmatrix} a_{1,1} & a_{1,2} & \cdots & a_{1,n} \\ a_{2,1} & a_{2,2} & \cdots & a_{2,n} \\ \vdots & \vdots & & \vdots \\ a_{m,1} & a_{m,2} & \cdots & a_{m,n} \end{bmatrix} \quad \text{and} \quad B = \begin{bmatrix} b_{1,1} & b_{1,2} & \cdots & b_{1,q} \\ b_{2,1} & b_{2,2} & \cdots & b_{2,q} \\ \vdots & \vdots & & \vdots \\ b_{p,1} & b_{p,2} & \cdots & b_{p,q} \end{bmatrix}$$

are equal if and only if $m = p$, $n = q$, and $a_{i,j} = b_{i,j}$ for $i = 1, 2, \ldots, m$ and $j = 1, 2, \ldots, n$.

The following theorem expresses the important relation between linear transformations and matrices.

(18.2) theorem Let $\mathcal{V}$ and $\mathcal{W}$ be vector spaces with bases $\{\mathbf{U}_1, \mathbf{U}_2, \ldots, \mathbf{U}_n\}$ and $\{\mathbf{W}_1, \mathbf{W}_2, \ldots, \mathbf{W}_m\}$, respectively, and let $\mathfrak{M}$ be the set of all linear transformations of $\mathcal{V}$ into $\mathcal{W}$. The correspondence that associates each linear transformation $L \in \mathfrak{M}$ with the matrix of L with respect to the given bases is a one-to-one correspondence of $\mathfrak{M}$ onto the set of all m by n real matrices.

PROOF Let $L \in \mathfrak{M}$. The matrix of L with respect to the given bases in $\mathcal{V}$ and $\mathcal{W}$ is an m by n matrix A, where the elements of the jth column of A are the coordinates $a_{1,j}, a_{2,j}, \ldots, a_{m,j}$ of $L(\mathbf{U}_j)$ with respect to $\{\mathbf{W}_1, \mathbf{W}_2, \ldots, \mathbf{W}_m\}$. Therefore, each L uniquely determines an m by n matrix (by Theorem 10.5). Conversely, let $A = [a_{i,j}]$ be any m by n matrix. Then it can be shown that the mapping L of $\mathcal{V}$ into $\mathcal{W}$ defined by

$$L(r_1 \cdot \mathbf{U}_1 + r_2 \cdot \mathbf{U}_2 + \cdots + r_n \cdot \mathbf{U}_n) = s_1 \cdot \mathbf{W}_1 + s_2 \cdot \mathbf{W}_2 + \cdots + s_m \cdot \mathbf{W}_m$$

where the r's and s's are related by equations (18–1), is a linear transformation of $\mathcal{V}$ into $\mathcal{W}$ with associated matrix A. [This portion of the proof is left as an exercise (Exercise 18.7).] Therefore, we have shown that the correspondence which associates the matrix of a linear transformation L with each $L \in \mathfrak{M}$ is a function from the set of all linear transformations *onto* the set of all m by n matrices. To prove that this correspondence is one-to-one, we must show that if a linear transformation L corresponds to an m by n matrix $A = [a_{i,j}]$ and a linear transformation M corresponds to an m by n matrix $B = [b_{i,j}]$, and $A = B$, then $L = M$. Let $\mathbf{V} = r_1 \cdot \mathbf{U}_1 + r_2 \cdot \mathbf{U}_2 + \cdots + r_n \cdot \mathbf{U}_n$ be any vector in $\mathcal{V}$. Since $A = [a_{i,j}]$ is the matrix of L, it follows from equations (18–1) that the ith coordinate of $L(\mathbf{V})$ is $a_{i,1}r_1 + a_{i,2}r_2 + \cdots + a_{i,n}r_n$ for $i = 1, 2, \ldots, m$. Similarly, since $B = [b_{i,j}]$ is the matrix of M, the ith coordinate of $M(\mathbf{V})$ is $b_{i,1}r_1 + b_{i,2}r_2 + \cdots + b_{i,n}r_n$ for $i = 1, 2, \ldots, m$. If $A = B$, then $a_{i,j} = b_{i,j}$ for all i, j. Therefore, $L(\mathbf{V}) = M(\mathbf{V})$ for all $\mathbf{V} \in \mathcal{V}$. Thus, $L = M$, and we have shown that the correspondence is one-to-one. ∎

We now consider the important special case where L is a linear transformation of the vector space $\mathcal{V}$ into itself. Let $\{\mathbf{U}_1, \mathbf{U}_2, \ldots, \mathbf{U}_n\}$ be a basis of $\mathcal{V}$. Then the image $L(\mathbf{V})$ of a vector $\mathbf{V} = r_1 \cdot \mathbf{U}_1 + r_2 \cdot \mathbf{U}_2 + \cdots + r_n \cdot \mathbf{U}_n$ can be expressed in terms of this same basis. That is,

$$\begin{aligned} L(\mathbf{V}) &= r_1 \cdot L(\mathbf{U}_1) + r_2 \cdot L(\mathbf{U}_2) + \cdots + r_n \cdot L(\mathbf{U}_n) \\ &= s_1 \cdot \mathbf{U}_1 + s_2 \cdot \mathbf{U}_2 + \cdots + s_n \cdot \mathbf{U}_n. \end{aligned}$$

In particular, if we write $L(\mathbf{U}_i) = a_{1,i} \cdot \mathbf{U}_1 + a_{2,i} \cdot \mathbf{U}_2 + \cdots + a_{n,i} \cdot \mathbf{U}_n$, for $i = 1, 2, \ldots, n$, then as before, we have L given by the linear equations (18–1) with $m = n$. By Theorem 18.2, there is a one-to-one correspondence between the set $\mathfrak{M}$ of all linear transformations of $\mathcal{V}$ into itself and the set of all square n by n matrices.

EXAMPLE 3 Let L be a linear transformation of $\mathcal{V}_3$ into itself. Suppose that

$$\begin{aligned} L[(1, 0, 0)] &= (a_{1,1}, a_{2,1}, a_{3,1}) \\ L[(0, 1, 0)] &= (a_{1,2}, a_{2,2}, a_{3,2}) \end{aligned}$$

and

$$L[(0, 0, 1)] = (a_{1,3}, a_{2,3}, a_{3,3}).$$

Then the matrix of L with respect to the basis $E^{(3)} = \{(1, 0, 0), (0, 1, 0), (0, 0, 1)\}$ is the 3 by 3 matrix

$$A = \begin{bmatrix} a_{1,1} & a_{1,2} & a_{1,3} \\ a_{2,1} & a_{2,2} & a_{2,3} \\ a_{3,1} & a_{3,2} & a_{3,3} \end{bmatrix}$$

If $(r_1, r_2, r_3) \in \mathcal{V}_3$, then $L[(r_1, r_2, r_3)] = (s_1, s_2, s_3)$, where

$$\begin{aligned} a_{1,1}r_1 + a_{1,2}r_2 + a_{1,3}r_3 &= s_1 \\ a_{2,1}r_1 + a_{2,2}r_2 + a_{2,3}r_3 &= s_2 \\ a_{3,1}r_1 + a_{3,2}r_2 + a_{3,3}r_3 &= s_3. \end{aligned}$$

Conversely, if

$$B = \begin{bmatrix} b_{1,1} & b_{1,2} & b_{1,3} \\ b_{2,1} & b_{2,2} & b_{2,3} \\ b_{3,1} & b_{3,2} & b_{3,3} \end{bmatrix}$$

is any 3 by 3 matrix, then the mapping M defined by $M[(r_1, r_2, r_3)] = (s_1, s_2, s_3)$, where

$$\begin{aligned} b_{1,1}r_1 + b_{1,2}r_2 + b_{1,3}r_3 &= s_1 \\ b_{2,1}r_1 + b_{2,2}r_2 + b_{2,3}r_3 &= s_2 \\ b_{3,1}r_1 + b_{3,2}r_2 + b_{3,3}r_3 &= s_3. \end{aligned}$$

is a linear transformation of $\mathcal{V}_3$ into itself.

exercises

18.1 Compute the matrices of the following linear transformations with respect to the given bases:

(a) The orthogonal projection of $\mathcal{E}_3$ into the subspace $\mathcal{S}$ of $\mathcal{E}_3$ spanned by the nonzero vector $\mathbf{V}_0$ (see Example 2, §15). The bases are $\{\mathbf{I}, \mathbf{J}, \mathbf{K}\}$ for $\mathcal{E}_3$ and $\{\mathbf{V}_0\}$ for $\mathcal{S}$.

(b) The linear transformation L of $\mathcal{E}_3$ into R defined by $L(\mathbf{V}) = (2 \cdot \mathbf{I} - 3 \cdot \mathbf{J} + \mathbf{K}) \circ \mathbf{V}$ for $\mathbf{V} \in \mathcal{E}_3$. The bases are $\{\mathbf{I}, \mathbf{J}, \mathbf{K}\}$ for $\mathcal{E}_3$ and $\{1\}$ for R.

(c) The linear transformation M of $\mathcal{V}_5$ into $\mathcal{V}_4$ defined by $M[(a_1, a_2, a_3, a_4, a_5)] = (a_1 + 2a_2, a_3 - a_5, 3a_2 + a_3, a_5)$. The bases are $E^{(5)}$ for $\mathcal{V}_5$ and $E^{(4)}$ for $\mathcal{V}_4$.

(d) The linear transformation L of $\mathcal{V}_4$ into $\mathcal{V}_5$ defined by $L[(a_1, a_2, a_3, a_4)] = (a_1 - a_2, a_3 + a_4, a_1 + a_2 + a_3, a_4, a_1 - a_3 + a_4)$. The bases are $E^{(4)}$ for $\mathcal{V}_4$ and $E^{(5)}$ for $\mathcal{V}_5$.

18.2 Compute the associated matrix for each of the following linear transformations of $\mathcal{V}$ into itself with respect to the given basis:

(a) The linear transformation D_k defined by $D_k(\mathbf{V}) = k \cdot \mathbf{V}$, $k \neq 0$, of a vector space $\mathcal{V}$ of dimension n into itself, with respect to any basis $\{\mathbf{U}_1, \mathbf{U}_2, \ldots, \mathbf{U}_n\}$.

(b) The linear transformation L defined by $L(\mathbf{V}) = (\mathbf{I} + \mathbf{J} - \mathbf{K}) \times \mathbf{V}$ of $\mathcal{E}_3$ into itself, with respect to the basis $\{\mathbf{I}, \mathbf{J}, \mathbf{K}\}$.

(c) The linear transformation $T_{30°}$ of $\mathcal{E}_2$ into itself, with respect to the basis $\{\mathbf{I}, \mathbf{J}\}$ (see Example 1, §15).

(d) The linear transformation L of $\mathcal{V}_5$ into itself given in Example 1, §17, with respect to the basis $E^{(5)}$.

(e) The linear transformation M of $\mathcal{V}_5$ into itself given in Example 1, §17, with respect to the basis $E^{(5)}$.

18.3 Let $\mathcal{S}$ be the subspace of the vector space $R[x]$ consisting of the polynomials in $R[x]$ of degree at most 4 (see Example 5, §9). The ordinary derivative D of a polynomial is a linear transformation of $\mathcal{S}$ into itself. Find the matrix of D with respect to the basis $\{1, x, x^2, x^3, x^4\}$ of $\mathcal{S}$. Show that $B = \{1 - x, x + x^2, 1 + x^3, x - x^2 + x^4, x^4\}$ is also a basis of $\mathcal{S}$. Compute the matrix of D with respect to the basis B.

18.4 Let M be the linear transformation of $\mathcal{V}_5$ into $\mathcal{V}_4$ given in Exercise 18.1(c), and let L be the linear transformation of $\mathcal{V}_4$ into $\mathcal{V}_5$ given in Exercise 18.1(d). Then LM is a linear transformation of $\mathcal{V}_5$ into itself. Compute the matrix of LM with respect to the basis $E^{(5)}$.

18.5 Let L be the linear transformation of $\mathcal{V}_3$ into given $\mathcal{V}_5$ given in Example 3, §16, and let M be the linear transformation of $\mathcal{V}_5$ into $\mathcal{V}_4$ given in Example 2, §17. The matrix of L with respect to the bases $E^{(3)}$ and $E^{(5)}$ was computed in Example 1 of this section. Compute the matrix of M with respect to the bases $E^{(5)}$ and $E^{(4)}$; also compute the matrix of ML with respect to the bases $E^{(3)}$ and $E^{(4)}$.

18.6 Let L and M be the linear transformations of $\mathcal{V}_5$ into itself given in Example 1, §17. With respect to the basis $E^{(5)}$ of $\mathcal{V}_5$, find the matrices of the following linear transformations of $\mathcal{V}_5$ into itself: L, M, $L + M$, $10 \cdot L$, ML, LM.

***18.7** Let $\mathcal{V}$ be a vector space of dimension n with basis $\{\mathbf{U}_1, \mathbf{U}_2, \ldots, \mathbf{U}_n\}$. Let $\mathcal{W}$ be a vector space of dimension m with basis $\{\mathbf{W}_1, \mathbf{W}_2, \ldots, \mathbf{W}_m\}$. Let L be the mapping of $\mathcal{V}$ into $\mathcal{W}$ defined by

$$L(r_1 \cdot \mathbf{U}_1 + r_2 \cdot \mathbf{U}_2 + \cdots + r_n \cdot \mathbf{U}_n) = s_1 \cdot \mathbf{W}_1 + s_2 \cdot \mathbf{W}_2 + \cdots + s_m \cdot \mathbf{W}_m,$$

where

$$\begin{aligned} a_{1,1}r_1 + a_{1,2}r_2 + \cdots + a_{1,n}r_n &= s_1 \\ a_{2,1}r_1 + a_{2,2}r_2 + \cdots + a_{2,n}r_n &= s_2 \\ &\vdots \\ a_{m,1}r_1 + a_{m,2}r_2 + \cdots + a_{m,n}r_n &= s_m. \end{aligned}$$

Prove that L is a linear transformation of $\mathcal{V}$ into $\mathcal{W}$ with matrix

$$A = \begin{bmatrix} a_{1,1} & a_{1,2} & \cdots & a_{1,n} \\ a_{2,1} & a_{2,2} & \cdots & a_{2,n} \\ \vdots & \vdots & & \vdots \\ a_{m,1} & a_{m,2} & \cdots & a_{m,n} \end{bmatrix}.$$

18.8 Let L be the linear transformation of $\mathcal{E}_2$ into $\mathcal{E}_2$ with matrix

$$\begin{bmatrix} \frac{1}{2} & \frac{-\sqrt{3}}{2} \\ \frac{\sqrt{3}}{2} & \frac{1}{2} \end{bmatrix}$$

with respect to the basis $\{\mathbf{I}, \mathbf{J}\}$ of $\mathcal{E}_2$. Show that $L = T_\varphi$, with $\varphi = 60°$, where T_φ is the linear transformation of $\mathcal{E}_2$ described in Example 1, §15.

18.9 Let L be the linear transformation of $\mathcal{E}_3$ into itself with matrix

$$\begin{bmatrix} 1/2 & 1/2 & 0 \\ 1/2 & 1/2 & 0 \\ 0 & 0 & 1 \end{bmatrix}$$

with respect to the basis $\{\mathbf{I}, \mathbf{J}, \mathbf{K}\}$. Show that L is the orthogonal projection of $\mathcal{E}_3$ into the subspace of $\mathcal{E}_3$ spanned by the vectors $\mathbf{U} = \mathbf{I} + \mathbf{J}$ and $\mathbf{V} = \mathbf{K}$. Find $L(\mathbf{V})$ for the following vectors $\mathbf{V} \in \mathcal{E}_3$: $\mathbf{V} = 2 \cdot \mathbf{I} - 3 \cdot \mathbf{J} + 4 \cdot \mathbf{K}$; $\mathbf{V} = \mathbf{I} - \mathbf{J}$; $\mathbf{V} = \mathbf{J} + 5 \cdot \mathbf{K}$; $\mathbf{V} = \mathbf{J}$; $\mathbf{V} = 5 \cdot \mathbf{I} + 5 \cdot \mathbf{J} + 10 \cdot \mathbf{K}$.

18.10 Let

$$A = \begin{bmatrix} a_{1,1} & a_{1,2} & a_{1,3} \\ a_{2,1} & a_{2,2} & a_{2,3} \\ a_{3,1} & a_{3,2} & a_{3,3} \end{bmatrix}, \quad B = \begin{bmatrix} b_{1,1} & b_{1,2} & b_{1,3} \\ b_{2,1} & b_{2,2} & b_{2,3} \\ b_{3,1} & b_{3,2} & b_{3,3} \end{bmatrix}$$

be the matrices of linear transformations L and M, respectively, of $\mathcal{V}_3$ into itself with respect to the basis $E^{(3)}$. Find the matrices of $-L$, $L + M$, $L - M$, $3 \cdot L$, $-5 \cdot M$, and $2 \cdot L + 5 \cdot M$ with respect to $E^{(3)}$.

18.11 Let L be the linear transformation of $\mathcal{V}_2$ into $\mathcal{V}_4$ with matrix

$$\begin{bmatrix} -1 & 3 \\ 0 & 2 \\ -4 & 3 \\ 1 & 0 \end{bmatrix}$$

with respect to the bases $E^{(2)}$ and $E^{(4)}$. Find $L(\mathbf{V})$ for the following vectors $\mathbf{V} \in \mathcal{V}_2$: $(-2, 3)$; $(0, 1)$; $(-1, -5)$; $(1, 0)$; $(\sqrt{2}, \sqrt{3})$.

19 the algebra of matrices

The definitions of the operations of addition, scalar multiplication, and multiplication of matrices are based on the corresponding operations for linear transformations.

(19.1) definition If $A = [a_{i,j}]$ and $B = [b_{i,j}]$ are m by n matrices, then the *sum* of A and B is the m by n matrix

$$A + B = C = \begin{bmatrix} a_{1,1} + b_{1,1} & a_{1,2} + b_{1,2} & \cdots & a_{1,n} + b_{1,n} \\ a_{2,1} + b_{2,1} & a_{2,2} + b_{2,2} & \cdots & a_{2,n} + b_{2,n} \\ \vdots & \vdots & & \vdots \\ a_{m,1} + b_{m,1} & a_{m,2} + b_{m,2} & \cdots & a_{m,n} + b_{m,n} \end{bmatrix}.$$

Thus, $C = A + B$ is an m by n matrix with elements $c_{i,j} = a_{i,j} + b_{i,j}$, for $i = 1, 2, \ldots, m$ and $j = 1, 2, \ldots, n$. (Note that two matrices can be added only if they have the same dimensions.)

It is easy to establish the connection between matrix addition and the addition of linear transformations. Let $\mathcal{V}$ and $\mathcal{W}$ be vector spaces of dimensions n and m, respectively, and let L and M be linear transformations of $\mathcal{V}$ into $\mathcal{W}$ such that $A = [a_{i,j}]$ is the matrix of L and $B = [b_{i,j}]$ is the matrix of M with respect to chosen bases in $\mathcal{V}$ and $\mathcal{W}$. Then A and B are m by n matrices, and the matrix of the linear transformation $L + M$ of $\mathcal{V}$ into $\mathcal{W}$ is the sum of the matrices, $A + B$, as defined in 19.1.

Since, by Definition 19.1, matrices are added element-by-element, it follows from the associative and commutative properties of addition of real numbers that matrix addition is associative and commutative. Thus, if A, B, and C are m by n matrices,

$$(A + B) + C = A + (B + C) \tag{19–1}$$

and

$$A + B = B + A. \tag{19–2}$$

It also follows from Definition 19.1 that, for any dimensions m and n, there is a *zero matrix* $\mathbf{O}$, which has the number 0 in every position. Then for any m by n matrix $A = [a_{i,j}]$, we have

$$A + \mathbf{O} = \mathbf{O} + A = A.$$

Let $A = [a_{i,j}]$ and $B = [b_{i,j}]$ be m by n matrices. The *negative* of A is the m by n matrix $-A = [-a_{i,j}]$, which has the negative of $a_{i,j}$ in every position. Then subtraction of m by n matrices is defined by

$$B - A = B + (-A).$$

In particular,

$$A - A = A + (-A) = (-A) + A = \mathbf{O}.$$

(19.2) definition If $A = [a_{i,j}]$ is an m by n matrix, then the *scalar product* of a real number r and the matrix A is the m by n matrix

$$r \cdot A = \begin{bmatrix} ra_{1,1} & ra_{1,2} & \cdots & ra_{1,n} \\ ra_{2,1} & ra_{2,2} & \cdots & ra_{2,n} \\ \vdots & \vdots & & \vdots \\ ra_{m,1} & ra_{m,2} & \cdots & ra_{m,n} \end{bmatrix}.$$

Thus, $r \cdot A$ is the m by n matrix obtained from A by multiplying each element of A by the real number r.

It follows that if L is a linear transformation of $\mathcal{V}$ into $\mathcal{W}$ with matrix $A = [a_{i,j}]$ with respect to chosen bases in $\mathcal{V}$ and $\mathcal{W}$, then $r \cdot A$ is the matrix of the linear transformation $r \cdot L$ with respect to these bases.

Let A and B be m by n matrices, and let r and s be real numbers. Then

$$r \cdot (A + B) = r \cdot A + r \cdot B \tag{19–3}$$

$$(r + s) \cdot A = r \cdot A + s \cdot A \tag{19–4}$$

$$(rs) \cdot A = r \cdot (s \cdot A) = s \cdot (r \cdot A). \tag{19–5}$$

These identities are easily derived from Definition 19.2, and their proof is left as an exercise (Exercise 19.5).

EXAMPLE 1 Let

$$A = \begin{bmatrix} -2 & 0 & 3 & 5 & -1 \\ 6 & 7 & -2 & 0 & 1 \\ 5 & -10 & 6 & -2 & 1 \end{bmatrix}, \quad B = \begin{bmatrix} 0 & -7 & 4 & 3 & 2 \\ -1 & 2 & 0 & 1 & 2 \\ 5 & 6 & -3 & 4 & -9 \end{bmatrix}.$$

Then

$$(\tfrac{1}{2}) \cdot (A + B) = (\tfrac{1}{2}) \cdot \begin{bmatrix} -2 & -7 & 7 & 8 & 1 \\ 5 & 9 & -2 & 1 & 3 \\ 10 & -4 & 3 & 2 & -8 \end{bmatrix}$$

$$= \begin{bmatrix} -1 & -\tfrac{7}{2} & \tfrac{7}{2} & 4 & \tfrac{1}{2} \\ \tfrac{5}{2} & \tfrac{9}{2} & -1 & \tfrac{1}{2} & \tfrac{3}{2} \\ 5 & -2 & \tfrac{3}{2} & 1 & -4 \end{bmatrix}$$

$$6 \cdot (A - B) - 3 \cdot (A - B) = 3 \cdot (A - B)$$

$$= 3 \cdot \begin{bmatrix} -2 & 7 & -1 & 2 & -3 \\ 7 & 5 & -2 & -1 & -1 \\ 0 & -16 & 9 & -6 & 10 \end{bmatrix} = \begin{bmatrix} -6 & 21 & -3 & 6 & -9 \\ 21 & 15 & -6 & -3 & -3 \\ 0 & -48 & 27 & -18 & 30 \end{bmatrix}.$$

The next theorem establishes a fundamental result.

(19.3) theorem Let ${}_m\mathfrak{A}_n$ be the set of all m by n real matrices. Then ${}_m\mathfrak{A}_n$ is a finite dimensional vector space with respect to the operations of addition and scalar multiplication defined by 19.1 and 19.2. Moreover, ${}_m\mathfrak{A}_n$ is isomorphic to the vector space $\mathfrak{M}$ of all linear transformations of a vector space $\mathcal{V}$ of dimension n into a vector space $\mathcal{W}$ of dimension m.

PROOF The fact that ${}_m\mathfrak{A}_n$ is a vector space is a consequence of the properties of addition and scalar multiplication given above. Let $E_{i,j}$ be the m by n matrix with the number 1 in the ith row and the jth column and zeros elsewhere. The set of mn matrices, $\{E_{i,j} | i = 1, 2, \ldots, m;\ j = 1, 2, \ldots, n\}$, is a linearly independent set, for, if

$$a_{1,1} \cdot E_{1,1} + \cdots + a_{1,n} \cdot E_{1,n} + a_{2,1} \cdot E_{2,1} + \cdots + a_{2,n} \cdot E_{2,n} + \cdots + a_{m,1} \cdot E_{m,1} + \cdots + a_{m,n} \cdot E_{m,n}$$

is the zero m by n matrix, then

$$\begin{bmatrix} a_{1,1} & a_{1,2} & \cdots & a_{1,n} \\ a_{2,1} & a_{2,2} & \cdots & a_{2,n} \\ \vdots & \vdots & & \vdots \\ a_{m,1} & a_{m,2} & \cdots & a_{m,n} \end{bmatrix} = \begin{bmatrix} 0 & 0 & \cdots & 0 \\ 0 & 0 & \cdots & 0 \\ \vdots & \vdots & & \vdots \\ 0 & 0 & \cdots & 0 \end{bmatrix}.$$

Therefore, $a_{i,j} = 0$ for all i, j. Moreover, if $A = [a_{i,j}]$ is any real m by n matrix, then

$$A = a_{1,1} \cdot E_{1,1} + \cdots + a_{1,n} \cdot E_{1,n} + a_{2,1} \cdot E_{2,1} + \cdots + a_{2,n} \cdot E_{2,n} + \cdots + a_{m,1} \cdot E_{m,1} + \cdots + a_{m,n} \cdot E_{m,n}.$$

Thus, $\{E_{i,j}\}$ spans ${}_m\mathfrak{A}_n$, and therefore is a basis. Hence, ${}_m\mathfrak{A}_n$ is a vector space of dimension mn.

Let $\mathcal{V}$ be a vector space with basis $\{\mathbf{U}_1, \mathbf{U}_2, \ldots, \mathbf{U}_n\}$, and let $\mathcal{W}$ be a vector space with basis $\{\mathbf{W}_1, \mathbf{W}_2, \ldots, \mathbf{W}_m\}$. By Theorem 18.2, the correspondence that associates each linear transformation L of $\mathcal{V}$ into $\mathcal{W}$ with the matrix of L with respect to these bases is a one-to-one correspondence of $\mathfrak{M}$ onto ${}_m\mathfrak{A}_n$. If A is the matrix of L and B is the matrix of a linear transformation M, then we have observed that $A + B$ is the matrix of $L + M$ and $r \cdot A$ is the matrix of $r \cdot L$. Thus, the correspondence of Theorem 18.2 is a one-to-one linear transformation of $\mathfrak{M}$ onto ${}_m\mathfrak{A}_n$. That is, $\mathfrak{M}$ is isomorphic to the vector space ${}_m\mathfrak{A}_n$. ∎

Let $\mathfrak{A}_n$ denote the set of all n by n square matrices. Then it follows from Theorem 19.3 that $\mathfrak{A}_n$ is a vector space that is isomorphic to the vector space $\mathfrak{M}$ of all linear transformations of a vector space $\mathcal{V}$ of dimension n into itself.

Let $\mathcal{V}$, $\mathcal{W}$, and $\mathcal{P}$ be vector spaces of dimensions n, m, and p, respectively, and let L be a linear transformation of $\mathcal{V}$ into $\mathcal{W}$ and M be a linear transformation of $\mathcal{W}$ into $\mathcal{P}$. Finally, suppose that the matrix of L is the m by n matrix $A = [a_{i,j}]$ and that the matrix of M is the p by m matrix $B = [b_{i,j}]$ with respect to chosen bases in $\mathcal{V}$, $\mathcal{W}$, and $\mathcal{P}$. If $\mathbf{V} \in \mathcal{V}$, equations (18–1),

$$\begin{aligned}
a_{1,1}r_1 + a_{1,2}r_2 + \cdots + a_{1,n}r_n &= s_1\\
a_{2,1}r_1 + a_{2,2}r_2 + \cdots + a_{2,n}r_n &= s_2\\
&\vdots\\
a_{m,1}r_1 + a_{m,2}r_2 + \cdots + a_{m,n}r_n &= s_m
\end{aligned}$$

give the coordinates of $L(\mathbf{V}) \in \mathcal{W}$ in terms of the coordinates of $\mathbf{V}$. Similarly, the equations

$$\begin{aligned}
b_{1,1}s_1 + b_{1,2}s_2 + \cdots + b_{1,m}s_m &= t_1\\
b_{2,1}s_1 + b_{2,2}s_2 + \cdots + b_{2,m}s_m &= t_2\\
&\vdots\\
b_{p,1}s_1 + b_{p,2}s_2 + \cdots + b_{p,m}s_m &= t_p
\end{aligned} \tag{19–6}$$

express the coordinates $t_1, t_2, \ldots, t_p$ of $M[L(\mathbf{V})] \in \mathcal{P}$ in terms of the coordinates $s_1, s_2, \ldots, s_m$ of $L(\mathbf{V})$. Substituting the expressions for $s_1, s_2, \ldots, s_m$ from (18–1) into (19–6), we obtain the equations

$$\begin{aligned}
&(b_{1,1}a_{1,1} + b_{1,2}a_{2,1} + \cdots + b_{1,m}a_{m,1})r_1 + \cdots\\
&\qquad + (b_{1,1}a_{1,n} + b_{1,2}a_{2,n} + \cdots + b_{1,m}a_{m,n})r_n = t_1\\
&(b_{2,1}a_{1,1} + b_{2,2}a_{2,1} + \cdots + b_{2,m}a_{m,1})r_1 + \cdots\\
&\qquad + (b_{2,1}a_{1,n} + b_{2,2}a_{2,n} + \cdots + b_{2,m}a_{m,n})r_n = t_2\\
&(b_{p,1}a_{1,1} + b_{p,2}a_{2,1} + \cdots + b_{p,m}a_{m,1})r_1 + \cdots \qquad \vdots\\
&\qquad + (b_{p,1}a_{1,n} + b_{p,2}a_{2,n} + \cdots + b_{p,m}a_{m,n})r_n = t_p
\end{aligned}$$

which give the coordinates of $ML(\mathbf{V})$ in terms of the coordinates of $\mathbf{V}$. Consequently, the matrix of the linear transformation ML of $\mathcal{V}$ into $\mathcal{P}$ is the p by n matrix $C = [c_{i,j}]$, where

$$c_{i,j} = b_{i,1}a_{1,j} + b_{i,2}a_{2,j} + \cdots + b_{i,m}a_{m,j} = \sum_{k=1}^{m} b_{i,k}a_{k,j}.$$

This rule for computing the matrix of the product of two linear transformations leads to the following definition of matrix multiplication.

(19.4) definition Let $A = [a_{i,j}]$ be an m by n matrix, and let $B = [b_{i,j}]$ be a p by m matrix. Then the *product* BA is the p by n matrix that has the number $\sum_{k=1}^{m} b_{i,k}a_{k,j}$ in the ith row and jth column for $i = 1, 2, \ldots, p$ and $j = 1, 2, \ldots, n$.

For example, in the case of 2 by 2 matrices, this definition says

$$\begin{bmatrix} a_{1,1} & a_{1,2}\\ a_{2,1} & a_{2,2}\end{bmatrix}\begin{bmatrix} b_{1,1} & b_{1,2}\\ b_{2,1} & b_{2,2}\end{bmatrix} = \begin{bmatrix} a_{1,1}b_{1,1} + a_{1,2}b_{2,1} & a_{1,1}b_{1,2} + a_{1,2}b_{2,2}\\ a_{2,1}b_{1,1} + a_{2,2}b_{2,1} & a_{2,1}b_{1,2} + a_{2,2}b_{2,2}\end{bmatrix}.$$

By Definition 19.4, two matrices can be multiplied only when the number of columns in the first matrix equals the number of rows in the second matrix. Definition 19.4 is called the *row by column rule* for multiplying matrices. With this definition, our previous computations show that the matrix of the product of two linear transformations is the product of their matrices. This fact, together with Theorem 18.2, will enable us to conclude that matrix multiplication satisfies the same algebraic identities as the multiplication of linear transformations.

EXAMPLE 2 Let

$$A = \begin{bmatrix} 0 & -1 \\ \frac{1}{2} & 3 \\ 7 & 4 \end{bmatrix}, \quad B = \begin{bmatrix} 5 & -3 & \frac{1}{3} \\ 0 & 2 & 1 \end{bmatrix}, \quad C = \begin{bmatrix} 4 & 1 & -2 \\ 1 & 0 & 10 \\ -1 & 6 & 0 \end{bmatrix}.$$

Then

$$AB = \begin{bmatrix} 0 & -2 & -1 \\ \frac{5}{2} & \frac{9}{2} & \frac{19}{6} \\ 35 & -13 & \frac{19}{3} \end{bmatrix}, \quad BA = \begin{bmatrix} \frac{5}{6} & -\frac{38}{3} \\ 8 & 10 \end{bmatrix}, \quad BC = \begin{bmatrix} \frac{50}{3} & 7 & -40 \\ 1 & 6 & 20 \end{bmatrix},$$

$$CA = \begin{bmatrix} -\frac{27}{2} & -9 \\ 70 & 39 \\ 3 & 19 \end{bmatrix}, \quad C^2 = \begin{bmatrix} 19 & -8 & 2 \\ -6 & 61 & -2 \\ 2 & -1 & 62 \end{bmatrix}.$$

The products AC and CB are not defined.

EXAMPLE 3 Let L be a linear transformation of a vector space $\mathcal{V}$ of dimension n into a vector space $\mathcal{W}$ of dimension m, and let $A = [a_{i,j}]$ be the matrix of L with respect to chosen bases in each space. Then the system of equations (18–1), which give the coordinates $s_1, s_2, \ldots, s_m$ of $L(\mathbf{V})$, for $\mathbf{V} \in \mathcal{V}$, in terms of the coordinates $r_1, r_2, \ldots, r_n$ of $\mathbf{V}$, can be written as the single matrix equation

$$\begin{bmatrix} a_{1,1} & a_{1,2} & \cdots & a_{1,n} \\ a_{2,1} & a_{2,2} & \cdots & a_{2,n} \\ \vdots & \vdots & & \vdots \\ a_{m,1} & a_{m,2} & \cdots & a_{m,n} \end{bmatrix} \begin{bmatrix} r_1 \\ r_2 \\ \vdots \\ r_n \end{bmatrix} = \begin{bmatrix} s_1 \\ s_2 \\ \vdots \\ s_m \end{bmatrix}.$$

By (17–4), multiplication of linear transformations is associative. Then, as indicated above, the one-to-one correspondence between matrices and linear transformations (Theorem 18.2) and the fact that the matrix of the product of two linear transformations is the product of their matrices imply the associative law for matrix multiplication:

(19.5) theorem Let C be a q by p matrix, B be a p by m matrix, and A be an m by n matrix. Then

$$C(BA) = (CB)A.$$

Alternately, it can be proved directly from Definition 19.4 that matrix multiplication is associative.

Since matrix multiplication is associative, we may use the notation CBA to stand for either product $C(BA)$ or $(CB)A$. More generally, it follows that parentheses may be dropped from any indicated product. In particular, if A is a square matrix, then

$$A^t = \underbrace{AA\cdots A}_{t \text{ terms}}$$

is defined for every positive integer t.

EXAMPLE 4 Let

$$A = \begin{bmatrix} -1 & 3 & 4 \\ 0 & -2 & 1 \\ 6 & 2 & -3 \end{bmatrix}, \qquad B = \begin{bmatrix} -4 & 1 \\ 2 & 0 \\ -3 & 5 \end{bmatrix}.$$

Then

$$A^2 = \begin{bmatrix} 25 & -1 & -13 \\ 6 & 6 & -5 \\ -24 & 8 & 35 \end{bmatrix},$$

$$A^3 = A^2A = \begin{bmatrix} -103 & 51 & 138 \\ -36 & -4 & 45 \\ 234 & -18 & -193 \end{bmatrix},$$

and

$$A^2B = \begin{bmatrix} -63 & -40 \\ 3 & -19 \\ 7 & 141 \end{bmatrix}.$$

The matrix of the identity transformation of a vector space $\mathcal{V}$ of dimension n into itself with respect to any basis of $\mathcal{V}$ is the n by n matrix

$$I_n = \begin{bmatrix} 1 & 0 & 0 & \cdots & 0 & 0 \\ 0 & 1 & 0 & \cdots & 0 & 0 \\ \vdots & \vdots & \vdots & & \vdots & \vdots \\ 0 & 0 & 0 & \cdots & 0 & 1 \end{bmatrix}$$

which has 1 in the ith row and ith column for $i = 1, 2, \ldots, n$, and zero elsewhere. Such a matrix is called an *identity matrix*. There is an identity matrix I_n for each positive integer n. If $A = [a_{i,j}]$ is an m by n matrix, then, from the definition of matrix multiplication,

$$A = I_mA = AI_n.$$

Both the sum and product of any pair of matrices in $\mathfrak{A}_n$, the set of all n by n square matrices, is defined. Let A, B, and C be in $\mathfrak{A}_n$, and let r be a real number. Then matrix multiplication in $\mathfrak{A}_n$ is related to addition and scalar multiplication by the following identities:

$$A(B + C) = AB + AC \tag{19-7}$$
$$(A + B)C = AC + BC \tag{19-8}$$
$$r \cdot (AB) = (r \cdot A)B = A(r \cdot B). \tag{19-9}$$

However, matrix multiplication in $\mathfrak{A}_n$ is not, in general, commutative.

EXAMPLE 5 Let

$$A = \begin{bmatrix} 0 & -1 & 2 \\ \frac{1}{2} & 6 & 0 \\ 5 & -3 & 1 \end{bmatrix}, \qquad B = \begin{bmatrix} 2 & 1 & 1 \\ -5 & 0 & 3 \\ -1 & 1 & 0 \end{bmatrix}.$$

Then

$$AB = \begin{bmatrix} 3 & 2 & -3 \\ -29 & \frac{1}{2} & \frac{37}{2} \\ 24 & 6 & -4 \end{bmatrix}, \qquad BA = \begin{bmatrix} \frac{11}{2} & 1 & 5 \\ 15 & -4 & -7 \\ \frac{1}{2} & 7 & -2 \end{bmatrix}.$$

If a and b are real numbers and $ab = 0$, then either $a = 0$ or $b = 0$. Hence, it follows that multiplication of real numbers satisfies a cancellation law. That is, if $ac = ad$, and $a \neq 0$, then $c = d$. Matrix multiplication, in general, does not have these properties.

EXAMPLE 6 Let

$$A = \begin{bmatrix} \frac{4}{5} & 1 & -1 \\ 0 & 5 & 3 \\ -1 & 0 & 2 \end{bmatrix}, \qquad B = \begin{bmatrix} 10 & \frac{2}{3} & 2 \\ -3 & \frac{1}{3} & -\frac{3}{5} \\ 3 & \frac{1}{3} & 1 \end{bmatrix}.$$

Then $A \neq \mathbf{O}$ and $B \neq \mathbf{O}$, but

$$AB = \begin{bmatrix} 0 & 0 & 0 \\ 0 & 0 & 0 \\ 0 & 0 & 0 \end{bmatrix} = \mathbf{O}.$$

Let

$$C = \begin{bmatrix} -4 & 1 & -1 \\ 2 & 0 & \frac{1}{5} \\ 1 & 1 & -2 \end{bmatrix}, \qquad D = \begin{bmatrix} -14 & \frac{1}{3} & -3 \\ 5 & -\frac{1}{3} & \frac{4}{5} \\ -4 & \frac{2}{3} & -3 \end{bmatrix}.$$

Then $B = C - D$, so that

$$AC - AD = A(C - D) = AB = \mathbf{O}.$$

Consequently, $AC = AD$, with $A \neq \mathbf{O}$ and $C \neq D$.

In §17, we proved that L is a nonsingular transformation of a finite dimensional vector space $\mathcal{V}$ into itself if and only if there exists a transformation M of $\mathcal{V}$ into $\mathcal{V}$ such that $ML = LM = I_{\mathcal{V}}$. Let $\mathcal{V}$ have dimension n, and let A and B be the matrices of L and M, respectively, for some given basis of $\mathcal{V}$. Then by the one-to-one correspondence between the set of linear transformations of $\mathcal{V}$ into $\mathcal{V}$ and the set $\mathfrak{A}_n$ of n by n matrices, ML, LM, and $I_{\mathcal{V}}$ correspond to BA, AB, and I_n, respectively. Therefore, $BA = AB = I_n$, which suggests the following definition:

(19.6) definition   A square n by n matrix A is *nonsingular* if there exists an n by n matrix B such that $BA = AB = I_n$. Otherwise, the matrix is *singular*.

Thus, the matrix of a nonsingular linear transformation L of $\mathcal{V}$ into itself is a nonsingular matrix. Conversely, a nonsingular matrix defines a nonsingular linear transformation. There is at most one matrix B that satisfies the condition of Definition 19.6. Indeed, if $BA = AB = I_n$ and $CA = AC = I_n$, then

$$C = CI_n = C(AB) = (CA)B = I_n B = B.$$

Thus, the matrix B is unique, and is called the *inverse* of A. It is customary to write $B = A^{-1}$. (In Chapter 5, we develop a method for computing the inverse of a nonsingular matrix.) Appealing once more to the correspondence between linear transformations and matrices, it is not difficult to show that if either of the equations $AB = I_n$ or $BA = I_n$ holds, then so does the other one. That is, an n by n matrix A is nonsingular if and only if there is an n by n matrix B such that either $AB = I_n$ or $BA = I_n$. The proof is left as an exercise (Exercise 19.13).

EXAMPLE 7   Show that the matrix

$$A = \begin{bmatrix} -2 & 1 \\ 4 & -3 \end{bmatrix}$$

is nonsingular and find A^{-1}. If

$$B = \begin{bmatrix} b_{1,1} & b_{1,2} \\ b_{2,1} & b_{2,2} \end{bmatrix}$$

is a matrix such that

$$AB = \begin{bmatrix} 1 & 0 \\ 0 & 1 \end{bmatrix},$$

then

$$\begin{aligned} -2b_{1,1} + b_{2,1} \qquad\qquad\qquad\qquad &= 1 \\ -2b_{1,2} + b_{2,2} &= 0 \\ 4b_{1,1} - 3b_{2,1} \qquad\qquad\qquad\qquad &= 0 \\ 4b_{1,2} - 3b_{2,2} &= 1. \end{aligned}$$

Solving this system of equations by the methods of Chapter 3, we find $b_{1,1} = -3/2$, $b_{2,1} = -2$, $b_{1,2} = -1/2$, $b_{2,2} = -1$. Therefore,

$$A^{-1} = B = \begin{bmatrix} -3/2 & -1/2 \\ -2 & -1 \end{bmatrix}$$

and

$$\begin{aligned} AB &= \begin{bmatrix} -2 & 1 \\ 4 & -3 \end{bmatrix} \begin{bmatrix} -3/2 & -1/2 \\ -2 & -1 \end{bmatrix} = \begin{bmatrix} 1 & 0 \\ 0 & 1 \end{bmatrix} \\ &= \begin{bmatrix} -3/2 & -1/2 \\ -2 & -1 \end{bmatrix} \begin{bmatrix} -2 & 1 \\ 4 & -3 \end{bmatrix} = BA. \end{aligned}$$

We conclude this section by stating two useful properties of nonsingular matrices, the first of which follows immediately from Definition 19.6.

(19.7) The inverse of a nonsingular matrix is nonsingular.

(19.8) The product of two nonsingular n by n matrices is a nonsingular n by n matrix.

PROOF Let A and B be nonsingular n by n matrices. Then A^{-1} and B^{-1} are n by n matrices such that

$$(B^{-1}A^{-1})(AB) = I_n = (AB)(B^{-1}A^{-1}).$$

Hence, AB is nonsingular and $(AB)^{-1} = B^{-1}A^{-1}$. ∎

exercises

19.1 Let L and M be linear transformations of $\mathcal{V}$ into $\mathcal{W}$ with matrices $A = [a_{i,j}]$ and $B = [b_{i,j}]$, respectively, with respect to given bases in $\mathcal{V}$ and $\mathcal{W}$. Prove that the matrix of the linear transformation $L + M$ of $\mathcal{V}$ into $\mathcal{W}$ is $A + B$. Also, prove that $-A$ is the matrix of $-L$.

19.2 Prove that matrix addition is associative and commutative.

19.3 Let

$$A = \begin{bmatrix} 2 & 0 & -1 & 5 \\ 7 & -6 & 3 & 1 \end{bmatrix}, \qquad B = \begin{bmatrix} 1/2 & 7/5 & -2 & -1/10 \\ -3/5 & 5/2 & 7/10 & 5 \end{bmatrix},$$

$$C = \begin{bmatrix} 1 & 1/3 & 2 & 0 \\ 0 & 3/2 & 1 & 17 \end{bmatrix}.$$

Compute the following matrices: $(A - B) + C$; $A - (B + C)$; $A + B$; $A + (C - B)$; $5 \cdot A + 10 \cdot B + \sqrt{2} \cdot C$; $20 \cdot (A - B)$.

19.4 Let L be a linear transformation of $\mathcal{V}$ into $\mathcal{W}$ with matrix $A = [a_{i,j}]$ with respect to given bases in $\mathcal{V}$ and $\mathcal{W}$. Prove that the matrix of $r \cdot L$ is $r \cdot A$.

19.5 Prove the identities (19–3), (19–4), and (19–5).

19.6 Find the 3 by 5 matrix X that is the solution of the following matrix equations:

(a)
$$\begin{bmatrix} 1 & -5 & 2 & 6 & -1 \\ -2 & 1 & 0 & 3 & 0 \\ 0 & 10 & 4 & 2 & -1 \end{bmatrix} - 5 \cdot X = \begin{bmatrix} 1 & 0 & 0 & 0 & 0 \\ 0 & 1 & 0 & 0 & 0 \\ 0 & 0 & 1 & 0 & 0 \end{bmatrix};$$

(b)
$$(\tfrac{1}{7}) \cdot X + \begin{bmatrix} 3 & -11 & 7 & 2 & 1 \\ 0 & 0 & 0 & 0 & 0 \\ -1 & 5 & 17 & 1 & 4 \end{bmatrix} = \begin{bmatrix} -6 & 6 & 3 & 10 & 2 \\ 1 & 4 & 0 & -1 & 7 \\ -16 & 3 & -1 & 0 & 4 \end{bmatrix}.$$

***19.7** Let $\mathfrak{M}$ be the vector space of all linear transformations of a vector space $\mathcal{V}$ of dimension n into a vector space $\mathcal{W}$ of dimension m. Prove that $\mathfrak{M}$ has dimension mn.

19.8 Let $A = [1 \quad -2 \quad 3 \quad 0]$,

$$B = \begin{bmatrix} 2 & 5 \\ -3 & -1 \\ 0 & 1 \\ 6 & 0 \end{bmatrix}, \qquad C = \begin{bmatrix} 0 & 1 & -1 & 10 \\ 2 & 6 & -1 & 5 \end{bmatrix}.$$

Compute the following matrices: $A(BC)$; BC; CB; $AB - 3 \cdot [2 \quad 1]$;

$$\begin{bmatrix} -3 & 1 \\ 2 & 5 \end{bmatrix} + 6 \cdot (CB).$$

19.9 Prove the associative law of matrix multiplication directly from Definition 19.4.

19.10 If A is a square matrix, and s and t are positive integers, show that

$$A^s A^t = A^{s+t} \quad \text{and} \quad (A^s)^t = A^{st}.$$

Show by example that the third law of exponents, $(AB)^s = A^s B^s$, is not satisfied in general.

***19.11** For a nonsingular n by n matrix A, define $A^0 = I_n$, and $A^{-s} = (A^{-1})^s$ for a positive integer s. Show that the first two laws of exponents, given in Exercise 19.10, are satisfied for a nonsingular matrix A and for all integers s and t.

19.12 Show that if A and B are n by n matrices such that $AB = \mathbf{O}$, then either A or B is singular.

19.13 Let A be an n by n matrix and C and D be n by p matrices. Show that if $AC = AD$ and A is nonsingular, then $C = D$.

19.14 Let s be a positive integer and let A be any matrix. Show that

$$s \cdot A = \underbrace{A + A + \cdots + A}_{s \text{ terms}}.$$

19.15 Show that $(A + B)^2 = A^2 + 2 \cdot AB + B^2$ if and only if A and B commute.

19.16 If A is an n by m matrix, B is an m by p matrix, and r and s are real numbers, show that $(r \cdot A)(s \cdot B) = (rs) \cdot AB$.

***19.17** A *diagonal matrix* is an m by n matrix $A = [a_{i,j}]$ such that $a_{i,j} = 0$ if $i \neq j$. Let $A = [a_{i,j}]$ be an m by n diagonal matrix, and let $B = [b_{i,j}]$ be any n by p matrix. Let $C = [c_{i,j}]$ be the m by p matrix AB. Prove that $c_{i,j} = a_{i,i}b_{i,j}$, for $i = 1, 2, \ldots, m$ and $j = 1, 2, \ldots, p$. Show that any pair of square diagonal matrices commutes.

***19.18** A *scalar matrix* is a square diagonal matrix with equal diagonal elements. Prove that an n by n scalar matrix is a scalar multiple of I_n. Let $A = [a_{i,j}]$ be the n by n scalar matrix such that $a_{i,i} = r$, for $i = 1, 2, \ldots, n$. Let $B = [b_{i,j}]$ be any n by p matrix. Prove that $AB = r \cdot B$.

***19.19** Let $A = [a_{i,j}]$ be an m by n matrix, and let $B = [b_{i,j}]$ and $C = [c_{i,j}]$ be n by p matrices. Prove that $A(B + C) = AB + AC$. Also, prove that $r \cdot (AB) = (r \cdot A)B = A(r \cdot B)$. If $A = [a_{i,j}]$ and $B = [b_{i,j}]$ are m by n matrices and $C = [c_{i,j}]$ is an n by p matrix, prove that $(A + B)C = AC + BC$. The identities (19–7), (19–8), and (19–9) for n by n square matrices are special cases of the results of this exercise.

19.20 Let A be an n by n matrix. Prove that A is nonsingular if and only if there exists an n by n matrix B such that either $AB = I_n$ or $BA = I_n$.

19.21 Prove that a nonzero scalar matrix is nonsingular.

19.22 Show that the following matrices are nonsingular and find their inverses:

(a) $\begin{bmatrix} -7 & 2 \\ 0 & 1 \end{bmatrix}$ (b) $\begin{bmatrix} 1/3 & 1/2 \\ 2/5 & 1/4 \end{bmatrix}$ (c) $\begin{bmatrix} 1 & 0 & 2 \\ -3 & 4 & 0 \\ 0 & 0 & 1 \end{bmatrix}$

(d) $\begin{bmatrix} a & b \\ -b & a \end{bmatrix}$, where not both a and b are 0.

***19.23** Let $A = [a_{i,j}]$ be an m by n matrix. The *transpose of A* is the n by m matrix $B = [b_{i,j}]$, where $b_{i,j} = a_{j,i}$ for $i = 1, 2, \ldots, n$ and $j = 1, 2, \ldots, m$. Thus, the rows of B are the columns of A and the columns of B are the rows of A. The transpose of a matrix A is denoted by A'. Prove the following statements:

(a) For any m by n matrix A, $(A')' = A$.
(b) If A is an m by n matrix and B is an n by p matrix, then $(AB)' = B'A'$.
(c) If A is a nonsingular n by n matrix, then A' is a nonsingular n by n matrix and $(A')^{-1} = (A^{-1})'$.
(d) If D is a square diagonal matrix, then $D' = D$.
(e) If A and B are m by n matrices and r and s are real numbers, then $(r \cdot A + s \cdot B)' = r \cdot A' + s \cdot B'$.

20 applications to systems of linear equations

A system of n linear equations in m unknowns,

$$\begin{aligned} a_{1,1}x_1 + a_{1,2}x_2 + \cdots + a_{1,m}x_m &= b_1 \\ a_{2,1}x_1 + a_{2,2}x_2 + \cdots + a_{2,m}x_m &= b_2 \\ &\vdots \\ a_{n,1}x_1 + a_{n,2}x_2 + \cdots + a_{n,m}x_m &= b_n \end{aligned} \tag{20–1}$$

can be written as a single matrix equation, using Definition 19.4 for matrix multiplication. Let

$$A = \begin{bmatrix} a_{1,1} & a_{1,2} & \cdots & a_{1,m} \\ a_{2,1} & a_{2,2} & \cdots & a_{2,m} \\ \vdots & \vdots & & \vdots \\ a_{n,1} & a_{n,2} & \cdots & a_{n,m} \end{bmatrix},$$

$$X = \begin{bmatrix} x_1 \\ x_2 \\ \vdots \\ x_m \end{bmatrix}, \quad \text{and} \quad B = \begin{bmatrix} b_1 \\ b_2 \\ \vdots \\ b_n \end{bmatrix}.$$

Then the system (20–1) can be written compactly as the matrix equation

$$AX = B.$$

Using this notation, a solution of the system is an m by 1 matrix

$$C = \begin{bmatrix} c_1 \\ c_2 \\ \vdots \\ c_m \end{bmatrix}$$

such that $AC = B$.

The n by m matrix $A = [a_{i,j}]$ of system (20–1) is called the *matrix of coefficients,* and the n by $(m + 1)$ matrix

$$A^* = \begin{bmatrix} a_{1,1} & a_{1,2} & \cdots & a_{1,m} & b_1 \\ a_{2,1} & a_{2,2} & \cdots & a_{2,m} & b_2 \\ \vdots & \vdots & & \vdots & \vdots \\ a_{n,1} & a_{n,2} & \cdots & a_{n,m} & b_n \end{bmatrix}$$

is called the *augmented matrix of coefficients*. Thus, the row space of the system is the subspace of $\mathcal{V}_{m+1}$ that is spanned by the vectors $(a_{i,1}, a_{i,2}, \ldots, a_{i,m}, b_i)$, $i = 1, 2, \ldots, n$, which are the rows of the augmented matrix of coefficients. According to the procedure developed in Chapter 3, the system of equations (20–1) can be replaced by an equivalent system in echelon form by performing a sequence of elementary transformations on the set of vectors $\{(a_{i,1}, a_{i,2}, \ldots, a_{i,m}, b_i) \mid i = 1, 2, \ldots, n\}$, which span the row space of the system. Since these transformations are performed on the rows of the augmented matrix A^*, they can be regarded as transformations of A^*. Accordingly, they are called *elementary row transformations* of the matrix A^*.

EXAMPLE 1 The system of equations

$$\begin{aligned} 2x_1 - x_2 + 3x_3 \qquad &= 1 \\ x_1 + 5x_2 \qquad + x_4 &= 0 \\ -3x_1 + 2x_2 - x_3 + 6x_4 &= 4 \end{aligned}$$

can be expressed as the single matrix equation

$$\begin{bmatrix} 2 & -1 & 3 & 0 \\ 1 & 5 & 0 & 1 \\ -3 & 2 & -1 & 6 \end{bmatrix} \begin{bmatrix} x_1 \\ x_2 \\ x_3 \\ x_4 \end{bmatrix} = \begin{bmatrix} 1 \\ 0 \\ 4 \end{bmatrix}.$$

The matrix of coefficients is

$$A = \begin{bmatrix} 2 & -1 & 3 & 0 \\ 1 & 5 & 0 & 1 \\ -3 & 2 & -1 & 6 \end{bmatrix}$$

and the augmented matrix of coefficients is

$$A^* = \begin{bmatrix} 2 & -1 & 3 & 0 & 1 \\ 1 & 5 & 0 & 1 & 0 \\ -3 & 2 & -1 & 6 & 4 \end{bmatrix}.$$

The following sequence of elementary row transformations carries A^* to echelon form:

$$\begin{bmatrix} 2 & -1 & 3 & 0 & 1 \\ 1 & 5 & 0 & 1 & 0 \\ -3 & 2 & -1 & 6 & 4 \end{bmatrix} \xrightarrow{\text{Type I}} \begin{bmatrix} 1 & 5 & 0 & 1 & 0 \\ 2 & -1 & 3 & 0 & 1 \\ -3 & 2 & -1 & 6 & 4 \end{bmatrix} \xrightarrow{\text{Type II}}$$

$$\begin{bmatrix} 1 & 5 & 0 & 1 & 0 \\ 0 & -11 & 3 & -2 & 1 \\ -3 & 2 & -1 & 6 & 4 \end{bmatrix} \xrightarrow{\text{Type II}} \begin{bmatrix} 1 & 5 & 0 & 1 & 0 \\ 0 & -11 & 3 & -2 & 1 \\ 0 & 17 & -1 & 9 & 4 \end{bmatrix} \xrightarrow{\text{Type III}}$$

$$\begin{bmatrix} 1 & 5 & 0 & 1 & 0 \\ 0 & 1 & -3/11 & 2/11 & -1/11 \\ 0 & 17 & -1 & 9 & 4 \end{bmatrix} \xrightarrow{\text{Type II}} \begin{bmatrix} 1 & 5 & 0 & 1 & 0 \\ 0 & 1 & -3/11 & 2/11 & -1/11 \\ 0 & 0 & 40/11 & 65/11 & 61/11 \end{bmatrix}$$

$$\xrightarrow{\text{Type III}} \begin{bmatrix} 1 & 5 & 0 & 1 & 0 \\ 0 & 1 & -3/11 & 2/11 & -1/11 \\ 0 & 0 & 1 & 13/8 & 61/40 \end{bmatrix}.$$

Thus, the original system of equations is equivalent to the system

$$\begin{aligned} x_1 + 5x_2 \qquad\qquad + \quad x_4 &= 0 \\ x_2 - 3/11\,x_3 + 2/11\,x_4 &= -1/11 \\ x_3 + 13/8\,x_4 &= 61/40. \end{aligned}$$

Therefore, every solution of the system is given by $x_4 = r$, $x_3 = 61/40 - 13/8\,r$, $x_2 = 143/440 - 5/8\,r$, $x_1 = -143/88 + 17/8\,r$, where r is any real number.

It will be useful to apply the terminology that we have introduced in the discussion of systems of linear equations to rectangular matrices in general.

(20.1) definition Let $A = [a_{i,j}]$ be an n by m matrix. The *row space* of A is the subspace of $\mathcal{V}_m$ spanned by the vectors $(a_{i,1}, a_{i,2}, \ldots, a_{i,m})$, $i = 1, 2, \ldots, n$, which are the rows of A. An *elementary row transformation* of A is an elementary transformation on the set of vectors in $\mathcal{V}_m$ that are the rows of A.

We can now translate the results of Chapter 3 into matrix language.

(20.2) theorem If B is a matrix obtained from the matrix A by a sequence of elementary row transformations, then B and A have the same row space.

PROOF Theorem 20.2 is an immediate consequence of Theorem 12.2. ∎

Since Definition 13.3 of a system of linear equations in echelon form involves only a description of the matrix $A = [a_{i,j}]$ of coefficients of the system, it can be used to define an n by m matrix in echelon form. Thus, an n by m matrix $A = [a_{i,j}]$ is in echelon form if there is an integer k, where $0 \le k \le n$, such that:

(a) Every element in the matrix A below the kth row is zero. (If $k = 0$, then $A = \mathbf{O}$.)

(b) If $k > 0$, there is an increasing sequence of k positive integers, $1 \le m_1 < m_2 < \cdots < m_k \le m$, such that the ith row of A, for $1 \le i \le k$, has zeros in the first $m_i - 1$ columns and 1 in the m_ith column.

The integer k described above is called the *associated integer* k of the echelon form. For example, the following matrices are in echelon form:

$$\begin{bmatrix} 0 & 1 & -3 & 1 \\ 0 & 0 & 1 & 0 \\ 0 & 0 & 0 & 1 \end{bmatrix}, \qquad k = 3,\ m_1 = 2,\ m_2 = 3,\ m_3 = 4;$$

$$\begin{bmatrix} 1 & 0 & 0 & -3 \\ 0 & 0 & 0 & 1 \\ 0 & 0 & 0 & 0 \\ 0 & 0 & 0 & 0 \end{bmatrix}, \qquad k = 2,\ m_1 = 1,\ m_2 = 4; \qquad \begin{bmatrix} 0 & 0 \\ 0 & 0 \end{bmatrix}, \qquad k = 0;$$

$$\begin{bmatrix} 0 & 1 & 6 & -5 & 0 \\ 0 & 0 & 1 & -2 & 1 \end{bmatrix}, \qquad k = 2,\ m_1 = 2,\ m_2 = 3.$$

The following theorem is the matrix form of Theorem 13.4.

(20.3) theorem An n by m matrix $A = [a_{i,j}]$ can be transformed into a matrix in echelon form by a sequence of elementary row transformations.

EXAMPLE 2 Transform the matrix

$$A = \begin{bmatrix} -2 & -5 & 0 \\ 3 & 1 & 2 \\ 0 & 6 & -7 \\ -4 & 3 & 1 \end{bmatrix}$$

into echelon form by elementary row transformations. We have

$$\begin{bmatrix} -2 & -5 & 0 \\ 3 & 1 & 2 \\ 0 & 6 & -7 \\ -4 & 3 & 1 \end{bmatrix} \xrightarrow{\text{Type III}} \begin{bmatrix} 1 & 5/2 & 0 \\ 3 & 1 & 2 \\ 0 & 6 & -7 \\ -4 & 3 & 1 \end{bmatrix} \xrightarrow{\text{Type II}} \begin{bmatrix} 1 & 5/2 & 0 \\ 0 & -13/2 & 2 \\ 0 & 6 & -7 \\ -4 & 3 & 1 \end{bmatrix}$$

$$\xrightarrow{\text{Type II}} \begin{bmatrix} 1 & 5/2 & 0 \\ 0 & -13/2 & 2 \\ 0 & 6 & -7 \\ 0 & 13 & 1 \end{bmatrix} \xrightarrow{\text{Type III}} \begin{bmatrix} 1 & 5/2 & 0 \\ 0 & 1 & -4/13 \\ 0 & 6 & -7 \\ 0 & 13 & 1 \end{bmatrix} \xrightarrow{\text{Type II}} \begin{bmatrix} 1 & 5/2 & 0 \\ 0 & 1 & -4/13 \\ 0 & 0 & -67/13 \\ 0 & 13 & 1 \end{bmatrix}$$

$$\xrightarrow{\text{Type II}} \begin{bmatrix} 1 & 5/2 & 0 \\ 0 & 1 & -4/13 \\ 0 & 0 & -67/13 \\ 0 & 0 & 5 \end{bmatrix} \xrightarrow{\text{Type III}} \begin{bmatrix} 1 & 5/2 & 0 \\ 0 & 1 & -4/13 \\ 0 & 0 & 1 \\ 0 & 0 & 5 \end{bmatrix} \xrightarrow{\text{Type II}} \begin{bmatrix} 1 & 5/2 & 0 \\ 0 & 1 & -4/13 \\ 0 & 0 & 1 \\ 0 & 0 & 0 \end{bmatrix}.$$

In this echelon form, $k = 3$, $m_1 = 1$, $m_2 = 2$, $m_3 = 3$.

The results of §14 on systems of linear equations can be restated conveniently in terms of the following concept:

(20.4) definition The *row rank* of an n by m matrix $A = [a_{i,j}]$ is the dimension of the row space of A.

Theorem 20.2 states that if B is a matrix obtained from a matrix A by a sequence of elementary row transformations, then B and A have the same row space. Thus, by Definition 20.4, the row rank of B is equal to the row rank of A. Hence, if an n by m matrix A is in echelon form, it follows from the proof of Theorem 14.1 that the k nonzero rows of A are linearly independent vectors in $\mathcal{V}_m$, the row space of A. That is, the row rank of A is the associated integer k of the echelon form.

In the following theorem we will use the notation $A^* = [A, B]$ for the augmented matrix of the system $AX = B$ of n linear equations in m unknowns. That is, $A^* = [A, B]$ is an n by $m + 1$ matrix obtained by adjoining the n by 1 matrix $B = [b_i]$ as an $(m + 1)$st column to the n by m matrix $A = [a_{i,j}]$.

(20.5) theorem Let $AX = B$ be a system of n linear equations in m unknowns. The system is consistent if and only if the row rank of the augmented matrix $A^* = [A, B]$ is equal to the row rank of the matrix of coefficients A. If the system is consistent, then the solution is unique if and only if the row rank of A is equal to m.

PROOF If the augmented matrix $A^* = [A, B]$ is carried into echelon form $A_1^* = [A_1, B_1]$ by elementary row transformations, then A_1 is an echelon form for A. Moreover, the system of equations $AX = B$ is carried into an equivalent system $A_1X = B_1$ in echelon form. By Theorem 20.2, row rank A_1^* = row rank A^*, and row rank A_1 = row rank A. Now since the systems $AX = B$ and $A_1X = B_1$ are equivalent, their augmented matrices A^* and A_1^* have the same row rank, and their matrices of coefficients A and A_1 have the same row rank, it is sufficient to prove the theorem for the system $A_1X = B_1$ in echelon form.

By Theorem 14.1, the system $A_1X = B_1$ is consistent if and only if the dimension of its row space is k, where k is the associated integer of this system in echelon form. We can now make the following identifications:

(a) The associated integer k of the system of equations $A_1X = B_1$ is the same as the associated integer k of the matrix A_1 in echelon form;
(b) by the discussion preceding the theorem, the row rank of A_1 is k;
(c) the row space of the system $A_1X = B_1$ is the same as the row space of the augmented matrix $A_1^* = [A_1, B_1]$;
(d) by Definition 20.4, the dimension of the row space of A_1^* is the row rank of A_1^*.

Because of the above identifications, in Theorem 14.1 we can replace "the dimension of its row space" by "the row rank of the augmented matrix

A_1^*," and "the associated integer of this system of equations" by "the row rank of A_1." Then Theorem 14.1 reads: The system $A_1X = B_1$ is consistent if and only if the row rank of the augmented matrix A_1^* is k, where k is the row rank of A_1. This proves the first statement of the theorem.

If $A_1X = B_1$ is consistent, then by Theorem 14.1, this system has a unique solution if and only if $k =$ row rank $A_1 = m$. ∎

EXAMPLE 3 In Example 1 above we found that the given system of three equations in four unknowns has

$$A = \begin{bmatrix} 2 & -1 & 3 & 0 \\ 1 & 5 & 0 & 1 \\ -3 & 2 & -1 & 6 \end{bmatrix}$$

as matrix of coefficients, and

$$A^* = \begin{bmatrix} 2 & -1 & 3 & 0 & 1 \\ 1 & 5 & 0 & 1 & 0 \\ -3 & 2 & -1 & 6 & 4 \end{bmatrix}$$

as augmented matrix. Since an echelon form of A^* is

$$A_1^* = \begin{bmatrix} 1 & 5 & 0 & 1 & 0 \\ 0 & 1 & -3/11 & 2/11 & -1/11 \\ 0 & 0 & 1 & 13/8 & 61/40 \end{bmatrix},$$

it follows that an echelon form of A, which consists of the first four columns of A^*, is

$$A_1 = \begin{bmatrix} 1 & 5 & 0 & 1 \\ 0 & 1 & -3/11 & 2/11 \\ 0 & 0 & 0 & 1 \end{bmatrix}.$$

Thus, $k = 3$ is the associated integer of this echelon form of A_1. We have

$$\text{row rank } A_1 = 3 = \text{row rank } A_1^*.$$

By Theorem 20.5, the system of equations is consistent. Here, $m = 4$. Since $k = 3 < 4 = m$, it follows from Theorem 20.5 that the system does not have a unique solution. These facts, of course, were evident from the solution of the system given in Example 1.

EXAMPLE 4 The system of equations

$$\begin{aligned} 2x_1 - x_2 + 3x_3 &= 4 \\ x_1 + 4x_2 - x_3 &= 2 \\ 4x_1 + 7x_2 + x_3 &= 1 \end{aligned}$$

has

$$A = \begin{bmatrix} 2 & -1 & 3 \\ 1 & 4 & -1 \\ 4 & 7 & 1 \end{bmatrix}$$

as matrix of coefficients, and

$$A^* = \begin{bmatrix} 2 & -1 & 3 & 4 \\ 1 & 4 & -1 & 2 \\ 4 & 7 & 1 & 1 \end{bmatrix}$$

as augmented matrix. The following sequence of elementary row transformations carries A^* to echelon form:

$$A^* = \begin{bmatrix} 2 & -1 & 3 & 4 \\ 1 & 4 & -1 & 2 \\ 4 & 7 & 1 & 1 \end{bmatrix} \xrightarrow{\text{Type I}} \begin{bmatrix} 1 & 4 & -1 & 2 \\ 2 & -1 & 3 & 4 \\ 4 & 7 & 1 & 1 \end{bmatrix} \xrightarrow{\text{Type II}}$$

$$\begin{bmatrix} 1 & 4 & -1 & 2 \\ 0 & -9 & 5 & 0 \\ 4 & 7 & 1 & 1 \end{bmatrix} \xrightarrow{\text{Type II}} \begin{bmatrix} 1 & 4 & -1 & 2 \\ 0 & -9 & 5 & 0 \\ 0 & -9 & 5 & -9 \end{bmatrix} \xrightarrow{\text{Type II}}$$

$$\begin{bmatrix} 1 & 4 & -1 & 2 \\ 0 & -9 & 5 & 0 \\ 0 & 0 & 0 & -9 \end{bmatrix} \xrightarrow{\text{Type III}} \begin{bmatrix} 1 & 4 & -1 & 2 \\ 0 & 1 & -5/9 & 0 \\ 0 & 0 & 0 & 1 \end{bmatrix} = A_1^*.$$

Thus, an echelon form for A is

$$A_1 = \begin{bmatrix} 1 & 4 & -1 \\ 0 & 1 & -5/9 \\ 0 & 0 & 0 \end{bmatrix}.$$

We have row rank $A_1 = 2$ and row rank $A_1^* = 3$. By Theorem 20.5, the given system of equations is inconsistent.

The system of equations $AX = B$ is homogeneous if

$$B = \begin{bmatrix} 0 \\ 0 \\ \vdots \\ 0 \end{bmatrix} = \mathbf{0}.$$

As a corollary of Theorem 20.5, we obtain the following version of Theorem 14.2.

(20.6) corollary A homogeneous system $AX = \mathbf{O}$ of n equations in m unknowns has a nontrivial solution if and only if the row rank of A is less than m.

The proof of this statement is left as an exercise (Exercise 20.6). Corollary 20.6 has the following application.

(20.7) theorem A square n by n matrix $A = [a_{i,j}]$ is nonsingular if and only if the row rank of A is n.

PROOF Suppose first that A is nonsingular; hence, A^{-1} exists. Let

$$C = \begin{bmatrix} c_1 \\ c_2 \\ \vdots \\ c_n \end{bmatrix}, \quad \text{such that} \quad AC = \begin{bmatrix} 0 \\ 0 \\ \vdots \\ 0 \end{bmatrix}.$$

Then

$$C = I_n C = (A^{-1}A)C = A^{-1}(AC) = A^{-1}\begin{bmatrix} 0 \\ 0 \\ \vdots \\ 0 \end{bmatrix} = \begin{bmatrix} 0 \\ 0 \\ \vdots \\ 0 \end{bmatrix}.$$

Therefore, the homogeneous system of equations

$$A\begin{bmatrix} x_1 \\ x_2 \\ \vdots \\ x_n \end{bmatrix} = \begin{bmatrix} 0 \\ 0 \\ \vdots \\ 0 \end{bmatrix} \tag{20–2}$$

has only the trivial solution. By Corollary 20.6, the row rank of A is n.

Conversely, suppose that the row rank of $A = n$. By Corollary 20.6, the homogeneous system of equations (20–2) has only the trivial solution $x_1 = x_2 = \cdots = x_n = 0$. Thus, the linear transformation L of $\mathcal{V}_n$ into $\mathcal{V}_n$ defined by the matrix A (for any fixed basis of $\mathcal{V}_n$) has kernel $\{\mathbf{0}\}$. Therefore L is nonsingular, and its matrix A is nonsingular. ∎

(20.8) theorem Let

$$A\begin{bmatrix} x_1 \\ x_2 \\ \vdots \\ x_n \end{bmatrix} = \begin{bmatrix} b_1 \\ b_2 \\ \vdots \\ b_n \end{bmatrix}$$

be a system of n linear equations in n unknowns. The system has a unique solution if and only if the n by n matrix A is nonsingular. When A is nonsingular, the solution is given by

$$\begin{bmatrix} c_1 \\ c_2 \\ \vdots \\ c_n \end{bmatrix} = A^{-1} \begin{bmatrix} b_1 \\ b_2 \\ \vdots \\ b_n \end{bmatrix}.$$

The proof of Theorem 20.8 is left as an exercise (Exercise 20.8).

EXAMPLE 5 By Example 7, §19, the system of equations

$$\begin{aligned} -2x_1 + x_2 &= 4 \\ 4x_1 - 3x_2 &= -7 \end{aligned}$$

has a nonsingular matrix

$$A = \begin{bmatrix} -2 & 1 \\ 4 & -3 \end{bmatrix}$$

with inverse

$$A^{-1} = \begin{bmatrix} -3/2 & -1/2 \\ -2 & -1 \end{bmatrix}.$$

Thus, the unique solution of the system is given by

$$\begin{bmatrix} c_1 \\ c_2 \end{bmatrix} = \begin{bmatrix} -3/2 & -1/2 \\ -2 & -1 \end{bmatrix} \begin{bmatrix} 4 \\ -7 \end{bmatrix} = \begin{bmatrix} -5/2 \\ -1 \end{bmatrix}.$$

That is, the solution is $x_1 = c_1 = -5/2$, $x_2 = c_2 = -1$.

Let L be a linear transformation of an n-dimensional vector space $\mathcal{V}$ into an m-dimensional space $\mathcal{W}$ that has matrix $A = [a_{i,j}]$ with respect to given bases in the two spaces. Then L is given by the matrix equation

$$A \begin{bmatrix} r_1 \\ r_2 \\ \vdots \\ r_n \end{bmatrix} = \begin{bmatrix} s_1 \\ s_2 \\ \vdots \\ s_m \end{bmatrix}$$

where the coordinates of $\mathbf{V} \in \mathcal{V}$ are $r_1, r_2, \ldots, r_n$ with respect to the given basis in $\mathcal{V}$, and the coordinates of $L(\mathbf{V}) = \mathbf{W} \in \mathcal{W}$ are $s_1, s_2, \ldots, s_m$ with respect to the given basis in $\mathcal{W}$ (see Example 3, §19). The set of all vectors in $\mathcal{V}$ that are sent into a fixed vector $\mathbf{W} \in \mathcal{W}$ by L is found by solving the system of linear equations $AX = S$, where

$$X = \begin{bmatrix} x_1 \\ x_2 \\ \vdots \\ x_n \end{bmatrix}, \quad \text{and} \quad S = \begin{bmatrix} s_1 \\ s_2 \\ \vdots \\ s_m \end{bmatrix}.$$

In particular, the null space of L is the solution space of the homogeneous system $AX = \mathbf{O}$. Therefore, it follows from the results of Chapter 3 that any vector in $\mathcal{V}$ that is sent into a given vector $\mathbf{W} \in \mathcal{W}$ is the sum of a fixed vector that is sent into $\mathbf{W}$ and a linear combination of the vectors in a basis for the null space of L.

exercises

20.1 Write down the matrix of coefficients and the augmented matrix of coefficients for each of the systems of equations given in Exercise 13.2. Transform these matrices into echelon form by elementary row transformations. What is the row rank of each matrix?

20.2 Transform the following matrices into echelon form by elementary row transformations:

(a) $\begin{bmatrix} 2 & -1 & 0 & 1 & 2 & 2 \\ 1 & 3 & 4 & -3 & -1 & 4 \\ 3 & 2 & 4 & 2 & 3 & 2 \end{bmatrix}$ (b) $\begin{bmatrix} \frac{1}{2} & \frac{2}{3} & -1 \\ 0 & 5 & \frac{3}{2} \\ 3 & 14 & -3 \end{bmatrix}$

(c) $\begin{bmatrix} 2 & -1 & 1 \\ 1 & 1 & -2 \\ 4 & -3 & 4 \\ 2 & -1 & -4 \end{bmatrix}$

(d) $\begin{bmatrix} -1 & 3 & 7 & 2 \\ 2 & 1 & -3 & 5 \\ 4 & 0 & -2 & 1 \\ 1 & 2 & -6 & 4 \end{bmatrix}$ (e) $\begin{bmatrix} 0 & 0 & 1 & 0 \\ 0 & 1 & 0 & 0 \\ 0 & -2 & 1 & 3 \\ 0 & 4 & 0 & 6 \\ 0 & 2 & 1 & 9 \end{bmatrix}$

What is the row rank of each matrix?

20.3 Solve the systems of equations that have the matrices of Exercise 20.2 as augmented matrices of coefficients. Solve the homogeneous systems of equations that have these matrices as matrices of coefficients.

20.4 Prove that the row rank of an n by m matrix A is less than or equal to the smaller of the dimensions.

20.5 Show that if the matrix of coefficients of a system of n linear equations in m unknowns has row rank n, then the system is consistent.

20.6 Prove Corollary 20.6.

20.7 Which of the following matrices are nonsingular?

(a) $\begin{bmatrix} 1 & -2 & 3 & 0 \\ 4 & -1 & \frac{7}{2} & 1 \\ 4 & 2 & 1 & 0 \\ 2 & 0 & \frac{8}{5} & 1 \end{bmatrix}$ (b) $\begin{bmatrix} \frac{1}{2} & -2 & 0 \\ 1 & 3 & 2 \\ 6 & 12 & -1 \end{bmatrix}$

(c) $\begin{bmatrix} -1 & 1 & -1 & 1 \\ 2 & -3 & 4 & 1 \\ 1 & -1 & 1 & -1 \\ 0 & 0 & 0 & 1 \end{bmatrix}$ (d) $\begin{bmatrix} 0 & 0 & 0 & 0 & 1 \\ 0 & 1 & 0 & 0 & 0 \\ 0 & 0 & 1 & 0 & 0 \\ 1 & 0 & 0 & 0 & 0 \\ 0 & 0 & 0 & 1 & 0 \end{bmatrix}$

20.8 Prove Theorem 20.8.

20.9 The following systems of equations have nonsingular matrices of coefficients. Solve each system by finding the inverse of the matrix of coefficients as in Example 3.

(a) $\begin{aligned} x_1 - 3x_2 &= 1 \\ 5x_1 + x_2 &= 0 \end{aligned}$ (b) $\begin{aligned} x_1 + x_2 + x_3 &= 1 \\ x_2 + x_3 &= 2 \\ x_3 &= 3 \end{aligned}$

20.10 Let $AX = B$ be a system of n linear equations in m unknowns. That is, A is an n by m matrix, B is an n by 1 matrix, and

$$X = \begin{bmatrix} x_1 \\ x_2 \\ \vdots \\ x_m \end{bmatrix}.$$

Let P be a nonsingular n by n matrix. Prove that the system of equations $(PA)X = PB$ is equivalent to the system $AX = B$.

20.11 Find the set of all vectors in $\mathcal{V}_6$ that are mapped into the vector $(1, 2, -3) \in \mathcal{V}_3$ by the linear transformation L that has the matrix of Exercise 20.2(a) with respect to the bases $E^{(6)}$ and $E^{(3)}$ in $\mathcal{V}_6$ and $\mathcal{V}_3$, respectively.

20.12 Let the matrix of Exercise 20.2(b) be the matrix of a linear transformation L of $\mathcal{V}_3$ into itself with respect to the basis $E^{(3)}$. Find the null space of L.

20.13 Let L be the linear transformation of $\mathcal{E}_3$ into $\mathcal{E}_3$ defined by $L(\mathbf{V}) = \mathbf{K} \times \mathbf{V}$ for $\mathbf{V} \in \mathcal{E}_3$ (see Example 2, §16). Recall that $\mathbf{K}$ is the unit vector along the Z axis, and that by the result of Example 2, §16, the range space of L is $\mathcal{E}_2$, the set of all vectors in the XY plane. Find the set of vectors in $\mathcal{E}_3$ which L sends into the vector $3 \cdot \mathbf{I} + 2 \cdot \mathbf{J}$ in $\mathcal{E}_2$.

5

equivalence of matrices

21 change of basis

The matrix of a linear transformation of a finite dimensional vector space $\mathcal{V}$ into a finite dimensional vector space $\mathcal{W}$ depends on the particular choice of bases in each vector space. Often, it is convenient to change bases to deal with a particular problem. In this chapter we study the relations between different matrices associated with the same linear transformation and learn how to accomplish this change of bases. We begin this discussion by finding the equations that connect the coordinates of a given vector relative to two different bases in a vector space $\mathcal{V}$.

(21.1) theorem Let $\{\mathbf{U}_1, \mathbf{U}_2, \ldots, \mathbf{U}_k\}$ be a set of vectors in a vector space $\mathcal{V}$ of dimension n. Let $\mathbf{U}_i$ have coordinates $(r_{i,1}, r_{i,2}, \ldots, r_{i,n})$ for $i = 1, 2, \ldots, k$ with respect to some basis of $\mathcal{V}$. Then the row rank of the k by n matrix $R = [r_{i,j}]$ is equal to the dimension of the subspace $\mathcal{S}$ spanned by $\{\mathbf{U}_1, \mathbf{U}_2, \ldots, \mathbf{U}_k\}$.

PROOF By the isomorphism between $\mathcal{V}$ and the n-tuple space $\mathcal{V}_n$, the vector $\mathbf{U}_i$ corresponds to the vector $(r_{i,1}, r_{i,2}, \ldots, r_{i,n})$, $i = 1, 2, \ldots, k$ (see Theorem 16.7). Therefore, the subspace $\mathcal{S}$ of $\mathcal{V}$ spanned by $\{\mathbf{U}_1, \mathbf{U}_2, \ldots, \mathbf{U}_k\}$ is isomorphic to the row space of the matrix $R = [r_{i,j}]$. Hence, the dimension of $\mathcal{S}$ is equal to the dimension of the row space of R, which is the row rank of R. ▮

(21.2) corollary Let $\{\mathbf{U}_1, \mathbf{U}_2, \ldots, \mathbf{U}_n\}$ be a set of n vectors in a vector space $\mathcal{V}$ of dimension n. Let $\mathbf{U}_i$ have coordinates $(r_{i,1}, r_{i,2}, \ldots, r_{i,n})$ for $i =$

1, 2, . . . , n with respect to some basis of $\mathcal{V}$. Then the n by n matrix $R = [r_{i,j}]$ is nonsingular if and only if $\{\mathbf{U}_1, \mathbf{U}_2, \ldots, \mathbf{U}_n\}$ is a basis of $\mathcal{V}$.

PROOF If $R = [r_{i,j}]$ is nonsingular, then, by Theorem 20.7, R has row rank n. By Theorem 21.1, the subspace $\mathcal{S}$ of $\mathcal{V}$ spanned by $\{\mathbf{U}_1, \mathbf{U}_2, \ldots, \mathbf{U}_n\}$ has dimension n. Since $\mathcal{V}$ has dimension n, it follows from Corollary 10.9 that $\mathcal{S} = \mathcal{V}$. Thus, $\{\mathbf{U}_1, \mathbf{U}_2, \ldots, \mathbf{U}_n\}$ is a set of n vectors that spans the n-dimensional vector space $\mathcal{V}$. By Corollary 10.8, $\{\mathbf{U}_1, \mathbf{U}_2, \ldots, \mathbf{U}_n\}$ is a basis of $\mathcal{V}$.

Conversely, if $\{\mathbf{U}_1, \mathbf{U}_2, \ldots, \mathbf{U}_n\}$ is a basis of $\mathcal{V}$, then, by Theorem 21.1, the row rank of R is n. By Theorem 20.7, R is nonsingular. ∎

EXAMPLE 1 Consider the set of vectors $\{(1, -2, 0, 6), (-1, 3, 2, -4), (5, -13, -6, 24)\}$ in $\mathcal{V}_4$. The components of a vector in $\mathcal{V}_4$ are the coordinates of the vector with respect to the basis $E^{(4)}$. Thus, we can take

$$R = \begin{bmatrix} 1 & -2 & 0 & 6 \\ -1 & 3 & 2 & -4 \\ 5 & -13 & -6 & 24 \end{bmatrix}$$

to be the matrix $[r_{i,j}]$ in Theorem 21.1. An echelon form for R is

$$R^* = \begin{bmatrix} 1 & -2 & 0 & 6 \\ 0 & 1 & 2 & 2 \\ 0 & 0 & 0 & 0 \end{bmatrix}.$$

Thus, R^* and, consequently, R have row rank 2. By Theorem 21.1, the subspace $\mathcal{S}$ of $\mathcal{V}_4$ spanned by $\{(1, -2, 0, 6), (-1, 3, 2, -4), (5, -13, -6, 24)\}$ has dimension two. The fact that $\mathcal{S}$ has dimension two can, of course, be determined directly. Since $(5, -13, -6, 24) = 2 \cdot (1, -2, 0, 6) - 3 \cdot (-1, 3, 2, -4)$, $\mathcal{S}$ is spanned by the set $T = \{(1, -2, 0, 6), (-1, 3, 2, -4)\}$. It is routine to check that T is a linearly independent set. Hence, T is a basis of $\mathcal{S}$, so that $\mathcal{S}$ has dimension two.

EXAMPLE 2 Consider the set of vectors $S = \{(-7, 1, 2, 10), (0, -3, 2, 1), (1, 1, -5, 4), (8, -1, 0, 2)\}$ of four vectors in $\mathcal{V}_4$. The matrix

$$R = \begin{bmatrix} -7 & 1 & 2 & 10 \\ 0 & -3 & 2 & 1 \\ 1 & 1 & -5 & 4 \\ 8 & -1 & 0 & 2 \end{bmatrix}$$

has

$$R^* = \begin{bmatrix} 1 & 1 & -5 & 4 \\ 0 & 1 & -2/3 & 1/3 \\ 0 & 0 & 1 & -106/83 \\ 0 & 0 & 0 & 1 \end{bmatrix}$$

as an echelon form. Therefore, the row rank of R is 4, so that R is nonsingular. According to Corollary 21.2, the set S is a basis of $\mathcal{V}_4$.

EXAMPLE 3 In $\mathcal{E}_3$, let the vectors **U**, **V**, **W** have coordinates $(2, -1, 3)$, $(4, 0, 1)$, $(1, -1, 8)$ with respect to the basis $\{\mathbf{I}, \mathbf{J}, \mathbf{K}\}$. Then

$$\begin{aligned}\mathbf{U} &= 2 \cdot \mathbf{I} + (-1) \cdot \mathbf{J} + 3 \cdot \mathbf{K}\\ \mathbf{V} &= 4 \cdot \mathbf{I} + 0 \cdot \mathbf{J} + 1 \cdot \mathbf{K}\\ \mathbf{W} &= 1 \cdot \mathbf{I} + (-1) \cdot \mathbf{J} + 8 \cdot \mathbf{K}.\end{aligned}$$

The matrix

$$P = \begin{bmatrix} 2 & -1 & 3 \\ 4 & 0 & 1 \\ 1 & -1 & 8 \end{bmatrix}$$

which has rows that are the coordinates of **U**, **V**, **W** with respect to the basis $\{\mathbf{I}, \mathbf{J}, \mathbf{K}\}$, is nonsingular. This follows from the fact that an echelon form for P is the matrix

$$P^* = \begin{bmatrix} 1 & -1 & 8 \\ 0 & 1 & -13 \\ 0 & 0 & 1 \end{bmatrix}$$

which has row rank 3. Thus, by Corollary 21.2, $\{\mathbf{U}, \mathbf{V}, \mathbf{W}\}$ is a basis of $\mathcal{E}_3$.

If **T** is any vector in $\mathcal{E}_3$, it is natural to ask for the relation between the coordinates of **T** with respect to the basis $\{\mathbf{I}, \mathbf{J}, \mathbf{K}\}$ and the coordinates of **T** with respect to the basis $\{\mathbf{U}, \mathbf{V}, \mathbf{W}\}$. Let **T** have coordinates (a_1, a_2, a_3) with respect to $\{\mathbf{I}, \mathbf{J}, \mathbf{K}\}$ and coordinates $(\bar{a}_1, \bar{a}_2, \bar{a}_3)$ with respect to $\{\mathbf{U}, \mathbf{V}, \mathbf{W}\}$. We have

$$\mathbf{T} = \bar{a}_1 \cdot \mathbf{U} + \bar{a}_2 \cdot \mathbf{V} + \bar{a}_3 \cdot \mathbf{W}$$

or, in terms of coordinates with respect to $\{\mathbf{I}, \mathbf{J}, \mathbf{K}\}$,

$$\begin{aligned}(a_1, a_2, a_3) &= \bar{a}_1 \cdot (2, -1, 3) + \bar{a}_2 \cdot (4, 0, 1) + \bar{a}_3 \cdot (1, -1, 8)\\ &= (2\bar{a}_1 + 4\bar{a}_2 + 1\bar{a}_3, -1\bar{a}_1 + 0\bar{a}_2 - 1\bar{a}_3, 3\bar{a}_1 + 1\bar{a}_2 + 8\bar{a}_3).\end{aligned}$$

Hence,

$$\begin{aligned}a_1 &= 2\bar{a}_1 + 4\bar{a}_2 + 1\bar{a}_3\\ a_2 &= -1\bar{a}_1 + 0\bar{a}_2 - 1\bar{a}_3\\ a_3 &= 3\bar{a}_1 + 1\bar{a}_2 + 8\bar{a}_3.\end{aligned}$$

The above equations give the coordinates of **T** with respect to $\{\mathbf{I}, \mathbf{J}, \mathbf{K}\}$ in terms of the coordinates of **T** with respect to $\{\mathbf{U}, \mathbf{V}, \mathbf{W}\}$. These equations can be written as a single matrix equation

$$[a_1 \quad a_2 \quad a_3] = [\bar{a}_1 \quad \bar{a}_2 \quad \bar{a}_3] \begin{bmatrix} 2 & -1 & 3 \\ 4 & 0 & 1 \\ 1 & -1 & 8 \end{bmatrix} = [\bar{a}_1 \quad \bar{a}_2 \quad \bar{a}_3]P.$$

Since P is nonsingular, this matrix equation can be multiplied by P^{-1} to obtain

$$[\bar{a}_1 \quad \bar{a}_2 \quad \bar{a}_3] = [a_1 \quad a_2 \quad a_3]P^{-1}.$$

Using the rule for the transpose of the product of two matrices (Exercise 19.22) the above matrix equations can be written in the forms

$$\begin{bmatrix} a_1 \\ a_2 \\ a_3 \end{bmatrix} = P' \begin{bmatrix} \bar{a}_1 \\ \bar{a}_2 \\ \bar{a}_3 \end{bmatrix} \quad \text{and} \quad \begin{bmatrix} \bar{a}_1 \\ \bar{a}_2 \\ \bar{a}_3 \end{bmatrix} = (P^{-1})' \begin{bmatrix} a_1 \\ a_2 \\ a_3 \end{bmatrix}.$$

To see how this method works in practice, let us suppose that **T** has coordinates $(-5, 7, 9)$ with respect to $\{\mathbf{I}, \mathbf{J}, \mathbf{K}\}$. Then

$$\begin{bmatrix} -5 \\ 7 \\ 9 \end{bmatrix} = \begin{bmatrix} 2 & 4 & 1 \\ -1 & 0 & -1 \\ 3 & 1 & 8 \end{bmatrix} \begin{bmatrix} \bar{a}_1 \\ \bar{a}_2 \\ \bar{a}_3 \end{bmatrix}.$$

Solving this system of equations by the methods of Chapter 3, we find that the coordinates of **T** with respect to $\{\mathbf{U}, \mathbf{V}, \mathbf{W}\}$ are

$$(\bar{a}_1 \quad \bar{a}_2 \quad \bar{a}_3) = \left(\frac{-86}{7}, \frac{25}{7}, \frac{37}{7}\right).$$

Following the ideas suggested in Example 3, we now give a general discussion of the relationship between the coordinates of a vector **V** with respect to different bases in an n-dimensional vector space $\mathcal{V}$.

Suppose that $\{\mathbf{U}_1, \mathbf{U}_2, \ldots, \mathbf{U}_n\}$ is a basis of the vector space $\mathcal{V}$ and that $\{\bar{\mathbf{U}}_1, \bar{\mathbf{U}}_2, \ldots, \bar{\mathbf{U}}_n\}$ is any set of n vectors in $\mathcal{V}$. Then, since $\{\mathbf{U}_1, \mathbf{U}_2, \ldots, \mathbf{U}_n\}$ is a basis of $\mathcal{V}$, we have

$$\begin{aligned} \bar{\mathbf{U}}_1 &= p_{1,1} \cdot \mathbf{U}_1 + p_{1,2} \cdot \mathbf{U}_2 + \cdots + p_{1,n} \cdot \mathbf{U}_n \\ \bar{\mathbf{U}}_2 &= p_{2,1} \cdot \mathbf{U}_1 + p_{2,2} \cdot \mathbf{U}_2 + \cdots + p_{2,n} \cdot \mathbf{U}_n \\ &\vdots \\ \bar{\mathbf{U}}_n &= p_{n,1} \cdot \mathbf{U}_1 + p_{n,2} \cdot \mathbf{U}_2 + \cdots + p_{n,n} \cdot \mathbf{U}_n. \end{aligned} \tag{21–1}$$

The coordinates of $\bar{\mathbf{U}}_i$ with respect to the basis $\{\mathbf{U}_1, \mathbf{U}_2, \ldots, \mathbf{U}_n\}$ are $(p_{i,1}, p_{i,2}, \ldots, p_{i,n})$ for $i = 1, 2, \ldots, n$. By Corollary 21.2, $\{\bar{\mathbf{U}}_1, \bar{\mathbf{U}}_2, \ldots, \bar{\mathbf{U}}_n\}$ is a basis of $\mathcal{V}$ if and only if the matrix $P = [p_{i,j}]$ is nonsingular.

Now assume that $\{\bar{\mathbf{U}}_1, \bar{\mathbf{U}}_2, \ldots, \bar{\mathbf{U}}_n\}$ is a basis of $\mathcal{V}$ related to the basis $\{\mathbf{U}_1, \mathbf{U}_2, \ldots, \mathbf{U}_n\}$ by the equations (21–1). In this situation, we call the nonsingular matrix $P = [p_{i,j}]$ the *matrix of the change of basis* from the **U**-basis to the $\bar{\mathbf{U}}$-basis. Let $\mathbf{V} \in \mathcal{V}$, and suppose that **V** has coordinates $(a_1, a_2, \ldots, a_n)$ with respect to the **U**-basis and coordinates $(\bar{a}_1, \bar{a}_2, \ldots, \bar{a}_n)$ with respect to the $\bar{\mathbf{U}}$-basis. Then

$$\mathbf{V} = \bar{a}_1 \cdot \bar{\mathbf{U}}_1 + \bar{a}_2 \cdot \bar{\mathbf{U}}_2 + \cdots + \bar{a}_n \cdot \bar{\mathbf{U}}_n.$$

By the remarks following Theorem 16.7,

$$\begin{aligned}(a_1, a_2, \ldots, a_n) &= \bar{a}_1 \cdot (p_{1,1}, p_{1,2}, \ldots, p_{1,n}) + \bar{a}_2 \cdot (p_{2,1}, p_{2,2}, \ldots, p_{2,n}) \\ &\quad + \cdots + \bar{a}_n \cdot (p_{n,1}, p_{n,2}, \ldots, p_{n,n}) \\ &= (\bar{a}_1 p_{1,1} + \bar{a}_2 p_{2,1} + \cdots + \bar{a}_n p_{n,1}, \bar{a}_1 p_{1,2} + \bar{a}_2 p_{2,2} \\ &\quad + \cdots + \bar{a}_n p_{n,2}, \ldots, \bar{a}_1 p_{1,n} + \bar{a}_2 p_{2,n} \\ &\qquad + \cdots + \bar{a}_n p_{n,n}).\end{aligned}$$

Hence,

$$\begin{aligned}a_1 &= \bar{a}_1 p_{1,1} + \bar{a}_2 p_{2,1} + \cdots + \bar{a}_n p_{n,1} \\ a_2 &= \bar{a}_1 p_{1,2} + \bar{a}_2 p_{2,2} + \cdots + \bar{a}_n p_{n,2} \\ &\vdots \\ a_n &= \bar{a}_1 p_{1,n} + \bar{a}_2 p_{2,n} + \cdots + \bar{a}_n p_{n,n}.\end{aligned} \tag{21–2}$$

Equations (21–2) can be written compactly as a single matrix equation in either of the two following ways:

$$[a_1 \quad a_2 \quad \ldots \quad a_n] = [\bar{a}_1 \quad \bar{a}_2 \quad \ldots \quad \bar{a}_n]P \tag{21–3}$$

or

$$\begin{bmatrix} a_1 \\ a_2 \\ \vdots \\ a_n \end{bmatrix} = P' \begin{bmatrix} \bar{a}_1 \\ \bar{a}_2 \\ \vdots \\ \bar{a}_n \end{bmatrix} \tag{21–4}$$

where P' is the transpose of the matrix P (see Exercise 19.22). Since $P = [p_{i,j}]$ the matrix of the change of basis, is nonsingular, equation (21–3) can be multiplied by P^{-1} to obtain

$$[\bar{a}_1 \quad \bar{a}_2 \quad \ldots \quad \bar{a}_n] = [a_1 \quad a_2 \quad \ldots \quad a_n]P^{-1} \tag{21–5}$$

which can be written in the form

$$\begin{bmatrix} \bar{a}_1 \\ \bar{a}_2 \\ \vdots \\ \bar{a}_n \end{bmatrix} = (P^{-1})' \begin{bmatrix} a_1 \\ a_2 \\ \vdots \\ a_n \end{bmatrix}. \tag{21–6}$$

EXAMPLE 4 Let $\{\mathbf{I}, \mathbf{J}\}$ be the set of unit vectors associated with a rectangular Cartesian coordinate system in $\mathcal{E}_2$. Define vectors $\bar{\mathbf{I}}$ and $\bar{\mathbf{J}}$ by the equations

$$\begin{aligned}\bar{\mathbf{I}} &= \cos\theta \cdot \mathbf{I} + (-\sin\theta) \cdot \mathbf{J} \\ \bar{\mathbf{J}} &= \sin\theta \cdot \mathbf{I} + \cos\theta \cdot \mathbf{J}\end{aligned}$$

for some fixed angle θ. It is easy to discover that the matrix

$$P = \begin{bmatrix} \cos\theta & -\sin\theta \\ \sin\theta & \cos\theta \end{bmatrix}$$

has an inverse

$$P^{-1} = \begin{bmatrix} \cos\theta & \sin\theta \\ -\sin\theta & \cos\theta \end{bmatrix}.$$

Thus, P is nonsingular, and $\{\bar{\mathbf{I}}, \bar{\mathbf{J}}\}$ is a basis of $\mathcal{E}_2$. We find

$$|\bar{\mathbf{I}}| = \sqrt{\cos^2\theta + \sin^2\theta} = 1, \qquad |\bar{\mathbf{J}}| = \sqrt{\sin^2\theta + \cos^2\theta} = 1$$

and

$$\bar{\mathbf{I}} \circ \bar{\mathbf{J}} = \cos\theta \sin\theta + (-\sin\theta)\cos\theta = 0.$$

Hence, the pair of orthogonal unit vectors $\{\bar{\mathbf{I}}, \bar{\mathbf{J}}\}$ determines a new rectangular Cartesian coordinate system for $\mathcal{E}_2$. We may associate with each vector $\mathbf{V} \in \mathcal{E}_2$ a coordinate pair $(\bar{x}, \bar{y})$ with respect to $\{\bar{\mathbf{I}}, \bar{\mathbf{J}}\}$. If the coordinates of $\mathbf{V}$ are (x, y) with respect to $\{\mathbf{I}, \mathbf{J}\}$, then the new coordinates $(\bar{x}, \bar{y})$ of $\mathbf{V}$ with respect to $\{\bar{\mathbf{I}}, \bar{\mathbf{J}}\}$ are given by the matrix equation

$$[\bar{x} \quad \bar{y}] = [x \quad y]\begin{bmatrix} \cos\theta & \sin\theta \\ -\sin\theta & \cos\theta \end{bmatrix}. \tag{21-7}$$

This equation is the matrix version of the standard equations of rotation of a rectangular Cartesian coordinate system through an angle θ.

exercises

21.1 Prove that $\{\mathbf{U}_1, \mathbf{U}_2, \mathbf{U}_3, \mathbf{U}_4\}$ is a basis of $\mathcal{V}_4$, where $\mathbf{U}_1 = (-2, 3, 4, 1)$, $\mathbf{U}_2 = (0, -1, 6, 3)$, $\mathbf{U}_3 = (1, 1, 4, 2)$, and $\mathbf{U}_4 = (-5, 6, -1, 0)$. Find the dimension of the subspace $\mathcal{S}$ of $\mathcal{V}_4$ spanned by the following sets of vectors:
(a) $\{\mathbf{U}_1 - 2\cdot\mathbf{U}_2 + \mathbf{U}_3, 3\cdot\mathbf{U}_1 - \mathbf{U}_3, -6\cdot\mathbf{U}_2 + 4\cdot\mathbf{U}_3\}$;
(a) $\{\mathbf{U}_1 - \mathbf{U}_2 + \mathbf{U}_3 - \mathbf{U}_4, \mathbf{U}_1 - \mathbf{U}_3 + \mathbf{U}_4, \mathbf{U}_2 - \mathbf{U}_3 + \mathbf{U}_4, \mathbf{U}_3 - \mathbf{U}_4\}$;
(c) $\{\sqrt{2}\cdot\mathbf{U}_1 + \sqrt{3}\cdot\mathbf{U}_2 + \sqrt{5}\cdot\mathbf{U}_3 + \sqrt{7}\cdot\mathbf{U}_4, \sqrt{7}\cdot\mathbf{U}_1 + \sqrt{5}\cdot\mathbf{U}_2 + \sqrt{3}\cdot\mathbf{U}_3 + \sqrt{2}\cdot\mathbf{U}_4\}$.

21.2 Prove that $\{(1, 0, 0), (1, 1, 0), (1, 1, 1)\}$ and $\{(1, 2, -1), (3, -1, 4), (2, 0, 1)\}$ are bases of $\mathcal{V}_3$. Find the matrix of the change of basis from the first basis to the second basis. Find the coordinates of the vector $\mathbf{U} = (7, -3, 4)$ with respect to each basis.

21.3 Prove that the following sets of vectors are bases of $\mathcal{E}_3$:
(a) $\{\mathbf{I} + \mathbf{J} + \mathbf{K}, \mathbf{J} + \mathbf{K}, \mathbf{K}\}$
(b) $\{2\cdot\mathbf{I} - 3\cdot\mathbf{J} + 5\cdot\mathbf{K}, \mathbf{I} - \mathbf{J} + 4\cdot\mathbf{K}, -3\cdot\mathbf{I} + 4\cdot\mathbf{J} - \mathbf{K}\}$
(c) $\{\tfrac{1}{2}\cdot\mathbf{I} - \sqrt{3}/2\cdot\mathbf{J}, \sqrt{3}/2\cdot\mathbf{I} + \tfrac{1}{2}\cdot\mathbf{J}, 4\cdot\mathbf{J} + 7\cdot\mathbf{K}\}$

Find the coordinates of the vector $-2\cdot\mathbf{I} + 3\cdot\mathbf{J} + 5\cdot\mathbf{K}$ with respect to each of the above bases.

21.4 The set $\{\mathbf{I}, \mathbf{J}\}$ of unit vectors is associated with a rectangular Cartesian coordinate system in $\mathcal{E}_2$. Define vectors $\bar{\mathbf{I}}, \bar{\mathbf{J}}$ by the equations

$$\bar{\mathbf{I}} = p_{1,1}\cdot\mathbf{I} + p_{2,2}\cdot\mathbf{J}$$
$$\bar{\mathbf{J}} = p_{2,1}\cdot\mathbf{I} + p_{2,2}\cdot\mathbf{J}$$

where the matrix

$$P = \begin{bmatrix} p_{1,1} & p_{1,2} \\ p_{2,1} & p_{2,2} \end{bmatrix}$$

is nonsingular and $P^{-1} = P'$. Prove that $\{\bar{\mathbf{I}}, \bar{\mathbf{J}}\}$ determines a new rectangular Cartesian coordinate system in $\mathcal{E}_2$.

21.5 Let $\{\mathbf{U}_1, \mathbf{U}_2, \ldots, \mathbf{U}_n\}$ be an orthonormal basis of a Euclidean vector space $\mathcal{E}$. Let $\{\bar{\mathbf{U}}_1, \bar{\mathbf{U}}_2, \ldots, \bar{\mathbf{U}}_n\}$ be a second basis of $\mathcal{E}$, and let $P = [p_{i,j}]$ be the matrix of the change of basis from the $\mathbf{U}$-basis to the $\bar{\mathbf{U}}$-basis. Prove that the $\bar{\mathbf{U}}$-basis is orthonormal if and only if $P^{-1} = P'$.

22 matrices of linear transformations with respect to different bases

In the following discussion, $\mathcal{V}$ is an n-dimensional vector space with basis $\{\mathbf{U}_1, \mathbf{U}_2, \ldots, \mathbf{U}_n\}$ and $\mathcal{W}$ is an m-dimensional space with basis $\{\mathbf{W}_1, \mathbf{W}_2, \ldots, \mathbf{W}_m\}$. We assume that L is a linear transformation of $\mathcal{V}$ into $\mathcal{W}$ that has matrix A with respect to the given bases, and we investigate the effect of a change of basis in each space on the matrix A.

Suppose first that $\{\bar{\mathbf{U}}_1, \bar{\mathbf{U}}_2, \ldots, \bar{\mathbf{U}}_n\}$ and $\{\bar{\mathbf{W}}_1, \bar{\mathbf{W}}_2, \ldots, \bar{\mathbf{W}}_m\}$ are second bases of $\mathcal{V}$ and $\mathcal{W}$, respectively. Let $\mathbf{V}$ be any vector in $\mathcal{V}$ that has coordinates $(r_1, r_2, \ldots, r_n)$ with respect to the basis $\{\mathbf{U}_1, \mathbf{U}_2, \ldots, \mathbf{U}_n\}$ and coordinates $(\bar{r}_1, \bar{r}_2, \ldots, \bar{r}_n)$ with respect to the basis $\{\bar{\mathbf{U}}_1, \bar{\mathbf{U}}_2, \ldots, \bar{\mathbf{U}}_n\}$. Denote the coordinates of $L(\mathbf{V})$ with respect to the basis $\{\mathbf{W}_1, \mathbf{W}_2, \ldots, \mathbf{W}_m\}$ by $(s_1, s_2, \ldots, s_m)$ and the coordinates of $L(\mathbf{V})$ with respect to $\{\bar{\mathbf{W}}_1, \bar{\mathbf{W}}_2, \ldots, \bar{\mathbf{W}}_m\}$ by $(\bar{s}_1, \bar{s}_2, \ldots, \bar{s}_m)$. Then

$$\begin{bmatrix} s_1 \\ s_2 \\ \vdots \\ s_m \end{bmatrix} = A \begin{bmatrix} r_1 \\ r_2 \\ \vdots \\ r_n \end{bmatrix}$$

where A is the m by n matrix of L with respect to the first bases in $\mathcal{V}$ and $\mathcal{W}$ (see Example 3, §19). Let P be the matrix of the change of basis from the $\mathbf{U}$-basis to the $\bar{\mathbf{U}}$-basis in $\mathcal{V}$, and let Q be the matrix of the change of basis from the $\mathbf{W}$-basis to the $\bar{\mathbf{W}}$-basis in $\mathcal{W}$. Then by the results of §21,

$$\begin{bmatrix} r_1 \\ r_2 \\ \vdots \\ r_n \end{bmatrix} = P' \begin{bmatrix} \bar{r}_1 \\ \bar{r}_2 \\ \vdots \\ \bar{r}_n \end{bmatrix} \quad \text{and} \quad \begin{bmatrix} \bar{s}_1 \\ \bar{s}_2 \\ \vdots \\ \bar{s}_m \end{bmatrix} = (Q^{-1})' \begin{bmatrix} s_1 \\ s_2 \\ \vdots \\ s_m \end{bmatrix}$$

Therefore,

$$\begin{bmatrix} \bar{s}_1 \\ \bar{s}_2 \\ \vdots \\ \bar{s}_m \end{bmatrix} = (Q^{-1})' \begin{bmatrix} s_1 \\ s_2 \\ \vdots \\ s_m \end{bmatrix} = (Q^{-1})'A \begin{bmatrix} r_1 \\ r_1 \\ \vdots \\ r_n \end{bmatrix} = (Q^{-1})'AP' \begin{bmatrix} \bar{r}_1 \\ \bar{r}_2 \\ \vdots \\ \bar{r}_n \end{bmatrix}.$$

Hence, the matrix of the linear transformation L with respect to the second bases in $\mathcal{V}$ and $\mathcal{W}$ is $(Q^{-1})'AP'$, where $(Q^{-1})'$ and P' are nonsingular matrices.

The following diagram may help to clarify the above discussion:

$$\begin{array}{ccc}
\overset{\mathcal{V}}{\{\mathbf{U}_1, \mathbf{U}_2, \ldots, \mathbf{U}_n\}} & \xrightarrow[A]{L} & \overset{\mathcal{W}}{\{\mathbf{W}_1, \mathbf{W}_2, \ldots, \mathbf{W}_m\}} \\
\downarrow P & & Q \downarrow\uparrow Q^{-1} \\
\{\bar{\mathbf{U}}_1, \bar{\mathbf{U}}_2, \ldots, \bar{\mathbf{U}}_n\} & \xrightarrow[(Q^{-1})'AP']{L} & \{\bar{\mathbf{W}}_1, \bar{\mathbf{W}}_2, \ldots, \bar{\mathbf{W}}_m\}
\end{array}$$

This diagram indicates that A is the matrix of L with respect to the original bases $\{\mathbf{U}_1, \mathbf{U}_2, \ldots, \mathbf{U}_n\}$ in $\mathcal{V}$ and $\{\mathbf{W}_1, \mathbf{W}_2, \ldots, \mathbf{W}_m\}$ in $\mathcal{W}$; that P is the matrix of the change of basis from $\{\mathbf{U}_1, \mathbf{U}_2, \ldots, \mathbf{U}_n\}$ to the basis $\{\bar{\mathbf{U}}_1, \bar{\mathbf{U}}_2, \ldots, \bar{\mathbf{U}}_n\}$ in $\mathcal{V}$; that Q is the matrix of the change of basis from $\{\mathbf{W}_1, \mathbf{W}_2, \ldots, \mathbf{W}_m\}$ to the basis $\{\bar{\mathbf{W}}_1, \bar{\mathbf{W}}_2, \ldots, \bar{\mathbf{W}}_m\}$ in $\mathcal{W}$ (and therefore Q^{-1} is the matrix of the change of basis from $\{\bar{\mathbf{W}}_1, \bar{\mathbf{W}}_2, \ldots, \bar{\mathbf{W}}_m\}$ to $\{\mathbf{W}_1, \mathbf{W}_2, \ldots, \mathbf{W}_m\}$); and finally that $(Q^{-1})'AP'$ is the matrix of L with respect to the new bases $\{\bar{\mathbf{U}}_1, \bar{\mathbf{U}}_2, \ldots, \bar{\mathbf{U}}_n\}$ in $\mathcal{V}$ and $\{\bar{\mathbf{W}}_1, \bar{\mathbf{W}}_2, \ldots, \bar{\mathbf{W}}_m\}$ in $\mathcal{W}$.

Now suppose that B is an m by n matrix and that there exist a nonsingular m by m matrix E and a nonsingular n by n matrix F such that $B = EAF$. By the results of §21, a new basis $\{\bar{\mathbf{U}}_1, \bar{\mathbf{U}}_2, \ldots, \bar{\mathbf{U}}_n\}$ can be defined in $\mathcal{V}$, using the nonsingular matrix F' as the matrix of the change of basis. Then the coordinates $(\bar{r}_1, \bar{r}_2, \ldots, \bar{r}_n)$ of $\mathbf{V}$ with respect to this new basis are related to the coordinates $(r_1, r_2, \ldots, r_n)$ of $\mathbf{V}$ with respect to the basis $\{\mathbf{U}_1, \mathbf{U}_2, \ldots, \mathbf{U}_n\}$ by

$$\begin{bmatrix} r_1 \\ r_2 \\ \vdots \\ r_n \end{bmatrix} = (F')' \begin{bmatrix} \bar{r}_1 \\ \bar{r}_2 \\ \vdots \\ \bar{r}_n \end{bmatrix} = F \begin{bmatrix} \bar{r}_1 \\ \bar{r}_2 \\ \vdots \\ \bar{r}_n \end{bmatrix}.$$

Similarly, a new basis $\{\bar{\mathbf{W}}_1, \bar{\mathbf{W}}_2, \ldots, \bar{\mathbf{W}}_m\}$ can be defined in $\mathcal{W}$, using the nonsingular matrix $(E')^{-1}$ as the matrix of the change of basis, such that the coordinates $(\bar{s}_1, \bar{s}_2, \ldots, \bar{s}_m)$ of $L(\mathbf{V})$ with respect to this basis are given by

$$\begin{bmatrix} \bar{s}_1 \\ \bar{s}_1 \\ \vdots \\ \bar{s}_m \end{bmatrix} = (((E')^{-1})^{-1})' \begin{bmatrix} s_1 \\ s_2 \\ \vdots \\ s_m \end{bmatrix} = E \begin{bmatrix} s_2 \\ s_2 \\ \vdots \\ s_m \end{bmatrix}$$

where $(s_1, s_2, \ldots, s_m)$ are the coordinates of $L(\mathbf{V})$ with respect to $\{\mathbf{W}_1, \mathbf{W}_2, \ldots, \mathbf{W}_m\}$. Hence,

$$\begin{bmatrix} \bar{s}_1 \\ \bar{s}_2 \\ \vdots \\ \bar{s}_m \end{bmatrix} = E \begin{bmatrix} s_1 \\ s_2 \\ \vdots \\ s_m \end{bmatrix} = EA \begin{bmatrix} r_1 \\ r_2 \\ \vdots \\ r_n \end{bmatrix} = EAF \begin{bmatrix} \bar{r}_1 \\ \bar{r}_2 \\ \vdots \\ \bar{r}_n \end{bmatrix} = B \begin{bmatrix} \bar{r}_1 \\ \bar{r}_2 \\ \vdots \\ \bar{r}_n \end{bmatrix}.$$

Therefore, there is a suitable change of basis in each space $\mathcal{V}$ and $\mathcal{W}$ such that B is the matrix of the linear transformation L with respect to these new bases.

The diagram related to this discussion should be interpreted in the same way as the previous one.

$$\begin{array}{ccc}
\overset{\mathcal{V}}{\{\mathbf{U}_1, \mathbf{U}_2, \ldots, \mathbf{U}_n\}} & \xrightarrow[A]{L} & \overset{\mathcal{W}}{\{\mathbf{W}_1, \mathbf{W}_2, \ldots, \mathbf{W}_m\}} \\
\big\downarrow F' & & (E')^{-1} \big\downarrow\!\big\uparrow E' \\
\{\bar{\mathbf{U}}_1, \bar{\mathbf{U}}_2, \ldots, \bar{\mathbf{U}}_n\} & \xrightarrow[B = EAF]{L} & \{\bar{\mathbf{W}}_1, \bar{\mathbf{W}}_2 \ldots, \bar{\mathbf{W}}_m\}
\end{array}$$

The results obtained so far in this section can be summarized in the following theorem:

(22.1) theorem Let A and B be m by n matrices, and let $\mathcal{V}$ be an n-dimensional vector space and $\mathcal{W}$ be an m-dimensional vector space. The matrices A and B are matrices of the same linear transformation L of $\mathcal{V}$ into $\mathcal{W}$ (with respect to suitable bases in these spaces) if and only if there exist non-singular matrices E and F such that $B = EAF$.

EXAMPLE 1 Let L be the linear transformation of $\mathcal{V}_3$ into $\mathcal{V}_2$ which has matrix

$$A = \begin{bmatrix} 2 & -1 & 3 \\ 4 & 0 & -6 \end{bmatrix}$$

with respect to the bases $E^{(3)} = \{(1, 0, 0), (0, 1, 0), (0, 0, 1)\}$ in $\mathcal{V}_3$ and $E^{(2)} = \{(1, 0), (0, 1)\}$ in $\mathcal{V}_2$. Then a vector $\mathbf{V} = (r_1, r_2, r_3) \in \mathcal{V}_3$ is sent into a vector $L(\mathbf{V}) = (s_1, s_2) \,\varepsilon\, \mathcal{V}_2$, where

$$\begin{bmatrix} s_1 \\ s_2 \end{bmatrix} = A \begin{bmatrix} r_1 \\ r_2 \\ r_3 \end{bmatrix} = \begin{bmatrix} 2 & -1 & 3 \\ 4 & 0 & -6 \end{bmatrix} \begin{bmatrix} r_1 \\ r_2 \\ r_3 \end{bmatrix} = \begin{bmatrix} 2r_1 - r_2 + 3r_3 \\ 4r_1 - 6r_3 \end{bmatrix}.$$

Thus, $L(\mathbf{V}) = L[(r_1, r_2, r_3)] = (2r_1 - r_2 + 3r_3, 4r_1 - 6r_3)$. Now, the set $S = \{(1, 0, 0), (1, 1, 0), (1, 1, 1)\}$ is a second basis of $\mathcal{V}_3$, and the set $T = \{(1, 0), (1, 1)\}$ is a second basis of $\mathcal{V}_2$. Since

$$\begin{aligned}
(1, 0, 0) &= 1 \cdot (1, 0, 0) + 0 \cdot (0, 1, 0) + 0 \cdot (0, 0, 1) \\
(1, 1, 0) &= 1 \cdot (1, 0, 0) + 1 \cdot (0, 1, 0) + 0 \cdot (0, 0, 1) \\
(1, 1, 1) &= 1 \cdot (1, 0, 0) + 1 \cdot (0, 1, 0) + 1 \cdot (0, 0, 1)
\end{aligned}$$

the matrix of the change of basis from the basis $E^{(3)}$ to the basis S in $\mathcal{V}_3$ is

$$P = \begin{bmatrix} 1 & 0 & 0 \\ 1 & 1 & 0 \\ 1 & 1 & 1 \end{bmatrix}.$$

Similarly, in $\mathcal{V}_2$,

$$\begin{aligned}
(1, 0) &= 1 \cdot (1, 0) + 0 \cdot (0, 1) \\
(1, 1) &= 1 \cdot (1, 0) + 1 \cdot (0, 1)
\end{aligned}$$

so that the matrix of the change of basis from the basis $E^{(2)}$ to the basis T is

$$Q = \begin{bmatrix} 1 & 0 \\ 1 & 1 \end{bmatrix}.$$

It is easy to see that both P and Q are nonsingular.

Let $\mathbf{V}$ have coordinates $(\bar{r}_1, \bar{r}_2, \bar{r}_3)$ with respect to the basis S in $\mathcal{V}_3$. That is,

$$\mathbf{V} = \bar{r}_1 \cdot (1, 0, 0) + \bar{r}_2 \cdot (1, 1, 0) + \bar{r}_3 \cdot (1, 1, 1).$$

Let $L(\mathbf{V})$ have coordinates $(\bar{s}_1, \bar{s}_2)$ with respect to the basis T in $\mathcal{V}_2$, so that

$$L(\mathbf{V}) = \bar{s}_1 \cdot (1, 0) + \bar{s}_2 \cdot (1, 1).$$

Then following the above discussion, we have

$$\begin{bmatrix} r_1 \\ r_2 \\ r_3 \end{bmatrix} = P' \begin{bmatrix} \bar{r}_1 \\ \bar{r}_2 \\ \bar{r}_3 \end{bmatrix} = \begin{bmatrix} 1 & 1 & 1 \\ 0 & 1 & 1 \\ 0 & 0 & 1 \end{bmatrix} \begin{bmatrix} \bar{r}_1 \\ \bar{r}_2 \\ \bar{r}_3 \end{bmatrix}.$$

Since

$$\begin{bmatrix} 1 & 0 \\ 1 & 1 \end{bmatrix} \begin{bmatrix} 1 & 0 \\ -1 & 1 \end{bmatrix} = \begin{bmatrix} 1 & 0 \\ 0 & 1 \end{bmatrix},$$

it follows that

$$Q^{-1} = \begin{bmatrix} 1 & 0 \\ -1 & 1 \end{bmatrix}, \qquad \text{and} \qquad (Q^{-1})' = \begin{bmatrix} 1 & -1 \\ 0 & 1 \end{bmatrix}.$$

Therefore,

$$\begin{bmatrix} \bar{s}_1 \\ \bar{s}_2 \end{bmatrix} = (Q^{-1})' \begin{bmatrix} s_1 \\ s_2 \end{bmatrix} = \begin{bmatrix} 1 & -1 \\ 0 & 1 \end{bmatrix} \begin{bmatrix} s_1 \\ s_2 \end{bmatrix}. \tag{22–1}$$

Now, substituting

$$\begin{bmatrix} s_1 \\ s_2 \end{bmatrix} = A \begin{bmatrix} r_1 \\ r_2 \\ r_3 \end{bmatrix} \quad \text{and} \quad \begin{bmatrix} r_1 \\ r_2 \\ r_3 \end{bmatrix} = P' \begin{bmatrix} \bar{r}_1 \\ \bar{r}_2 \\ \bar{r}_3 \end{bmatrix}$$

in the matrix equation (22–1) we have

$$\begin{bmatrix} \bar{s}_1 \\ \bar{s}_2 \end{bmatrix} = (Q^{-1})' \begin{bmatrix} s_1 \\ s_2 \end{bmatrix} = (Q^{-1})'A \begin{bmatrix} r_1 \\ r_2 \\ r_3 \end{bmatrix} = (Q^{-1})'AP' \begin{bmatrix} \bar{r}_1 \\ \bar{r}_2 \\ \bar{r}_3 \end{bmatrix}$$

$$= \begin{bmatrix} 1 & -1 \\ 0 & 1 \end{bmatrix} \begin{bmatrix} 2 & -1 & 3 \\ 4 & 0 & -6 \end{bmatrix} \begin{bmatrix} 1 & 1 & 1 \\ 0 & 1 & 1 \\ 0 & 0 & 1 \end{bmatrix} \begin{bmatrix} \bar{r}_1 \\ \bar{r}_2 \\ \bar{r}_3 \end{bmatrix}$$

$$= \begin{bmatrix} -2 & -3 & 6 \\ 4 & 4 & -2 \end{bmatrix} \begin{bmatrix} \bar{r}_1 \\ \bar{r}_2 \\ \bar{r}_3 \end{bmatrix}.$$

Thus,

$$(Q^{-1})'AP' = \begin{bmatrix} -2 & -3 & 6 \\ 4 & 4 & -2 \end{bmatrix}$$

is the matrix of the linear transformation L with respect to the bases S and T in $\mathcal{V}_3$ and $\mathcal{V}_2$, respectively.

For example, if $\mathbf{V} = 3 \cdot (1, 0, 0) - 4 \cdot (1, 1, 0) + 5 \cdot (1, 1, 1)$, then

$$\begin{bmatrix} \bar{s}_1 \\ \bar{s}_2 \end{bmatrix} = \begin{bmatrix} -2 & -3 & 6 \\ 4 & 4 & -2 \end{bmatrix} \begin{bmatrix} 3 \\ -4 \\ 5 \end{bmatrix} = \begin{bmatrix} 36 \\ -14 \end{bmatrix}.$$

That is, $L(\mathbf{V}) = \bar{s}_1 \cdot (1, 0) + \bar{s}_2 \cdot (1, 1) = 36 \cdot (1, 0) - 14 \cdot (1, 1) = (22, -14)$. Since $\mathbf{V} = 3 \cdot (1, 0, 0) - 4 \cdot (1, 1, 0) + 5 \cdot (1, 1, 1) = (4, 1, 5)$,

$$\begin{bmatrix} s_1 \\ s_2 \end{bmatrix} = \begin{bmatrix} 2 & -1 & 3 \\ 4 & 0 & -6 \end{bmatrix} \begin{bmatrix} 4 \\ 1 \\ 5 \end{bmatrix} = \begin{bmatrix} 22 \\ -14 \end{bmatrix}$$

and we confirm the fact that $L(\mathbf{V}) = (s_1, s_2) = (22, -14)$.

Theorem 22.1 leads to the following definition:

(22.2) definition Let A and B be m by n matrices. The matrix B is *equivalent* to the matrix A if there exists a nonsingular m by m matrix E and a nonsingular n by n matrix F such that $B = EAF$.

The relation between m by n matrices given in Definition 22.2 is called *equivalence of rectangular matrices*. When regarded as a relation on the set ${}_m\mathfrak{M}_n$ of all real m by n matrices, equivalence of matrices satisfies three basic conditions:

(22.3) (a) A is equivalent to A, for every $A \in {}_m\mathfrak{M}_n$.

(b) If B is equivalent to A, then A is equivalent to B.

(c) If A is equivalent to B, and B is equivalent to C, then A is equivalent to C.

Let A be any matrix in ${}_m\mathfrak{A}_n$. Since the identity matrices I_m and I_n are nonsingular, and $A = I_m A I_n$, it follows from Definition 22.2 that A is equivalent to A. Thus, condition (a) of 22.3 is satisfied. If $A \in {}_m\mathfrak{A}_n$ and $B \in {}_m\mathfrak{A}_n$, and B is equivalent to A, then by Definition 22.2, there exist a nonsingular m by m matrix E and a nonsingular n by n matrix F such that $B = EAF$. Multiplying the latter equation by E^{-1} on the left and F^{-1} on the right, we obtain $A = E^{-1}BF^{-1}$, where E^{-1} is a nonsingular m by m matrix and F^{-1} is a nonsingular n by n matrix. Therefore, by Definition 22.2, A is equivalent to B. This proves part (b) of 22.3. Finally, suppose that A, B, and C are in ${}_m\mathfrak{A}_n$, A is equivalent to B, and B is equivalent to C. Then $A = E_1BF_1$ and $B = E_2CF_2$, where E_1 and E_2 are nonsingular m by m matrices, and F_1 and F_2 are nonsingular n by n matrices. Substituting the second of these equations into the first, we have $A = E_1(E_2CF_2)F_1 = (E_1E_2)C(F_1F_2)$, where E_1E_2 is a nonsingular m by m matrix and F_1F_2 is a nonsingular n by n matrix. Thus, A is equivalent to C by Definition 22.2. Therefore, condition (c) of 22.3 holds.

The conditions (a), (b), and (c) of 22.3 are called the *reflexive, symmetric,* and *transitive* laws, respectively. Any relation on an arbitrary set S that satisfies these three laws is called an *equivalence relation* on S. Ordinary equality of elements of a set has these properties, and an equivalence relation can be regarded as a generalization of equality. Equivalence relations occur, significantly, in all areas of contemporary mathematics. Besides the equivalence of rectangular matrices defined above, we will encounter other important relations on sets of matrices which are equivalence relations. The principal facts concerning equivalence relations are developed in the exercises at the end of this section.

It follows from the symmetry property that if a matrix B is equivalent to a matrix A, we can simply say that A and B are equivalent. In terms of the concept of equivalence of matrices, Theorem 22.1 states that two m by n matrices are matrices of the same linear transformation if and only if they are equivalent.

Suppose that a given m by n matrix A is equivalent to an m by n matrix B, which has a particularly simple form. Let L be the linear transformation of $\mathcal{V}$ into $\mathcal{W}$ defined by the matrix A relative to chosen bases in $\mathcal{V}$ and $\mathcal{W}$. Then the matrix A of L can be simplified by referring L to new bases in $\mathcal{V}$ and $\mathcal{W}$ for which the matrix of L is B. Moreover, certain properties of L that are reflected in its corresponding matrices become apparent.

EXAMPLE 2 Let L be the linear transformation of $\mathcal{V}_4$ into $\mathcal{V}_3$ that has the matrix

$$A = \begin{bmatrix} 0 & -3 & -1 & 1 \\ 1 & 5 & 4 & -1 \\ 1 & 2 & 3 & 0 \end{bmatrix}$$

with respect to the bases $E^{(4)}$ and $E^{(3)}$ in $\mathcal{V}_4$ and $\mathcal{V}_3$, respectively. Then any vector $(r_1, r_2, r_3, r_4) \in \mathcal{V}_4$ is sent into the vector $(s_1, s_2, s_3) \in \mathcal{V}_3$, where

$$\begin{bmatrix} s_1 \\ s_2 \\ s_3 \end{bmatrix} = \begin{bmatrix} 0 & -3 & -1 & 1 \\ 1 & 5 & 4 & -1 \\ 1 & 2 & 3 & 0 \end{bmatrix} \begin{bmatrix} r_1 \\ r_2 \\ r_3 \\ r_4 \end{bmatrix} = \begin{bmatrix} -3r_2 - r_3 + r_4 \\ r_1 + 5r_2 + 4r_3 - r_4 \\ r_1 + 2r_2 + 3r_3 \end{bmatrix}.$$

Thus,

$$L[(r_1, r_2, r_3, r_4)] = (-3r_2 - r_3 + r_4, r_1 + 5r_2 + 4r_3 - r_4, r_1 + 2r_2 + 3r_3).$$

The matrix

$$E = \begin{bmatrix} 0 & 1 & 0 \\ -1/3 & 0 & 0 \\ -1 & -1 & 1 \end{bmatrix}$$

is nonsingular with

$$E^{-1} = \begin{bmatrix} 0 & -3 & 0 \\ 1 & 0 & 0 \\ 1 & -3 & 1 \end{bmatrix}$$

The matrix

$$F = \begin{bmatrix} 1 & -5 & -7/3 & -2/3 \\ 0 & 1 & -1/3 & 1/3 \\ 0 & 0 & 1 & 0 \\ 0 & 0 & 0 & 1 \end{bmatrix}$$

is nonsingular with

$$F^{-1} = \begin{bmatrix} 1 & 5 & 4 & -1 \\ 0 & 1 & 1/3 & -1/3 \\ 0 & 0 & 1 & 0 \\ 0 & 0 & 0 & 1 \end{bmatrix}$$

Define a new basis in $\mathcal{V}_4$, using the matrix F':

$$\begin{aligned} \bar{\mathbf{U}}_1 &= 1 \cdot \mathbf{E}_1^{(4)} \\ \bar{\mathbf{U}}_2 &= (-5) \cdot \mathbf{E}_1^{(4)} + 1 \cdot \mathbf{E}_2^{(4)} \\ \bar{\mathbf{U}}_3 &= (-7/3) \cdot \mathbf{E}_1^{(4)} + (-1/3) \cdot \mathbf{E}_2^{(4)} + 1 \cdot \mathbf{E}_3^{(4)} \\ \bar{\mathbf{U}}_4 &= (-2/3) \cdot \mathbf{E}_1^{(4)} + (1/3) \cdot \mathbf{E}_2^{(4)} + 1 \cdot \mathbf{E}_4^{(4)}. \end{aligned}$$

Thus, $\bar{\mathbf{U}}_1 = (1, 0, 0, 0)$, $\bar{\mathbf{U}}_2 = (-5, 1, 0, 0)$, $\bar{\mathbf{U}}_3 = (-7/3, -1/3, 1, 0)$ and $\bar{\mathbf{U}}_4 = (-2/3, 1/3, 0, 1)$. Define a new basis in $\mathcal{V}_3$, using the matrix $(E')^{-1} = (E^{-1})'$:

$$\begin{aligned} \bar{\mathbf{W}}_1 &= 1 \cdot \mathbf{E}_2^{(3)} + 1 \cdot \mathbf{E}_3^{(3)} \\ \bar{\mathbf{W}}_2 &= (-3) \cdot \mathbf{E}_1^{(3)} + (-3) \cdot \mathbf{E}_3^{(3)} \\ \bar{\mathbf{W}}_3 &= 1 \cdot \mathbf{E}_3^{(3)}. \end{aligned}$$

Hence, $\bar{\mathbf{W}}_1 = (0, 1, 1)$, $\bar{\mathbf{W}}_2 = (-3, 0, -3)$, and $\bar{\mathbf{W}}_3 = (0, 0, 1)$. With respect to the bases $\{\bar{\mathbf{U}}_1, \bar{\mathbf{U}}_2, \bar{\mathbf{U}}_3, \bar{\mathbf{U}}_4\}$ and $\{\bar{\mathbf{W}}_1, \bar{\mathbf{W}}_2, \bar{\mathbf{W}}_3\}$ in $\mathcal{V}_4$ and $\mathcal{V}_3$, the linear transformation L has matrix

$$B = EAF = \begin{bmatrix} 0 & 1 & 0 \\ -\tfrac{1}{3} & 0 & 0 \\ -1 & -1 & 1 \end{bmatrix} \begin{bmatrix} 0 & -3 & -1 & 1 \\ 1 & 5 & 4 & -1 \\ 1 & 2 & 3 & 0 \end{bmatrix} \begin{bmatrix} 1 & -5 & -\tfrac{7}{3} & -\tfrac{2}{3} \\ 0 & 1 & -\tfrac{1}{3} & \tfrac{1}{3} \\ 0 & 0 & 1 & 0 \\ 0 & 0 & 0 & 1 \end{bmatrix}$$

$$= \begin{bmatrix} 1 & 0 & 0 & 0 \\ 0 & 1 & 0 & 0 \\ 0 & 0 & 0 & 0 \end{bmatrix}.$$

A vector $\mathbf{V} \in \mathcal{V}_4$ with coordinates $(\bar{r}_1, \bar{r}_2, \bar{r}_3, \bar{r}_4)$ with respect to the basis $\{\bar{\mathbf{U}}_1, \bar{\mathbf{U}}_2, \bar{\mathbf{U}}_3, \bar{\mathbf{U}}_4\}$ is sent into the vector $L(\mathbf{V})$ with coordinates $(\bar{s}_1, \bar{s}_2, \bar{s}_3)$ with respect to the basis $\{\bar{\mathbf{W}}_1, \bar{\mathbf{W}}_2, \bar{\mathbf{W}}_3\}$, where

$$\begin{bmatrix} \bar{s}_1 \\ \bar{s}_2 \\ \bar{s}_3 \end{bmatrix} = \begin{bmatrix} 1 & 0 & 0 & 0 \\ 0 & 1 & 0 & 0 \\ 0 & 0 & 0 & 0 \end{bmatrix} \begin{bmatrix} \bar{r}_1 \\ \bar{r}_2 \\ \bar{r}_3 \\ \bar{r}_4 \end{bmatrix} = \begin{bmatrix} \bar{r}_1 \\ \bar{r}_2 \\ 0 \end{bmatrix}.$$

For example, if

$$\mathbf{V} = 3 \cdot \bar{\mathbf{U}}_1 + (-6) \cdot \bar{\mathbf{U}}_2 + (\tfrac{1}{2}) \cdot \bar{\mathbf{U}}_3 + (-5) \cdot \bar{\mathbf{U}}_4,$$

then

$$\begin{aligned} L(\mathbf{V}) &= 3 \cdot \bar{\mathbf{W}}_1 + (-6) \cdot \bar{\mathbf{W}}_2 + 0 \cdot \bar{\mathbf{W}}_3 \\ &= 3 \cdot (0, 1, 1) + (-6) \cdot (-3, 0, -3) = (18, 3, 21). \end{aligned}$$

It is evident from the simple form of the matrix B of L that the range space $L(\mathcal{V}_4)$ is the subspace of $\mathcal{V}_3$ of dimension two spanned by $\bar{\mathbf{W}}_1$ and $\bar{\mathbf{W}}_2$. The null space $L^{-1}(\{\mathbf{0}\})$ of L is the subspace of $\mathcal{V}_4$ of dimension two spanned by $\bar{\mathbf{U}}_3$ and $\bar{\mathbf{U}}_4$.

We turn next to the situation where L is a linear transformation of an n-dimensional vector space $\mathcal{V}$ into itself. The matrix of L with respect to a basis $\{\mathbf{U}_1, \mathbf{U}_2, \ldots, \mathbf{U}_n\}$ of $\mathcal{V}$ is an n by n matrix A. If $\mathbf{V} \in \mathcal{V}$, and if $\mathbf{V}$ has coordinates $(r_1, r_2, \ldots, r_n)$ with respect to this basis, then $L(\mathbf{V})$ has coordinates $(s_1, s_2, \ldots, s_n)$ with respect to this same basis, where

$$\begin{bmatrix} s_1 \\ s_2 \\ \vdots \\ s_n \end{bmatrix} = A \begin{bmatrix} r_1 \\ r_2 \\ \vdots \\ r_n \end{bmatrix}.$$

IF $\{\bar{\mathbf{U}}_1, \bar{\mathbf{U}}_2, \ldots, \bar{\mathbf{U}}_n\}$ is a second basis of $\mathcal{V}$, and if P is the matrix of the change of basis from the **U**-basis to the $\bar{\mathbf{U}}$-basis, then

$$\begin{bmatrix} r_1 \\ r_2 \\ \vdots \\ r_n \end{bmatrix} = P' \begin{bmatrix} \bar{r}_1 \\ \bar{r}_2 \\ \vdots \\ \bar{r}_n \end{bmatrix} \quad \text{and} \quad \begin{bmatrix} s_1 \\ s_2 \\ \vdots \\ s_n \end{bmatrix} = P' \begin{bmatrix} \bar{s}_1 \\ \bar{s}_2 \\ \vdots \\ \bar{s}_n \end{bmatrix}$$

where **V** has coordinates $(\bar{r}_1, \bar{r}_2, \ldots, \bar{r}_n)$ and $L(\mathbf{V})$ has coordinates $(\bar{s}_1, \bar{s}_2, \ldots, \bar{s}_n)$ with respect to the second basis. Thus, we have

$$\begin{bmatrix} \bar{s}_1 \\ \bar{s}_2 \\ \vdots \\ \bar{s}_n \end{bmatrix} = (P')^{-1} \begin{bmatrix} s_1 \\ s_2 \\ \vdots \\ s_n \end{bmatrix} = (P')^{-1}A \begin{bmatrix} r_1 \\ r_2 \\ \vdots \\ r_n \end{bmatrix} = (P')^{-1}AP' \begin{bmatrix} \bar{r}_1 \\ \bar{r}_2 \\ \vdots \\ \bar{r}_n \end{bmatrix}.$$

Thus, the matrix of L with respect to the basis $\{\bar{\mathbf{U}}_1, \bar{\mathbf{U}}_2, \ldots, \bar{\mathbf{U}}_n\}$ is $(P')^{-1}AP'$. This situation is indicated in the following diagram:

$$\begin{array}{ccc}
\overset{\mathcal{V}}{\{\mathbf{U}_1, \mathbf{U}_2, \ldots, \mathbf{U}_n\}} & \xrightarrow[A]{L} & \overset{\mathcal{V}}{\{\mathbf{U}_1, \mathbf{U}_2, \ldots, \mathbf{U}_n\}} \\
\downarrow P & & P \downarrow \uparrow P^{-1} \\
\{\bar{\mathbf{U}}_1, \bar{\mathbf{U}}_2, \ldots, \bar{\mathbf{U}}_n\} & \xrightarrow[(P')^{-1}AP']{L} & \{\bar{\mathbf{U}}_1, \bar{\mathbf{U}}_2, \ldots, \bar{\mathbf{U}}_n\}
\end{array}$$

Conversely, if B is an n by n matrix, and if there exists a nonsingular n by n matrix Q such that $B = Q^{-1}AQ$, then a new basis can be defined in $\mathcal{V}$ by means of Q' so that B is the matrix of L with respect to this new basis.

(22.4) theorem Let A and B be n by n matrices, and let $\mathcal{V}$ be an n-dimensional vector space. The matrices A and B are matrices of the same linear transformation L of $\mathcal{V}$ into itself (with respect to suitable bases in $\mathcal{V}$) if and only if there exists a nonsingular matrix Q such that $B = Q^{-1}AQ$.

An important relation between square matrices is suggested by Theorem 22.4:

(22.5) definition Let A and B be n by n matrices. The matrix B is *similar* to the matrix A if there exists a nonsingular matrix Q such that $B = Q^{-1}AQ$.

Similarity is an equivalence relation on the set $\mathfrak{M}_n$ of all n by n matrices. By Theorem 22.4, two n by n matrices are similar if and only if they are matrices of the same linear transformation of an n-dimensional vector space $\mathcal{V}$ into itself. The analysis of a linear transformation of $\mathcal{V}$ into itself is accomplished by finding a simple form for an n by n matrix under the relation of similarity.

exercises

22.1 Let L be the linear transformation of $\mathcal{V}_3$ into $\mathcal{V}_2$ that has matrix

$$A = \begin{bmatrix} -1 & 2 & 0 \\ 1 & 0 & 1 \end{bmatrix}$$

with respect to the bases $\{(1, -2, 3), (0, 1, -1), (0, 0, 2)\}$ in $\mathcal{V}_3$ and $\{(5, 1), (2, 3)\}$ in $\mathcal{V}_2$. Find the matrix of L with respect to the bases $E^{(3)}$ in $\mathcal{V}_3$ and $E^{(2)}$ in $\mathcal{V}_2$.

22.2 Let L be the linear transformation of $\mathcal{V}_3$ into $\mathcal{V}_4$ that has matrix

$$A = \begin{bmatrix} -1 & 2 & 1 \\ 0 & 1 & -3 \\ 7 & 1 & 0 \\ -6 & 2 & 1 \end{bmatrix}$$

with respect to the bases $E^{(3)}$ in $\mathcal{V}_3$ and $E^{(4)}$ in $\mathcal{V}_4$. Let E and F be the nonsingular matrices given in Example 1. Find the bases in $\mathcal{V}_3$ and $\mathcal{V}_4$ such that L has matrix FAE with respect to these bases.

22.3 Let L be the linear transformation of $\mathcal{V}_2$ and $\mathcal{V}_3$ that has matrix

$$A = \begin{bmatrix} -4 & 1 \\ 0 & 2 \\ -1 & 3 \end{bmatrix}$$

with respect to the bases $E^{(2)}$ in $\mathcal{V}_2$ and $E^{(3)}$ in $\mathcal{V}_3$. Let

$$E = \begin{bmatrix} 0 & 0 & -1 \\ 0 & \frac{1}{2} & 0 \\ 1 & \frac{11}{2} & -4 \end{bmatrix}, \quad \text{and} \quad F = \begin{bmatrix} 1 & 3 \\ 0 & 1 \end{bmatrix}.$$

Define a new basis in $\mathcal{V}_2$, using the matrix F', and define a new basis in $\mathcal{V}_3$ using the matrix $(E')^{-1}$. Find the matrix of L with respect to these new bases.

***22.4** Let $S = \{a, b, c, \ldots\}$ be any set. A *relation* on S is a set R of ordered pairs (a, b) of elements of S. For example, less than ($<$) is a relation on the set of all real numbers, and this relation is the set R of all ordered pairs (x, y) of real numbers such that $x < y$. For the relation $<$, which of the following ordered pairs are elements of R? $(2, 5)$, $(-1, -6)$, $(5.2, 3.7)$; $(-1, \pi)$; $(\frac{2}{3}, \frac{2}{3})$; $(\frac{1}{2}, \sqrt{2})$.

***22.5** Let S be any set, and let R be a relation on S. Then R is an *equivalence relation* (a) if $(a, a) \in R$ for all $a \in S$; (b) if $(a, b) \in R$, then $(b, a) \in R$; (c) if $(a, b) \in R$ and $(b, c) \in R$, then $(a, c) \in R$. For a given relation R, it is customary to use a notation like $a\,R\,b$ or $a \sim b$ to stand for $(a, b) \in R$. Which of the following relations are equivalence relations?

(a) Congruence of triangles ($\cong$) on the set of all triangles in the plane.
(b) Less than or equal to ($\leq$) on the set of all real numbers.
(c) $a - b$ is divisible by 5 on the set of all integers.
(d) a divides b on the set of all positive integers.

***22.6** Let S be a set and let $\sim$ be an equivalence relation on S. For $a \in S$, denote by $[a]$ the set of all $x \in S$ such that $x \sim a$. In set notation,

$$[a] = \{x \in S \mid x \sim a\}.$$

The set $[a]$ is called the *equivalence class* of the element a with respect to the relation $\sim$. Prove the following properties of equivalence classes. Let $a \in S$ and $b \in S$. Then (a) $a \in [a]$; (b) $[a] = [b]$ if and only if $a \sim b$; (c) either $[a] = [b]$ or else $[a]$ and $[b]$ have no elements in common.

22.7 Let Z be the set of all integers, and let m be a fixed positive integer. For $a, b \in Z$, define $(a, b) \in R$ if and only if $a - b$ is divisible by m. Prove that R is an equivalence relation. Describe the equivalence classes of this relation.

22.8 Show that in the set $\mathfrak{M}_n$ of all n by n matrices, every nonsingular matrix P is equivalent to the identity matrix I_n.

22.9 Let $A \in {}_m\mathfrak{M}_n$, and let $\mathbf{O}$ be the zero matrix in ${}_m\mathfrak{M}_n$. Prove that A is equivalent to $\mathbf{O}$ if and only if $A = \mathbf{O}$.

22.10 Let A and B be in ${}_m\mathfrak{M}_n$. Prove that A is equivalent to B if and only if A' is equivalent to B' in ${}_n\mathfrak{M}_m$.

22.11 The matrix of the change of basis from the basis $E^{(3)}$ in $\mathcal{V}_3$ to the basis $\{(1, 2, 3), (3, 2, 1), (1, 2, 1)\}$ in $\mathcal{V}_3$ is

$$P = \begin{bmatrix} 1 & 2 & 3 \\ 3 & 2 & 1 \\ 1 & 2 & 1 \end{bmatrix}$$

where

$$P^{-1} = \begin{bmatrix} 0 & \frac{1}{2} & -\frac{1}{2} \\ -\frac{1}{4} & -\frac{1}{4} & 1 \\ \frac{1}{2} & 0 & -\frac{1}{2} \end{bmatrix}.$$

Let L be the linear transformation of $\mathcal{V}_3$ into itself that has matrix

$$A = \begin{bmatrix} 3 & 1 & 3 \\ 1 & -1 & 0 \\ 0 & 0 & 1 \end{bmatrix}$$

with respect to the basis $E^{(3)}$. Find the matrix of L with respect to the basis $\{(1, 2, 3), (3, 2, 1), (1, 2, 1)\}$. Find the coordinates of $L(\mathbf{E}_1^{(3)})$, $L(\mathbf{E}_2^{(3)})$, and $L(\mathbf{E}_3^{(3)})$ with respect to each basis.

22.12 Prove that similarity is an equivalence relation on $\mathfrak{M}_n$.

22.13 If two matrices in $\mathfrak{M}_n$ are similar, then they are equivalent (Definition 22.2). Show by an example that the converse of this statement is false.

22.14 Suppose that

$$P^{-1}AP = \begin{bmatrix} k_1 & 0 & \cdots & 0 \\ 0 & k_2 & & \vdots \\ \vdots & & \ddots & 0 \\ 0 & \cdots & 0 & k_n \end{bmatrix}$$

for $A \in \mathfrak{A}_n$ and $P = [p_{i,j}]$ is a nonsingular matrix in $\mathfrak{A}_n$. Prove that

$$A \begin{bmatrix} p_{1,j} \\ p_{2,j} \\ \vdots \\ p_{n,j} \end{bmatrix} = k_j \cdot \begin{bmatrix} p_{1,j} \\ p_{2,j} \\ \vdots \\ p_{n,j} \end{bmatrix}$$

for $j = 1, 2, \ldots, n$.

23 elementary transformation matrices

In the next three sections, we systematically investigate the simplification of m by n matrices under the relation of equivalence. The first important result, which we obtain in this section, is that two m by n matrices are equivalent if and only if one can be carried into the other by a sequence of elementary transformations. To get this result, we must show that an elementary transformation on a matrix can be accomplished by matrix multiplication, and that the multiplication of a matrix by a nonsingular matrix is equivalent to performing a sequence of elementary transformations on the matrix.

The three types of elementary row transformations that are performed on an m by n matrix A can be accomplished by multiplying the matrix A by certain simple matrices that we will now describe.

Let $I_m^{(i,j)}$ be the m by m matrix obtained from the identity matrix I_m by interchanging the ith and jth rows of I_m. Thus,

$$I_m^{(i,j)} = \begin{array}{cc}
 & \begin{array}{ccccc} \phantom{\begin{matrix}1&&\\&\ddots&\\&&1\end{matrix}} & i & \phantom{\begin{matrix}1&&\\&\ddots&\\&&1\end{matrix}} & j & \phantom{\begin{matrix}1&&\\&\ddots&\\&&1\end{matrix}} \end{array} \\
\begin{array}{c} \\ \\ \\ i \\ \\ \\ \\ j \\ \\ \\ \\ \end{array} &
\left[\begin{array}{ccccc}
\begin{matrix}1 & & \\ & \ddots & \\ & & 1\end{matrix} & \vdots & & \vdots & \\
\cdots & 0 & \cdots & 1 & \\
 & \vdots & \begin{matrix}1 & & \\ & \ddots & \\ & & 1\end{matrix} & \vdots & \\
\cdots & 1 & \cdots & 0 & \\
 & & & & \begin{matrix}1 & & \\ & \ddots & \\ & & 1\end{matrix}
\end{array}\right]
\end{array} . \tag{23–1}$$

Note that $I_m^{(i,j)}$ can also be obtained from I_m by interchanging the ith and jth columns. The matrix $I_m^{(i,j)}$ is called an *elementary transformation matrix of Type I*. It follows from the definition of matrix multiplication that an elementary transformation matrix of Type I has the following property:

(23.1) Let A be an m by n matrix. The matrix $I_m^{(i,j)}A$ is the matrix obtained from A by interchanging the ith and jth rows of A. The matrix $AI_n^{(i,j)}$ is the matrix obtained from A by interchanging the ith and jth columns of A.

The interchange of two columns of a matrix A is called an *elementary column transformation of Type I*. By 23.1, such a transformation can be accomplished on an m by n matrix A by performing this same column transformation on I_n to obtain $I_n^{(i,j)}$, and then postmultiplying A by $I_n^{(i,j)}$.

As a corollary of 23.1, we find that a matrix $I_m^{(i,j)}$ is nonsingular, and that it is its own inverse. Indeed, $I_m^{(i,j)}I_m^{(i,j)}$ is the matrix obtained from $I_m^{(i,j)}$ by interchanging the ith and jth rows of $I_m^{(i,j)}$, yielding the identity matrix I_m.

Let $I_m^{(i,j,c)}$ be the m by m matrix obtained from I_m by multiplying each element of the ith row of I_m by the number c and adding it to the corresponding element of the jth row ($i \neq j$). Then,

$$I_m^{(i,j,c)} = \begin{array}{c} \\ \\ \\ i \\ \\ j \\ \\ \\ \end{array}\overset{\begin{array}{ccccccc} & & i & & j & & \end{array}}{\begin{bmatrix} 1 & & \cdot & & \cdot & & \\ & \ddots & \vdots & & \vdots & & \\ \cdots & & 1 & & & & \\ & & \vdots & \ddots & & & \\ \cdots & & c & \cdots & 1 & & \\ & & & & & \ddots & \\ & & & & & & 1 \end{bmatrix}}. \qquad (23\text{–}2)$$

The matrix $I_m^{(i,j,c)}$ can also be described as the matrix obtained from I_m by multiplying each element of the jth column of I_m by the number c and adding it to the corresponding element of the ith column. The matrix $I_m^{(i,j,c)}$ is called an *elementary transformation matrix of Type II*, because of the following result:

(23.2) Let A be an m by n matrix. The matrix $I_m^{(i,j,c)}A$ is the matrix obtained from A by multiplying each element of the ith row of A by the number c and adding the product to the corresponding element of the jth row. The matrix $AI_n^{(i,j,c)}$ is the matrix obtained from A by multiplying each element of the jth column of A by the number c and adding the product to the corresponding element of the ith column.

The proof of 23.2 is an exercise in matrix multiplication. The transformation of a matrix A accomplished by postmultiplying A by an elementary transformation matrix of Type II is called an *elementary column transformation of Type II*. As an example of 23.2, we have

$$I_4^{(1,3,c)}A = \begin{bmatrix} 1 & 0 & 0 & 0 \\ 0 & 1 & 0 & 0 \\ c & 0 & 1 & 0 \\ 0 & 0 & 0 & 1 \end{bmatrix} \begin{bmatrix} a_{1,1} & a_{1,2} & a_{1,3} \\ a_{2,1} & a_{2,2} & a_{2,3} \\ a_{3,1} & a_{3,2} & a_{3,3} \\ a_{4,1} & a_{4,2} & a_{4,3} \end{bmatrix}$$

$$= \begin{bmatrix} a_{1,1} & a_{1,2} & a_{1,3} \\ a_{2,1} & a_{2,2} & a_{2,3} \\ ca_{1,1} + a_{3,1} & ca_{1,2} + a_{3,2} & ca_{1,3} + a_{3,3} \\ a_{4,1} & a_{4,2} & a_{4,3} \end{bmatrix}$$

and

$$AI_3^{(1,2,c)} = \begin{bmatrix} a_{1,1} & a_{1,2} & a_{1,3} \\ a_{2,1} & a_{2,2} & a_{2,3} \\ a_{3,1} & a_{3,2} & a_{3,3} \\ a_{4,1} & a_{4,2} & a_{4,3} \end{bmatrix} \begin{bmatrix} 1 & 0 & 0 \\ c & 1 & 0 \\ 0 & 0 & 1 \end{bmatrix} = \begin{bmatrix} a_{1,1} + ca_{1,2} & a_{1,2} & a_{1,3} \\ a_{2,1} + ca_{2,2} & a_{2,2} & a_{2,3} \\ a_{3,1} + ca_{3,2} & a_{3,2} & a_{3,3} \\ a_{4,1} + ca_{4,2} & a_{4,2} & a_{4,3} \end{bmatrix}.$$

The elementary transformation matrices of Type II are nonsingular, and the inverse of $I_m^{(i,j,c)}$ is $I_m^{(i,j,-c)}$, which is a matrix of the same type. To see this, note that by 23.2 $I_m^{(i,j,-c)}I_m^{(i,j,c)}$ is a matrix obtained from $I_m^{(i,j,c)}$ by multiplying each element of the ith row of $I_m^{(i,j,c)}$ by $-c$ and adding the result to the corresponding element of the jth row. Thus, $I_m^{(i,j,-c)}I_m^{(i,j,c)} = I_m$.

Let ${}^{(c)}I_m^{(i)}$ be the matrix obtained from I_m by multiplying each element of the ith row of I_m by the number $c \neq 0$ (or multiplying each element of the ith column of I_m by $c \neq 0$). The matrix ${}^{(c)}I_m^{(i)}$ is an *elementary transformation matrix of Type III*. We have,

$$ {}^{(c)}I_m^{(i)} = \begin{array}{c} \\ \\ \\ \\ \\ i \\ \\ \\ \\ \\ \end{array} \overset{\displaystyle i}{\begin{bmatrix} 1 & & & & \cdot & & & & \\ & \cdot & & & \cdot & & & & \\ & & \cdot & & \cdot & & & & \\ & & & \cdot & & & & & \\ & & & 1 & & & & & \\ & \cdots & & & c & & & & \\ & & & & & 1 & & & \\ & & & & & & \cdot & & \\ & & & & & & & \cdot & \\ & & & & & & & & \cdot \\ & & & & & & & & & 1 \end{bmatrix}} . \qquad (23\text{–}3)$$

(23.3) Let A be an m by n matrix. The matrix ${}^{(c)}I_m^{(i)}A$ is the matrix obtained from A by multiplying each element of the ith row of A by c. The matrix $A\,{}^{(c)}I_n^{(i)}$ is the matrix obtained from A by multiplying each element of ith column of A by c.

The proof of 23.3 is left as an exercise (Exercise 23.3). The multiplication of each element of a column of a matrix A by the same nonzero number c is called an *elementary column transformation of Type III.*

A matrix ${}^{(c)}I_m^{(i)}$ is nonsingular, since it follows from 23.3 that ${}^{(1/c)}I_m^{(i)}\,{}^{(c)}I_m^{(i)} = I_m$.

Summarizing the results 23.1, 23.2, and 23.3, we find that an elementary row or column transformation of any type can be performed on a matrix A as follows: Perform the transformation on an identity matrix of the proper size, obtaining an elementary transformation matrix. Then, for a row transformation, premultiply A by this elementary transformation matrix; for a column transformation, postmultiply A by this matrix. We have also discovered that each type of elementary transformation matrix is nonsingular and has an inverse that is a matrix of the same type.

As an application of these results, we will give a method for calculating the inverse of a nonsingular matrix. By Theorem 20.3, a matrix P can be transformed into a matrix in echelon form by a sequence of elementary row transformations. Thus, by the discussion of the preceding paragraph, there is a sequence of elementary transformation matrices, which we will denote by $E_1, E_2, \ldots, E_k$, such that $E_kE_{k-1} \cdots E_2E_1P = P^*$, where P^* is in echelon form. If P is an n by n nonsingular matrix, then, by Theorem 20.7, the row rank of P is n. Since P^* is obtained from P by elementary row transformations, the row rank of P^* is n. Moreover, the row rank of the echelon matrix P^* is the associated integer k of the echelon form. Hence, $k = n$ and

$$P^* = \begin{bmatrix} 1 & p^*_{1,2} & p^*_{1,3} & \cdots & & p^*_{1,n} \\ 0 & 1 & p^*_{2,3} & \cdots & & p^*_{2,n} \\ \cdot & 0 & \cdot & & & \cdot \\ \cdot & & \cdot & \cdot & & \cdot \\ \cdot & & & \cdot & \cdot & \cdot \\ & & & \cdot & 1 & p^*_{n-1,n} \\ 0 & \cdot & \cdot & \cdot & 0 & 1 \end{bmatrix}.$$

has 1 in each diagonal position and 0 in every position below the diagonal. Now it is possible to transform P^* into I_n by a sequence of elementary row transformations. First multiply each element of the second row of P^* by $-p^*_{1,2}$ and add the product to the corresponding element of the first row. This puts 0 in the (1, 2)-position, and the elements $p^{**}_{1,j} = p^*_{1,j} - p^*_{1,2}p^*_{2,j}$ in the $(1, j)$-position for $j = 3, \ldots, n$. Next, multiply each element of the third row of the resulting matrix by $-p^{**}_{1,3}$, and add the product to the corresponding element of the first row; multiply each element of the third row by $-p^*_{2,3}$ and add the product to the corresponding element of the second row. We now have a matrix with 1 on the diagonal in the first three columns and 0 elsewhere in these columns. Clearly, this process can be continued until we arrive at the identity matrix I_n. Thus, there is a sequence of elementary transformation matrices, $E_{k+1}, E_{k+2}, \ldots, E_t$, such that $E_tE_{t-1} \cdots E_{k+2}E_{k+1}P^* = I_n$. Hence,

$$(E_tE_{t-1} \cdots E_{k+1}E_k \cdots E_2E_1)P = I_n. \tag{23–4}$$

Therefore, $P^{-1} = E_tE_{t-1} \cdots E_{k+1}E_k \cdots E_2E_1$. Now E_1 is the matrix obtained from I_n by performing the first elementary row transformation on I_n, E_2E_1 is the matrix obtained from E_1 by performing the second elementary row transformation on E_1, and so forth. Finally, P^{-1} is the matrix obtained from $E_{t-1} \cdots E_{k+1}E_k \cdots E_2E_1$ by performing the last elementary row transformation on $E_{t-1} \cdots E_{k+1}E_k \cdots E_2E_1$. Thus, by this application of 23.1, 23.2, and 23.3, we have proved the following:

(23.4) The inverse of a nonsingular n by n matrix P is obtained by performing the same sequence of elementary row transformations on the identity matrix I_n as that which transforms the matrix P into I_n.

EXAMPLE 1 Let

$$P = \begin{bmatrix} 0 & 0 & 2 & -2 \\ 0 & 0 & 5 & 3 \\ 3 & -1 & 0 & 0 \\ 1 & 4 & 0 & 0 \end{bmatrix}.$$

The matrix P is transformed into I_4 by the following sequence of elementary row transformations:

$$P \xrightarrow{\text{Type I}} \begin{bmatrix} 1 & 4 & 0 & 0 \\ 0 & 0 & 5 & 3 \\ 3 & -1 & 0 & 0 \\ 0 & 0 & 2 & -2 \end{bmatrix} \xrightarrow{\text{Type II}} \begin{bmatrix} 1 & 4 & 0 & 0 \\ 0 & 0 & 5 & 3 \\ 0 & -13 & 0 & 0 \\ 0 & 0 & 2 & -2 \end{bmatrix} \xrightarrow{\text{Type I}}$$

$$\begin{bmatrix} 1 & 4 & 0 & 0 \\ 0 & -13 & 0 & 0 \\ 0 & 0 & 5 & 3 \\ 0 & 0 & 2 & -2 \end{bmatrix} \xrightarrow{\text{Type III}} \begin{bmatrix} 1 & 4 & 0 & 0 \\ 0 & 1 & 0 & 0 \\ 0 & 0 & 5 & 3 \\ 0 & 0 & 2 & -2 \end{bmatrix} \xrightarrow{\text{Type III}} \begin{bmatrix} 1 & 4 & 0 & 0 \\ 0 & 1 & 0 & 0 \\ 0 & 0 & 1 & 3/5 \\ 0 & 0 & 2 & -2 \end{bmatrix}$$

$$\xrightarrow{\text{Type II}} \begin{bmatrix} 1 & 4 & 0 & 0 \\ 0 & 1 & 0 & 0 \\ 0 & 0 & 1 & 3/5 \\ 0 & 0 & 0 & -16/5 \end{bmatrix} \xrightarrow{\text{Type III}} \begin{bmatrix} 1 & 4 & 0 & 0 \\ 0 & 1 & 0 & 0 \\ 0 & 0 & 1 & 3/5 \\ 0 & 0 & 0 & 1 \end{bmatrix} \xrightarrow{\text{Type II}}$$

$$\begin{bmatrix} 1 & 0 & 0 & 0 \\ 0 & 1 & 0 & 0 \\ 0 & 0 & 1 & 3/5 \\ 0 & 0 & 0 & 1 \end{bmatrix} \xrightarrow{\text{Type II}} \begin{bmatrix} 1 & 0 & 0 & 0 \\ 0 & 1 & 0 & 0 \\ 0 & 0 & 1 & 0 \\ 0 & 0 & 0 & 1 \end{bmatrix}$$

We now perform these same elementary row transformations on the identity matrix I_4:

$$I_4 \xrightarrow{\text{Type I}} \begin{bmatrix} 0 & 0 & 0 & 1 \\ 0 & 1 & 0 & 0 \\ 0 & 0 & 1 & 0 \\ 1 & 0 & 0 & 0 \end{bmatrix} \xrightarrow{\text{Type II}} \begin{bmatrix} 0 & 0 & 0 & 1 \\ 0 & 1 & 0 & 0 \\ 0 & 0 & 1 & -3 \\ 1 & 0 & 0 & 0 \end{bmatrix} \xrightarrow{\text{Type I}} \begin{bmatrix} 0 & 0 & 0 & 1 \\ 0 & 0 & 1 & -3 \\ 0 & 1 & 0 & 0 \\ 1 & 0 & 0 & 0 \end{bmatrix}$$

$$\xrightarrow{\text{Type III}} \begin{bmatrix} 0 & 0 & 0 & 1 \\ 0 & 0 & -1/13 & 3/13 \\ 0 & 1 & 0 & 0 \\ 1 & 0 & 0 & 0 \end{bmatrix} \xrightarrow{\text{Type III}} \begin{bmatrix} 0 & 0 & 0 & 1 \\ 0 & 0 & -1/13 & 3/13 \\ 0 & 1/5 & 0 & 0 \\ 1 & 0 & 0 & 0 \end{bmatrix} \xrightarrow{\text{Type II}}$$

$$\begin{bmatrix} 0 & 0 & 0 & 1 \\ 0 & 0 & -1/13 & 3/13 \\ 0 & 1/5 & 0 & 0 \\ 1 & -2/5 & 0 & 0 \end{bmatrix} \xrightarrow{\text{Type III}} \begin{bmatrix} 0 & 0 & 0 & 1 \\ 0 & 0 & -1/13 & 3/13 \\ 0 & 1/5 & 0 & 0 \\ -5/16 & 1/8 & 0 & 0 \end{bmatrix} \xrightarrow{\text{Type II}}$$

$$\begin{bmatrix} 0 & 0 & 4/13 & 1/13 \\ 0 & 0 & -1/13 & 3/13 \\ 0 & 1/5 & 0 & 0 \\ -5/16 & 1/8 & 0 & 0 \end{bmatrix} \xrightarrow{\text{Type II}} \begin{bmatrix} 0 & 0 & 4/13 & 1/13 \\ 0 & 0 & -1/13 & 3/13 \\ 3/16 & 1/8 & 0 & 0 \\ -5/16 & 1/8 & 0 & 0 \end{bmatrix}$$

Thus,

$$P^{-1} = \begin{bmatrix} 0 & 0 & 4/13 & 1/13 \\ 0 & 0 & -1/13 & 3/13 \\ 3/16 & 1/8 & 0 & 0 \\ -5/16 & 1/8 & 0 & 0 \end{bmatrix}$$

which can be checked directly by matrix multiplication. In practice, the work is shortened by performing each elementary transformation on I_4 as it is performed on the matrix P.

EXAMPLE 2 Find the inverse of the matrix

$$P = \begin{bmatrix} 1 & 2 & 3 \\ 3 & 2 & 1 \\ 1 & 2 & 1 \end{bmatrix}$$

$$I_3 = \begin{bmatrix} 1 & 0 & 0 \\ 0 & 1 & 0 \\ 0 & 0 & 1 \end{bmatrix} \qquad \begin{bmatrix} 1 & 2 & 3 \\ 3 & 2 & 1 \\ 1 & 2 & 1 \end{bmatrix} = P$$

$$\downarrow \quad \text{Type II} \quad \downarrow$$

$$\begin{bmatrix} 1 & 0 & 0 \\ -3 & 1 & 0 \\ 0 & 0 & 1 \end{bmatrix} \qquad \begin{bmatrix} 1 & 2 & 3 \\ 0 & -4 & -8 \\ 1 & 2 & 1 \end{bmatrix}$$

$$\downarrow \quad \text{Type II} \quad \downarrow$$

$$\begin{bmatrix} 1 & 0 & 0 \\ -3 & 1 & 0 \\ -1 & 0 & 1 \end{bmatrix} \qquad \begin{bmatrix} 1 & 2 & 3 \\ 0 & -4 & -8 \\ 0 & 0 & -2 \end{bmatrix}$$

$$\downarrow \quad \text{Type III} \quad \downarrow$$

$$\begin{bmatrix} 1 & 0 & 0 \\ 3/4 & -1/4 & 0 \\ -1 & 0 & 1 \end{bmatrix} \qquad \begin{bmatrix} 1 & 2 & 3 \\ 0 & 1 & 2 \\ 0 & 0 & -2 \end{bmatrix}$$

$$\downarrow \quad \text{Type III} \quad \downarrow$$

$$\begin{bmatrix} 1 & 0 & 0 \\ 3/4 & -1/4 & 0 \\ 1/2 & 0 & -1/2 \end{bmatrix} \qquad \begin{bmatrix} 1 & 2 & 3 \\ 0 & 1 & 2 \\ 0 & 0 & 1 \end{bmatrix}$$

$$\downarrow \quad \text{Type II} \quad \downarrow$$

$$\begin{bmatrix} -1/2 & 1/2 & 0 \\ 3/4 & -1/4 & 0 \\ 1/2 & 0 & -1/2 \end{bmatrix} \qquad \begin{bmatrix} 1 & 0 & -1 \\ 0 & 1 & 2 \\ 0 & 0 & 1 \end{bmatrix}$$

$$\downarrow \quad \text{Type II} \quad \downarrow$$

$$\begin{bmatrix} 0 & 1/2 & -1/2 \\ 3/4 & -1/4 & 0 \\ 1/2 & 0 & -1/2 \end{bmatrix} \qquad \begin{bmatrix} 1 & 0 & 0 \\ 0 & 1 & 2 \\ 0 & 0 & 1 \end{bmatrix}$$

$$\downarrow \quad \text{Type II} \quad \downarrow$$

$$P^{-1} = \begin{bmatrix} 0 & \frac{1}{2} & -\frac{1}{2} \\ -\frac{1}{4} & -\frac{1}{4} & 1 \\ \frac{1}{2} & 0 & -\frac{1}{2} \end{bmatrix} \qquad \begin{bmatrix} 1 & 0 & 0 \\ 0 & 1 & 0 \\ 0 & 0 & 1 \end{bmatrix} = I_3$$

This discussion yields the following important theorem and corollary:

(23.5) theorem A nonsingular matrix P is a product of elementary transformation matrices.

PROOF By equation (23–4),

$$\begin{aligned} P &= (E_t E_{t-1} \cdots E_{k+1} E_k \cdots E_2 E_1)^{-1} \\ &= E_1^{-1} E_2^{-1} \cdots E_k^{-1} E_{k+1}^{-1} \cdots E_{t-1}^{-1} E_t^{-1}. \end{aligned}$$

Theorem 23.5 follows from the fact that the inverse of each elementary transformation matrix E_i is again an elementary transformation matrix. ∎

(23.6) corollary An m by n matrix B is equivalent to an m by n matrix A if and only if A can be transformed into B by a sequence of elementary row and column transformations.

PROOF Suppose first that B is equivalent to A. By Definition 22.2, $B = PAQ$, where P is a nonsingular m by m matrix and Q is a nonsingular n by n matrix. By Theorem 23.5, P and Q can be expressed as products of elementary transformation matrices. Thus,

$$B = E_t E_{t-1} \cdots E_1 A \bar{E}_1 \bar{E}_2 \cdots \bar{E}_s \tag{23–5}$$

where the E_i and $\bar{E}_j$ are elementary transformation matrices. By 23.1, 23.2, and 23.3, premultiplication by each E_i performs a row transformation and postmultiplication by each $\bar{E}_j$ performs a column transformation. Thus, A is transformed into B by a sequence of elementary transformations.

Conversely, if B is obtained from A by a sequence of elementary row and column transformations, then each of these transformations can be accomplished by multiplication by an elementary transformation matrix. Thus, equation (23–5) holds. Since each E_i and $\bar{E}_j$ is nonsingular, it follows from 19.8 that $P = E_t E_{t-1} \cdots E_1$ and $Q = \bar{E}_1 \bar{E}_2 \cdots \bar{E}_s$ are nonsingular. Hence, $B = PAQ$, and B is equivalent to A. ∎

EXAMPLE 3 The matrix

$$A = \begin{bmatrix} 0 & -3 & -1 & 1 \\ 1 & 5 & 4 & -1 \\ 1 & 2 & 3 & 0 \end{bmatrix}$$

can be transformed into the matrix

$$B = \begin{bmatrix} 1 & 0 & 0 & 0 \\ 0 & 1 & 0 & 0 \\ 0 & 0 & 0 & 0 \end{bmatrix}$$

by a sequence of elementary row and column transformations:

$$A = \begin{bmatrix} 0 & -3 & -1 & 1 \\ 1 & 5 & 4 & -1 \\ 1 & 2 & 3 & 0 \end{bmatrix} \xrightarrow[\text{Type I}]{\text{Row}} \begin{bmatrix} 1 & 5 & 4 & -1 \\ 0 & -3 & -1 & 1 \\ 1 & 2 & 3 & 0 \end{bmatrix} \xrightarrow[\text{Type II}]{\text{Row}}$$

$$\begin{bmatrix} 1 & 5 & 4 & -1 \\ 0 & -3 & -1 & 1 \\ 0 & -3 & -1 & 1 \end{bmatrix} \xrightarrow[\text{Type II}]{\text{Column}} \begin{bmatrix} 1 & 0 & 4 & -1 \\ 0 & -3 & -1 & 1 \\ 0 & -3 & -1 & 1 \end{bmatrix} \xrightarrow[\text{Type II}]{\text{Column}}$$

$$\begin{bmatrix} 1 & 0 & 0 & -1 \\ 0 & -3 & -1 & 1 \\ 0 & -3 & -1 & 1 \end{bmatrix} \xrightarrow[\text{Type II}]{\text{Column}} \begin{bmatrix} 1 & 0 & 0 & 0 \\ 0 & -3 & -1 & 1 \\ 0 & -3 & -1 & 1 \end{bmatrix} \xrightarrow[\text{Type III}]{\text{Row}}$$

$$\begin{bmatrix} 1 & 0 & 0 & 0 \\ 0 & 1 & 1/3 & -1/3 \\ 0 & -3 & -1 & 1 \end{bmatrix} \xrightarrow[\text{Type II}]{\text{Row}} \begin{bmatrix} 1 & 0 & 0 & 0 \\ 0 & 1 & 1/3 & -1/3 \\ 0 & 0 & 0 & 0 \end{bmatrix} \xrightarrow[\text{Type II}]{\text{Column}}$$

$$\begin{bmatrix} 1 & 0 & 0 & 0 \\ 0 & 1 & 0 & -1/3 \\ 0 & 0 & 0 & 0 \end{bmatrix} \xrightarrow[\text{Type II}]{\text{Column}} \begin{bmatrix} 1 & 0 & 0 & 0 \\ 0 & 1 & 0 & 0 \\ 0 & 0 & 0 & 0 \end{bmatrix} = B.$$

Therefore, by Corollary 23.6, the matrices A and B are equivalent.

exercises

23.1 Construct the following elementary transformation matrices: $I_4^{(1,3)}$; $I_5^{(2,4,1/2)}$; ${}^{(\sqrt{2})}I_3^{(2)}$; $I_3^{(1,2)}$; $I_4^{(1,4,10)}$. What is the inverse of each matrix?

23.2 Let

$$A = \begin{bmatrix} 2 & -1 & 0 \\ 7 & 5 & 1 \\ -3 & 4 & 2 \\ 1 & 3 & 1 \end{bmatrix} \qquad B = \begin{bmatrix} 1/2 & 2/3 & -1 & 4 & 3/2 \\ -5 & 0 & 1 & 1/2 & 6 \\ 5/3 & 2 & -3 & 0 & 1 \end{bmatrix}.$$

Find the following products without multiplying the matrices: $I_4^{(1,2)}A$; $A\,{}^{(1/3)}I_3^{(2)}$; ${}^{(1/3)}I_3^{(2)}B$; $BI_5^{(1,3,6)}$; $I_4^{(1,3)}AI_3^{(1,2)}$; $I_3^{(1,2)}B\,{}^{(-1)}I_5^{(3)}$.

23.3 Prove statements 23.1, 23.2, and 23.3.

23.4 Let A and B be the matrices of Exercise 23.2. Find nonsingular matrices P_1 and P_2 such that P_1A and P_2B are in echelon form.

23.5 Find the inverses of the following nonsingular matrices:

(a) $\begin{bmatrix} 0 & 0 & 0 & 1 \\ 0 & 0 & 2 & 0 \\ 0 & 3 & 0 & 0 \\ 4 & 0 & 0 & 0 \end{bmatrix}$ (b) $\begin{bmatrix} 2 & -1 & 3 \\ 1 & 4 & 2 \\ 0 & 1 & -2 \end{bmatrix}$

(c) $\begin{bmatrix} 1 & 0 & 0 & 0 \\ 1/16 & 1/3 & 0 & 0 \\ -2 & 0 & 1 & 0 \\ 1/6 & 0 & 0 & 1/2 \end{bmatrix}$ (d) $\begin{bmatrix} 1 & 1 & 1 \\ 1 & 0 & 1 \\ 0 & 0 & 1 \end{bmatrix}$

23.6 Express the matrices of Exercise 23.5 as products of elementary transformation matrices.

23.7 Show that the following pairs of matrices are equivalent:

(a) $\begin{bmatrix} 2 & 2 & 2 & 2 \\ 2 & 2 & 2 & 2 \\ 2 & 2 & 2 & 2 \\ 2 & 2 & 2 & 2 \end{bmatrix}$ and $\begin{bmatrix} 1 & 0 & 0 & 0 \\ 0 & 0 & 0 & 0 \\ 0 & 0 & 0 & 0 \\ 0 & 0 & 0 & 0 \end{bmatrix}$

(b) $\begin{bmatrix} -2 & 3 & 1 \\ 1 & 2 & 4 \\ 1/2 & 9/2 & 13/2 \end{bmatrix}$ and $\begin{bmatrix} 1 & 0 & 0 \\ 0 & 1 & 0 \\ 0 & 0 & 0 \end{bmatrix}$

(c) $\begin{bmatrix} -2 & 1 \\ 6 & -3 \\ 0 & 1 \\ 7 & 5 \end{bmatrix}$ and $\begin{bmatrix} 1 & 0 \\ 0 & 1 \\ 0 & 0 \\ 0 & 0 \end{bmatrix}$

(d) $\begin{bmatrix} 1 & 2 & 3 & 4 \\ 2 & 3 & 4 & 1 \\ 3 & 4 & 1 & 2 \\ 4 & 1 & 2 & 3 \end{bmatrix}$ and $\begin{bmatrix} 1 & 0 & 0 & 0 \\ 0 & 1 & 0 & 0 \\ 0 & 0 & 1 & 0 \\ 0 & 0 & 0 & 1 \end{bmatrix}$

23.8 Show that the transpose of an elementary transformation matrix is an elementary transformation matrix of the same type.

23.9 Prove that a square matrix is nonsingular if and only if it is equivalent to the identity matrix.

24 rank of a matrix

The concept of the row rank of a matrix A has been useful in several previous discussions. Now we examine its relation to two additional definitions of *rank*.

(24.1) definition Let $\mathcal{V}$ and $\mathcal{W}$ be finite dimensional vector spaces, and let L be a linear transformation of $\mathcal{V}$ into $\mathcal{W}$. The *rank* of L is the dimension of the range space $L(\mathcal{V})$ of L.

(24.2) definition Let $A = [a_{i,j}]$ be an m by n matrix. The *column rank of A* is the dimension of the *column space of A*, which is the subspace of $\mathcal{V}_m$ spanned by the vectors $(a_{1,j}, a_{2,j}, \ldots, a_{m,j})$, $j = 1, 2, \ldots, n$, which are the columns of A.

It is not difficult to discover the relation between these two definitions.

(24.3) theorem Let L be a linear transformation of an n-dimensional vector space $\mathcal{V}$ into an m-dimensional vector space $\mathcal{W}$. The rank of L is equal to the column rank of any matrix A of L.

PROOF Suppose that A is the matrix of L with respect to bases $\{\mathbf{U}_1, \mathbf{U}_2, \ldots, \mathbf{U}_n\}$ in $\mathcal{V}$ and $\{\mathbf{W}_1, \mathbf{W}_2, \ldots, \mathbf{W}_m\}$ in $\mathcal{W}$. It follows from the definition of A (see §18) that the jth column of A, for $j = 1, 2, \ldots, n$, gives the coordinates $(a_{1,j}, a_{2,j}, \ldots, a_{m,j})$ of $L(\mathbf{U}_j)$ with respect to the basis $\{\mathbf{W}_1, \mathbf{W}_2, \ldots, \mathbf{W}_m\}$. The range space $L(\mathcal{V})$ of L is spanned by the vectors $L(\mathbf{U}_1), L(\mathbf{U}_2), \ldots, L(\mathbf{U}_n)$. The isomorphism between $\mathcal{W}$ and the m-tuple space $\mathcal{V}_m$ pairs each vector $\mathbf{W} \in \mathcal{W}$ with the m-tuple of coordinates of $\mathbf{W}$ with respect to the basis $\{\mathbf{W}_1, \mathbf{W}_2, \ldots, \mathbf{W}_m\}$. Thus, each $L(\mathbf{U}_j)$ corresponds to the m-tuple $(a_{1,j}, a_{2,j}, \ldots, a_{m,j})$, $j = 1, 2, \ldots, n$. Therefore, since the range space $L(\mathcal{V})$ of L is spanned by the vectors $L(\mathbf{U}_1), L(\mathbf{U}_2), \ldots, L(\mathbf{U}_n)$ in $\mathcal{W}$, it corresponds to the column space of the matrix A, which is spanned by the vectors $(a_{1,1}, a_{2,1}, \ldots, a_{m,1})$, $(a_{1,2}, a_{2,2}, \ldots, a_{m,2}), \ldots, (a_{1,n}, a_{2,n}, \ldots, a_{m,n})$ in $\mathcal{V}_m$. That is, $L(\mathcal{V})$ is isomorphic to the column space of A (see Exercise 24.6). Hence, these vector spaces have the same dimension. By Definitions 24.1 and 24.2, the rank of L is equal to the column rank of A. ▮

Theorem 24.3 provides a relatively simple method for selecting a basis for $L(\mathcal{V})$, the range space of a linear transformation L. If r is the column rank of a matrix A of L, we select r linearly independent columns of A. These columns of A are the coordinates of the vectors forming a basis for $L(\mathcal{V})$.

(24.4) corollary If an m by n matrix B is equivalent to an m by n matrix A, then A and B have the same column rank.

PROOF If B is equivalent to A, then, by Theorem 22.1, A and B are matrices of the same linear transformation L of an n-dimensional space $\mathcal{V}$ into an m-dimensional space $\mathcal{W}$ with respect to suitable bases in these spaces. By Theorem 24.3, column rank of A = rank of L = column rank of B. ▮

We wish to show next that the column rank of a matrix is equal to the row rank. To do this, we first consider a matrix that is in echelon form.

(24.5) theorem Let A be an m by n matrix in echelon form. The column rank of A is equal to the row rank of A.

PROOF We have already observed that the row rank of a matrix $A = [a_{i,j}]$ in echelon form is the associated integer k of the echelon form. If $i > k$, then $a_{i,j} = 0$ for all j. That is, the columns of A as vectors in $\mathcal{V}_m$ are of the form $(a_{1,j}, a_{2,j}, \ldots, a_{k,j}, 0, 0, \ldots, 0)$ for $j = 1, 2, \ldots, n$. Thus, the subspace of $\mathcal{V}_m$ spanned by these vectors is isomorphic to a subspace of $\mathcal{V}_k$ (see Exercise 24.7). Therefore, the dimension of the column space of A is at most k. The columns $m_1, m_2, \ldots, m_k$ of A as vectors in $\mathcal{V}_m$ are $(1, 0, \ldots, 0)$, $(a_{1,m_2}, 1, 0, \ldots, 0), \ldots$, $(a_{1,m_k}, a_{2,m_k}, \ldots, a_{k-1,m_k}, 1, 0, \ldots, 0)$. It is easy to verify that these k vectors are linearly independent. Hence, the dimension of the column space of A is at least k. Therefore, this dimension is k, so that the column rank of $A = k =$ row rank of A. ∎

EXAMPLE 1 The 4 by 7 matrix

$$A = \begin{bmatrix} 0 & 1 & -7 & \tfrac{1}{2} & 2 & -3 & 6 \\ 0 & 0 & 0 & 1 & -5 & 4 & \tfrac{1}{3} \\ 0 & 0 & 0 & 0 & 0 & 1 & -1 \\ 0 & 0 & 0 & 0 & 0 & 0 & 0 \end{bmatrix}$$

is in echelon form with $k = 3$, $m_1 = 2$, $m_2 = 4$, $m_3 = 6$. The vectors $(0, 1, -7, -\tfrac{1}{2}, 2, -3, 6)$, $(0, 0, 0, 1, -5, 4, \tfrac{1}{3})$, $(0, 0, 0, 0, 0, 1, -1)$ are a basis of the row space of A. Thus, the row rank of A is 3. The vectors $(1, 0, 0, 0)$, $(\tfrac{1}{2}, 1, 0, 0)$, $(-3, 4, 1, 0)$ are a basis of the column space of A, so that the column rank of A is 3.

(24.6) corollary The column rank of an m by n matrix A is equal to the row rank of A.

PROOF Carry the matrix A into echelon form A^* by a sequence of elementary row transformations. Then row rank of $A =$ row rank of A^*. By Corollary 23.6, the matrix A is equivalent to the matrix A^*. Hence, by Corollary 24.4, column rank of $A =$ column rank of A^*. Since A^* is in echelon form, it follows from Theorem 24.5 that column rank of $A^* =$ row rank of A^*. Thus, row rank of $A =$ row rank of $A^* =$ column rank of $A^* =$ column rank of A. ∎

EXAMPLE 2 As an example of the result stated in Corollary 24.6, we check directly that the row rank of the 4 by 5 matrix

$$A = \begin{bmatrix} -6 & 1 & 0 & 3 & 2 \\ 2 & -4 & 3 & -7 & 0 \\ 0 & 1 & -2 & -1 & 5 \\ -4 & -1 & -1 & -6 & 12 \end{bmatrix}$$

is equal to the column rank of A. The row space of A is the subspace $\mathcal{S}$ of $\mathcal{V}_5$ spanned by the set $\{(-6, 1, 0, 3, 2), (2, -4, 3, -7, 0), (0, 1, -2, -1, 5), (-4, -1, -1, -6, 12)\}$. Since

$$(-4, -1, -1, -6, 12) = 1 \cdot (-6, 1, 0, 3, 2) + 1 \cdot (2, -4, 3, -7, 0) + 2 \cdot (0, 1, -2, -1, 5),$$

it follows that $\mathcal{S}$ is spanned by the set

$$S = \{(-6, 1, 0, 3, 2), (2, -4, 3, -7, 0), (0, 1, -2, -1, 5)\}.$$

It is routine to verify that S is a linearly independent set. Hence, S is a basis of $\mathcal{S}$, so that $\mathcal{S}$ has dimension three. That is, the row rank of A is 3.

The column space of A is the subspace $\mathcal{T}$ of $\mathcal{V}_4$ spanned by the set $\{(-6, 2, 0, -4), (1, -4, 1, -1), (0, 3, -2, -1), (3, -7, -1, -6), (2, 0, 5, 12)\}$. Since

$$(3, -7, -1, 6) = 1/13 \cdot (-6, 2, 0, -4) + 45/13 \cdot (1, -4, 1, -1) + 29/13 \cdot (0, 3, -2, -1)$$

and

$$(2, 0, 5, 12) = 25/26 \cdot (-6, 2, 0, -4) - 49/13 \cdot (1, -4, 1, -1) - 57/13 \cdot (0, 3, -2, -1),$$

it follows that $\mathcal{T}$ is spanned by the set

$$T = \{(-6, 2, 0, -4), (1, -4, 1, -1), (0, 3, -2, -1)\}.$$

Again, it is easy to show that T is a linearly independent set. Thus, T is a basis of $\mathcal{T}$, so that $\mathcal{T}$ has dimension three. Hence, the column rank of $A = 3 =$ row rank of A.

Since the row rank of any matrix A is the same as its column rank, we will refer to this number simply as the *rank of A*. By Theorem 24.3, the rank of A is the same as the rank of the linear transformation L defined by A with respect to chosen bases in two finite dimensional vector spaces, $\mathcal{V}$ and $\mathcal{W}$. By Corollary 24.4, equivalent matrices have the same rank. Earlier results that involved the row rank of a matrix can now be stated simply in terms of rank. For example, an n by n matrix A is nonsingular if and only if the rank of A is n (Theorem 20.7).

exercises

24.1 Let L be a linear transformation of $\mathcal{V}$ into $\mathcal{W}$, and let M be a linear transformation of $\mathcal{W}$ into $\mathcal{P}$. Prove that rank of $ML \leq$ rank of M and rank of $ML \leq$ rank of L.

24.2 Let A be an m by n matrix and B be an n by p matrix. Use Exercise 24.1 to prove that the rank of AB does not exceed the rank of either A or B.

24.3 Let A be an m by n matrix. Prove that A and A' have the same rank.

24.4 Find the rank of the following matrices:

(a) $\begin{bmatrix} -1 & 2 \\ 3 & 1 \\ 4 & 6 \\ 2 & 9 \end{bmatrix}$ (b) $\begin{bmatrix} -2 & 1 & 0 & 3 \\ 1 & 5 & -1 & 6 \\ -1 & 15 & -3 & 18 \end{bmatrix}$ (c) $\begin{bmatrix} 3 & 3 & 3 & 3 & 3 \\ 2 & 2 & 2 & 2 & 2 \\ 1 & 1 & 1 & 1 & 1 \\ 2 & 2 & 2 & 2 & 2 \\ 3 & 3 & 3 & 3 & 3 \end{bmatrix}$

(d) $[2 \quad -1 \quad 3 \quad 7 \quad 6]$ (e) $\begin{bmatrix} 2 & -2 & 2 & 5 & 1 \\ 1 & 1 & 1 & 3 & -1 \\ 3 & 5 & 4 & 2 & 4 \end{bmatrix}$

(f) $\begin{bmatrix} 3 & 2 & 1 & 0 & 0 \\ 0 & 1 & 2 & 0 & 0 \\ 1 & 4 & 7 & 2 & 3 \\ 5 & 3 & -1 & 1 & 2 \\ 6 & 1 & -2 & 3 & 6 \end{bmatrix}$ (g) $\begin{bmatrix} 2 & 6 & -5 & 8 \\ 4 & 3 & -1 & 7 \\ -1 & 1 & 6 & 0 \\ 5 & 2 & 4 & 7 \end{bmatrix}$

24.5 Find bases for the row space and the column space of each of the matrices given in Exercise 24.4.

24.6 Let L be a linear transformation of an n-dimensional vector space $\mathcal{V}$ into an m-dimensional vector space $\mathcal{W}$, and let A be any matrix of L. Prove that the range space $L(\mathcal{V})$ of L is isomorphic to the column space of A (see the proof of Theorem 24.3).

24.7 Let $\mathcal{S}$ be the set of all m-tuples of real numbers of the form $(a_1, a_2, \ldots, a_k, 0, 0, \ldots, 0)$, where $m \geq k$. Prove that $\mathcal{S}$ is a subspace of $\mathcal{V}_m$ which is isomorphic to $\mathcal{V}_k$.

25 canonical form of a matrix under equivalence

In this section we solve the problem of simplifying a linear transformation L of an n-dimensional space $\mathcal{V}$ into an m-dimensional space $\mathcal{W}$ by obtaining the so-called "rational canonical form" for an m by n matrix under equivalence.

Let D_r be the m by n matrix (for any m and n) that has 1 in the first r positions on the diagonal and zeros elsewhere. It is clear that $r \geq 0$ and that r is less than or equal to the smaller of the two dimensions m or n. For example, if $m = 3$, $n = 4$,

$$D_0 = \begin{bmatrix} 0 & 0 & 0 & 0 \\ 0 & 0 & 0 & 0 \\ 0 & 0 & 0 & 0 \end{bmatrix}, \quad D_1 = \begin{bmatrix} 1 & 0 & 0 & 0 \\ 0 & 0 & 0 & 0 \\ 0 & 0 & 0 & 0 \end{bmatrix}, \quad D_2 = \begin{bmatrix} 1 & 0 & 0 & 0 \\ 0 & 1 & 0 & 0 \\ 0 & 0 & 0 & 0 \end{bmatrix},$$

$$D_3 = \begin{bmatrix} 1 & 0 & 0 & 0 \\ 0 & 1 & 0 & 0 \\ 0 & 0 & 1 & 0 \end{bmatrix}.$$

We already know that an m by n matrix A can be carried into echelon form A^* by elementary row transformations. The following example shows how a matrix A^* in echelon form can be carried into a D_r matrix by elementary column transformations.

EXAMPLE 1 Let

$$A^* = \begin{bmatrix} 0 & 1 & 2/3 & -5 & 0 \\ 0 & 0 & 0 & 1 & 4 \\ 0 & 0 & 0 & 0 & 0 \\ 0 & 0 & 0 & 0 & 0 \end{bmatrix}$$

Then A^* is in echelon form with $k = 2$, $m_1 = 2$, $m_2 = 4$. Perform the following sequence of elementary column transformations on A^*.

$$A^* \xrightarrow{\text{Type II}} \begin{bmatrix} 0 & 1 & 0 & -5 & 0 \\ 0 & 0 & 0 & 1 & 4 \\ 0 & 0 & 0 & 0 & 0 \\ 0 & 0 & 0 & 0 & 0 \end{bmatrix} \xrightarrow{\text{Type II}} \begin{bmatrix} 0 & 1 & 0 & 0 & 0 \\ 0 & 0 & 0 & 1 & 4 \\ 0 & 0 & 0 & 0 & 0 \\ 0 & 0 & 0 & 0 & 0 \end{bmatrix} \xrightarrow{\text{Type II}}$$

$$\begin{bmatrix} 0 & 1 & 0 & 0 & 0 \\ 0 & 0 & 0 & 1 & 0 \\ 0 & 0 & 0 & 0 & 0 \\ 0 & 0 & 0 & 0 & 0 \end{bmatrix} \xrightarrow{\text{Type I}} \begin{bmatrix} 1 & 0 & 0 & 0 & 0 \\ 0 & 0 & 0 & 1 & 0 \\ 0 & 0 & 0 & 0 & 0 \\ 0 & 0 & 0 & 0 & 0 \end{bmatrix} \xrightarrow{\text{Type I}} \begin{bmatrix} 1 & 0 & 0 & 0 & 0 \\ 0 & 1 & 0 & 0 & 0 \\ 0 & 0 & 0 & 0 & 0 \\ 0 & 0 & 0 & 0 & 0 \end{bmatrix} = D_2.$$

The method employed in Example 1 can be used to carry any matrix A^* in echelon form into a D_r matrix. The nonzero elements other than the leading 1's in each row can be replaced by 0's by elementary column transformations of Type II. Then the columns can be rearranged by Type I elementary column transformations to put the 1's on the diagonal. By 23.1 and 23.2, these column transformations can be accomplished by postmultiplying A^* by elementary transformation matrices, and the product of these elementary transformation matrices is a nonsingular matrix Q. Hence, $A^*Q = D_r$. The matrix Q is obtained by performing the same column transformations on the identity matrix I_n as are performed on A^*. In Example 1,

$$Q = \begin{bmatrix} 0 & 0 & 0 & 1 & 0 \\ 1 & 5 & -2/3 & 0 & -20 \\ 0 & 0 & 1 & 0 & 0 \\ 0 & 1 & 0 & 0 & -4 \\ 0 & 0 & 0 & 0 & 1 \end{bmatrix}$$

Our results enable us to characterize the relation of equivalence of rectangular matrices in terms of the concept of rank.

(25.1) theorem An m by n matrix A is equivalent to a D_r matrix, where r is the rank of A.

PROOF The matrix A can be carried to echelon form A^* by elementary row transformations. Hence, there is a nonsingular matrix P such that $PA = A^*$. By the remarks above, there exists a nonsingular matrix Q such that $A^*Q = D_r$. Therefore, $PAQ = A^*Q = D_r$, so that A is equivalent to D_r. The rank of a matrix in the form D_r is obviously r. By Corollary 24.4, the rank of A is equal to rank of D_r, which is r. ∎

In practice, the recipe of Theorem 25.1 for carrying a matrix to the diagonal form D_r can be varied as is indicated in the following example.

EXAMPLE 2 Let

$$A = \begin{bmatrix} -3 & \frac{1}{2} & 0 & 2 \\ 6 & -4 & 1 & 7 \\ \frac{2}{5} & 5 & 9 & 0 \end{bmatrix}.$$

Then,

$$A \xrightarrow[\text{Type I}]{\text{Row}} \begin{bmatrix} 6 & -4 & 1 & 7 \\ -3 & \frac{1}{2} & 0 & 2 \\ \frac{2}{5} & 5 & 9 & 0 \end{bmatrix} \xrightarrow[\text{Type I}]{\text{Column}} \begin{bmatrix} 1 & -4 & 6 & 7 \\ 0 & \frac{1}{2} & -3 & 2 \\ 9 & 5 & \frac{2}{5} & 0 \end{bmatrix} \xrightarrow[\text{Type II}]{\text{Row}}$$

$$\begin{bmatrix} 1 & -4 & 6 & 7 \\ 0 & \frac{1}{2} & -3 & 2 \\ 0 & 41 & -\frac{268}{5} & -63 \end{bmatrix} \xrightarrow[\text{Type II}]{\text{Column}} \begin{bmatrix} 1 & 0 & 6 & 7 \\ 0 & \frac{1}{2} & -3 & 2 \\ 0 & 41 & -\frac{268}{5} & -63 \end{bmatrix} \xrightarrow[\text{Type II}]{\text{Column}}$$

$$\begin{bmatrix} 1 & 0 & 0 & 7 \\ 0 & \frac{1}{2} & -3 & 2 \\ 0 & 41 & -\frac{268}{5} & -63 \end{bmatrix} \xrightarrow[\text{Type II}]{\text{Column}} \begin{bmatrix} 1 & 0 & 0 & 0 \\ 0 & \frac{1}{2} & -3 & 2 \\ 0 & 41 & -\frac{268}{5} & -63 \end{bmatrix} \xrightarrow[\text{Type III}]{\text{Row}}$$

$$\begin{bmatrix} 1 & 0 & 0 & 0 \\ 0 & 1 & -6 & 4 \\ 0 & 41 & -\frac{268}{5} & -63 \end{bmatrix} \xrightarrow[\text{Type II}]{\text{Row}} \begin{bmatrix} 1 & 0 & 0 & 0 \\ 0 & 1 & -6 & 4 \\ 0 & 0 & \frac{962}{5} & -227 \end{bmatrix} \xrightarrow[\text{Type II}]{\text{Column}}$$

$$\begin{bmatrix} 1 & 0 & 0 & 0 \\ 0 & 1 & 0 & 4 \\ 0 & 0 & \frac{962}{5} & -227 \end{bmatrix} \xrightarrow[\text{Type II}]{\text{Column}} \begin{bmatrix} 1 & 0 & 0 & 0 \\ 0 & 1 & 0 & 0 \\ 0 & 0 & \frac{962}{5} & -227 \end{bmatrix} \xrightarrow[\text{Type III}]{\text{Row}}$$

$$\begin{bmatrix} 1 & 0 & 0 & 0 \\ 0 & 1 & 0 & 0 \\ 0 & 0 & 1 & -1135/962 \end{bmatrix} \xrightarrow[\text{Type II}]{\text{Column}} \begin{bmatrix} 1 & 0 & 0 & 0 \\ 0 & 1 & 0 & 0 \\ 0 & 0 & 1 & 0 \end{bmatrix} = D_3.$$

The nonsingular matrices P and Q such that $PAQ = D_3$ are found by performing the row transformations on I_3 and the column transformations on I_4, respectively. In this example,

$$P = \begin{bmatrix} 0 & 1 & 0 \\ 2 & 0 & 0 \\ -205/481 & -45/962 & 5/962 \end{bmatrix}, \qquad Q = \begin{bmatrix} 0 & 0 & 1 & 1135/962 \\ 0 & 1 & 6 & 1481/481 \\ 1 & 4 & 18 & -848/481 \\ 0 & 0 & 0 & 1 \end{bmatrix}.$$

The matrices P and Q such that $PAQ = D_r$ are not unique, since a matrix A can be carried into a D_r matrix by different sequences of elementary transformations. However, it is a consequence of the following theorem that each matrix A is equivalent to exactly one diagonal matrix D_r.

(25.2) theorem Let A and B be m by n matrices. Then A and B are equivalent if and only if the rank of A = rank of B.

PROOF If A and B are equivalent, then the rank of A = rank of B, by Corollary 24.4. Conversely, suppose that the rank of A = rank of $B = r$. By Theorem 25.1, there exist nonsingular matrices P, P_1, Q, Q_1, such that $PAQ = D_r = P_1BQ_1$. Therefore, $(P_1^{-1}P)A(QQ_1^{-1}) = B$, where $P_1^{-1}P$ and QQ_1^{-1} are nonsingular matrices. Hence, A and B are equivalent. ∎

The matrix D_r to which a matrix A is equivalent is called the *rational canonical form* for A under the relation of equivalence. It is "rational" because the elements of D_r are rational functions of the elements of A. That is, only the rational operations of addition, subtraction, multiplication and division are performed on the elements of A to obtain D_r. Also, the matrices P and Q, such that $PAQ = D_r$, have elements which are rational functions of A. For example, if the elements of A are all rational numbers, then the elements of P and Q are rational numbers. The matrix D_r is called "canonical," since it is the unique matrix of this form to which A is equivalent.

EXAMPLE 3 Let the matrix A of Example 2 be the matrix of a linear transformation L of $\mathcal{V}_4$ into $\mathcal{V}_3$ with respect to the bases $E^{(4)}$ and $E^{(3)}$. Define new bases in $\mathcal{V}_4$ and $\mathcal{V}_3$ by means of the nonsingular matrices Q' and $(P^{-1})'$, where P and Q are the matrices determined in Example 2 such that $PAQ = D_3$. Using the method 23.4,

$$P^{-1} = \begin{bmatrix} 0 & 1/2 & 0 \\ 1 & 0 & 0 \\ 9 & 41 & 962/5 \end{bmatrix}.$$

Then,

$$\begin{aligned}
\bar{\mathbf{U}}_1 &= & & & 1 \cdot \mathbf{E}_3^{(4)} & &= (0, 0, 1, 0) \\
\bar{\mathbf{U}}_2 &= & & 1 \cdot \mathbf{E}_2^{(4)} + & 4 \cdot \mathbf{E}_3^{(4)} & &= (0, 1, 4, 0) \\
\bar{\mathbf{U}}_3 &= & 1 \cdot \mathbf{E}_1^{(4)} + & 6 \cdot \mathbf{E}_2^{(4)} + & 18 \cdot \mathbf{E}_3^{(4)} & &= (1, 6, 18, 0)
\end{aligned}$$

$$\bar{\mathbf{U}}_4 = \left(\frac{1135}{962}\right) \cdot \mathbf{E}_1^{(4)} + \left(\frac{1481}{481}\right) \cdot \mathbf{E}_2^{(4)} + \left(-\frac{848}{481}\right) \cdot \mathbf{E}_3^{(4)} + 1 \cdot \mathbf{E}_4^{(4)}$$

$$= \left(\frac{1135}{962}, \frac{1481}{481}, -\frac{848}{481}, 1\right)$$

$$\begin{aligned}
\bar{\mathbf{W}}_1 &= & 1 \cdot \mathbf{E}_2^{(3)} + 9 \cdot \mathbf{E}_3^{(3)} &= (0, 1, 9) \\
\bar{\mathbf{W}}_2 &= 1/2 \cdot \mathbf{E}_1^{(3)} & + 41 \cdot \mathbf{E}_3^{(3)} &= (1/2, 0, 41) \\
\bar{\mathbf{W}}_3 &= & \left(\frac{962}{5}\right) \cdot \mathbf{E}_3^{(3)} &= \left(0, 0, \frac{962}{5}\right).
\end{aligned}$$

With respect to these new bases, the matrix of L is D_3. Therefore, L is given by

$$\begin{bmatrix} \bar{s}_1 \\ \bar{s}_2 \\ \bar{s}_3 \end{bmatrix} = \begin{bmatrix} 1 & 0 & 0 & 0 \\ 0 & 1 & 0 & 0 \\ 0 & 0 & 1 & 0 \end{bmatrix} \begin{bmatrix} \bar{r}_1 \\ \bar{r}_2 \\ \bar{r}_3 \\ \bar{r}_4 \end{bmatrix} = \begin{bmatrix} \bar{r}_1 \\ \bar{r}_2 \\ \bar{r}_3 \end{bmatrix}.$$

That is, if $\mathbf{V} \in \mathcal{V}_4$ has coordinates $(\bar{r}_1, \bar{r}_2, \bar{r}_3, \bar{r}_4)$ with respect to the new basis in $\mathcal{V}_4$, then $L(\mathbf{V})$ has coordinates $(\bar{r}_1, \bar{r}_2, \bar{r}_3)$ with respect to the new basis in $\mathcal{V}_3$. It is clear that the range space $L(\mathcal{V}_4)$ of L is $\mathcal{V}_3$, and that the null space $L^{-1}(\{\mathbf{0}\})$ of L is the subspace of $\mathcal{V}_4$ spanned by the single vector $\bar{\mathbf{U}}_4$.

exercises

25.1 Carry the following matrices to rational canonical form D_r by elementary row and column transformations. Find the matrices P and Q such that $PAQ = D_r$ in each case:

(a) $\begin{bmatrix} 2 & -2 & 2 & 5 & 1 \\ 1 & 1 & 1 & 3 & -1 \\ 3 & 5 & 4 & 2 & 4 \end{bmatrix}$

(b) $\begin{bmatrix} 3 & 2 & 1 & 0 & 0 \\ 0 & 1 & 2 & 0 & 0 \\ 1 & 4 & 7 & 2 & 3 \\ 5 & 3 & -1 & 1 & 2 \\ 6 & 1 & -2 & 3 & 6 \end{bmatrix}$

(c) $\begin{bmatrix} -1 & 2 \\ 0 & 5 \\ 3 & 4 \\ 6 & 2 \end{bmatrix}$

(d) $\begin{bmatrix} \sqrt{2} & \sqrt{2} & \sqrt{2} & \sqrt{2} \\ \sqrt{6} & \sqrt{6} & \sqrt{6} & \sqrt{6} \\ \sqrt{30} & \sqrt{30} & \sqrt{30} & \sqrt{30} \end{bmatrix}$

25.2 How many equivalence classes are there for the relation of equivalence on the set ${}_m\mathfrak{A}_n$ of all m by n matrices?

25.3 As in Example 3, let each m by n matrix of Exercise 25.1 be the matrix of a linear transformation L from $\mathcal{V}_n$ into $\mathcal{V}_m$ with respect to the bases $E^{(n)}$ and $E^{(m)}$. Define new bases in $\mathcal{V}_n$ and $\mathcal{V}_m$ so that L has matrix D_r with respect to these new bases. Determine the null space and the range space of L in each case.

6

determinants

26 definition of a determinant

In elementary algebra, the discussion of systems of linear equations is usually limited to systems of n equations in n unknowns, where n is small, say $n = 2$, 3, or 4. The reader is probably familiar with the use of determinants to compute the solutions of such systems.

We recall that the determinant of a 2 by 2 matrix

$$\begin{bmatrix} a & b \\ c & d \end{bmatrix}$$

is the number $ad - bc$. In the system of two linear equations in two unknowns

$$\begin{aligned} a_{1,1}x + a_{1,2}y &= b_1 \\ a_{2,1}x + a_{2,2}y &= b_2 \end{aligned}$$

successive elimination of the unknowns y and x yields a simpler system

$$\begin{aligned} (a_{1,1}a_{2,2} - a_{1,2}a_{2,1})x &= (b_1a_{2,2} - a_{1,2}b_2) \\ (a_{1,1}a_{2,2} - a_{1,2}a_{2,1})y &= (a_{1,1}b_2 - b_1a_{2,1}). \end{aligned}$$

The numbers that appear in these latter equations are simply the determinants of the 2 by 2 matrices

$$\begin{bmatrix} a_{1,1} & a_{1,2} \\ a_{2,1} & a_{2,2} \end{bmatrix}, \quad \begin{bmatrix} b_1 & a_{1,2} \\ b_2 & a_{2,2} \end{bmatrix}, \quad \text{and} \quad \begin{bmatrix} a_{1,1} & b_1 \\ a_{2,1} & b_2 \end{bmatrix}.$$

If $a_{1,1}a_{2,2} - a_{1,2}a_{2,1} \neq 0$, then the system has a unique solution, which is given in terms of these determinants:

$$x = \frac{b_1a_{2,2} - a_{1,2}b_2}{a_{1,1}a_{2,2} - a_{1,2}a_{2,1}}$$

$$y = \frac{a_{1,1}b_2 - b_1a_{2,1}}{a_{1,1}a_{2,2} - a_{1,2}a_{2,1}}.$$

When a system of three linear equations in three unknowns,

$$\begin{aligned} a_{1,1}x + a_{1,2}y + a_{1,3}z &= b_1 \\ a_{2,1}x + a_{2,2}y + a_{2,3}z &= b_2 \\ a_{3,1}x + a_{3,2}y + a_{3,3}z &= b_3 \end{aligned} \tag{26–1}$$

has a unique solution, this solution can also be given in terms of determinants of 3 by 3 matrices. The determinant of a 3 by 3 matrix

$$\begin{bmatrix} a & b & c \\ d & e & f \\ i & j & k \end{bmatrix}$$

is the number $aek + bfi + cdj - cei - bdk - afj$. The 3 by 3 matrices involved in the solution of the system (26–1) are: The matrix of coefficients,

$$A = \begin{bmatrix} a_{1,1} & a_{1,2} & a_{1,3} \\ a_{2,1} & a_{2,2} & a_{2,3} \\ a_{3,1} & a_{3,2} & a_{3,3} \end{bmatrix};$$

the matrix A_1 obtained from A by replacing the first column of A by b_1, b_2, b_3,

$$A_1 = \begin{bmatrix} b_1 & a_{1,2} & a_{1,3} \\ b_2 & a_{2,2} & a_{2,3} \\ b_3 & a_{3,2} & a_{3,3} \end{bmatrix};$$

the matrix A_2 obtained from A by replacing the second column of A by b_1, b_2, b_3,

$$A_2 = \begin{bmatrix} a_{1,1} & b_1 & a_{1,3} \\ a_{2,1} & b_2 & a_{2,3} \\ a_{3,1} & b_3 & a_{3,3} \end{bmatrix};$$

and the matrix A_3 obtained from A replacing the third column of A by b_1, b_2, b_3,

$$A_3 = \begin{bmatrix} a_{1,1} & a_{1,2} & b_1 \\ a_{2,1} & a_{2,2} & b_2 \\ a_{3,1} & a_{3,2} & b_3 \end{bmatrix}.$$

In cases where the determinant of A is not zero, the system (26–1) has a unique solution. If we denote the determinants of the matrices A, A_1, A_2, and A_3 by $|A|$, $|A_1|$, $|A_2|$, and $|A_3|$, respectively, then the solution of (26–1) is given by

$$x = \frac{|A_1|}{|A|}, \; y = \frac{|A_2|}{|A|}, \; z = \frac{|A_3|}{|A|}.$$

The methods given above for computing the solutions of systems of two linear equations in two unknowns and systems of three linear equations in three unknowns are special cases of a general rule for solving systems of n linear equations in n unknowns. We will discuss this rule, called *Cramer's Rule*, in the final section of this chapter, as one application of the definition of the determinant of an n by n matrix.

Although our introductory remarks indicate that determinants are important as computational devices, this aspect of their usefulness should not be exaggerated. Determinants of square matrices are of theoretical importance in many branches of mathematics. For example, in Chapter 7, we use several of the properties of determinants in studying similarity of matrices.

Let $j_1, j_2, \ldots, j_n$ be an ordering of the positive integers $1, 2, \ldots, n$. An *inversion* occurs in this ordering whenever a greater integer precedes a smaller one. The *number of inversions* of $j_1, j_2, \ldots, j_n$ is the sum $k = \sum_{s=1}^{n-1} k_s$, where k_s is the number of integers greater than s that precede s in the given ordering. For example, if $n = 6$, then in the ordering 3, 2, 6, 1, 5, 4, we have $k_1 = 3$, $k_2 = 1$, $k_3 = 0$, $k_4 = 2$, $k_5 = 1$, and $k = 7$.

(26.1) definition The *determinant* of an n by n matrix $A = [a_{i,j}]$ is the number

$$|A| = \sum (-1)^k a_{1,j_1} a_{2,j_2} \cdots a_{n,j_n}$$

where in each term the column (second) subscripts $j_1, j_2, \ldots, j_n$ are some ordering of $1, 2, \ldots, n$, and the sum is taken over all possible orderings of the column subscripts. For each term, the exponent k of $(-1)^k$ is the number of inversions of the column subscripts $j_1, j_2, \ldots, j_n$.

In addition to using the notation $|A|$ for the determinant of $A = [a_{i,j}]$, we also write

$$|a_{i,j}| \quad \text{and} \quad \begin{vmatrix} a_{1,1} & a_{1,2} & \cdots & a_{1,n} \\ a_{2,1} & a_{2,2} & \cdots & a_{2,n} \\ \vdots & \vdots & & \vdots \\ a_{n,1} & a_{n,2} & \cdots & a_{n,n} \end{vmatrix}$$

The following example shows that for $n = 2$ and $n = 3$, Definition 26.1 agrees with the familiar definitions of determinants of 2 by 2 and 3 by 3 matrices.

EXAMPLE 1 If $n = 2$, then

$$A = \begin{bmatrix} a_{1,1} & a_{1,2} \\ a_{2,1} & a_{2,2} \end{bmatrix}$$

and the possible orderings of the column subscripts are 1, 2 and 2, 1. The first ordering has no inversions and the second ordering has one inversion. Therefore, by Definition 26.1,

$$\begin{vmatrix} a_{1,1} & a_{1,2} \\ a_{2,1} & a_{2,2} \end{vmatrix} = (-1)^0 a_{1,1}a_{2,2} + (-1)^1 a_{1,2}a_{2,1} = a_{1,1}a_{2,2} - a_{1,2}a_{2,1}.$$

If $n = 3$,

$$A = \begin{bmatrix} a_{1,1} & a_{1,2} & a_{1,3} \\ a_{2,1} & a_{2,2} & a_{2,3} \\ a_{3,1} & a_{3,2} & a_{3,3} \end{bmatrix}$$

and there are six possible orderings for the column subscripts: 1, 2, 3; 1, 3, 2; 2, 1, 3; 2, 3, 1; 3, 1, 2; and 3, 2, 1. By Definition 26.1,

$$\begin{aligned} \begin{vmatrix} a_{1,1} & a_{1,2} & a_{1,3} \\ a_{2,1} & a_{2,2} & a_{2,3} \\ a_{3,1} & a_{3,2} & a_{3,3} \end{vmatrix} &= (-1)^0 a_{1,1}a_{2,2}a_{3,3} + (-1)^1 a_{1,1}a_{2,3}a_{3,2} \\ &\quad + (-1)^1 a_{1,2}a_{2,1}a_{3,3} + (-1)^2 a_{1,2}a_{2,3}a_{3,1} \\ &\quad + (-1)^2 a_{1,3}a_{2,1}a_{3,2} + (-1)^3 a_{1,3}a_{2,2}a_{3,1} \\ &= a_{1,1}a_{2,2}a_{3,3} + a_{1,2}a_{2,3}a_{3,1} + a_{1,3}a_{2,1}a_{3,2} \\ &\quad - a_{1,3}a_{2,2}a_{3,1} - a_{1,1}a_{2,3}a_{3,2} - a_{1,2}a_{2,1}a_{3,3}. \end{aligned}$$

In the sum $\Sigma\, (-1)^k a_{1,j_1}a_{2,j_2} \cdots a_{n,j_n}$ of Definition 26.1, the row (first) subscripts of each term are in natural order 1, 2, . . . , n and the column (second) subscripts are these same numbers in some order. Therefore, each term is a signed product of n elements of $A = [a_{i,j}]$, with exactly one element from each row and one element from each column. Every product of n elements of A with exactly one element from each row and column can be written with the row subscripts in natural order. Since the sum is taken over all possible orderings of the column subscripts, the determinant $|A|$ is the sum of all products of n elements of A, with one element from each row and column, where the sign of each term is determined by the number of inversions of the column subscripts when the row subscripts are in natural order. There are $n(n-1)(n-2) \ldots (2)(1) = n!$ possible orderings for the column subscripts. Therefore, there are $n!$ terms in the sum.

For each square matrix A with real numbers as elements, $|A|$ is a real number. Thus, we may speak of the *determinant function* which maps a real square matrix into a real number. Moreover, the determinant function is an *integral function* of the elements $a_{i,j}$ of A since the evaluation of $|A|$

involves only the integral operations of addition, subtraction, and multiplication. Thus, if the elements of A are integers, $|A|$ is an integer, and if the elements of A are rational numbers, $|A|$ is a rational number. Although we have restricted our attention to matrices with real number elements, it is evident that Definition 26.1 would apply to matrices with elements in any algebraic system in which the operations of addition, subtraction, and multiplication can be performed, subject to the usual algebraic rules. For example, the elements of A can be polynomials in a variable x, in which case $|A|$ is a polynomial in x.

In order to establish a result that can be used in deriving the properties of determinants, we first note the following fact.

(26.2) Let l be the number of inversions of the ordering $i_1, i_2, \ldots, i_n$ of $1, 2, \ldots, n$. The ordering $i_1, i_2, \ldots, i_n$ can be carried into the natural ordering $1, 2, \ldots, n$ by l interchanges of adjacent integers.

PROOF The number of inversions l is equal to $\Sigma_{s=1}^{n-1} l_s$, where l_s is the number of integers greater than s that precede s in the ordering $i_1, i_2, \ldots, i_n$. Since l_1 numbers precede 1, it follows that 1 can be put in first position by l_1 adjacent interchanges. Moving 1 to first position does not change l_s for $s \geq 2$. Hence, the number of inversions of the new ordering is $\Sigma_{s=2}^{n-1} l_s$. Now 2 can be moved into second position by l_2 adjacent interchanges, producing a new ordering with $\Sigma_{s=3}^{n-1} l_s$ inversions. This process can be continued until the ordering $1, 2, \ldots, n$ is obtained by $l = \Sigma_{s=1}^{n-1} l_s$ adjacent interchanges. ∎

Consider the sum $S = \Sigma (-1)^k a_{i_1,j_1} a_{i_2,j_2} \cdots a_{i_n,j_n}$, which is described exactly as the sum appearing in Definition 26.1, except that the row subscripts are in a fixed order $i_1, i_2, \ldots, i_n$ which is not necessarily the natural order $1, 2, \ldots, n$. That is, the row subscripts are in the same order $i_1, i_2, \ldots, i_n$ in each term of S. There is a simple relation between S and $|A|$:

(26.3) theorem

$$S = \sum (-1)^k a_{i_1,j_1} a_{i_2,j_2} \cdots a_{i_n,j_n} = (-1)^l |A|$$

where l is the number of inversions of the fixed ordering $i_1, i_2, \ldots, i_n$.

PROOF Except possibly for differences in sign, the terms of S are the $n!$ distinct terms of $|A|$, since every product of n elements of $A = [a_{i,j}]$ with one element from each row and one from each column appears exactly once as a term of S. Thus, it is sufficient to show that each term of S is $(-1)^l$ times the term of $|A|$ that is the product of the same n elements. Let $(-1)^k a_{i_1,j_1} a_{i_2,j_2} \cdots a_{i_n,j_n}$ be a term of S. Rearranging the factors $a_{i,j}$ in this term, we have

$$(-1)^k a_{i_1,j_1} a_{i_2,j_2} \cdots a_{i_n,j_n} = (-1)^k a_{1,m_1} a_{2,m_2} \cdots a_{n,m_n}$$

where the latter product is a term of $|A|$, except possibly for sign. The corresponding term of $|A|$ is $(-1)^m a_{1,m_1}a_{2,m_2}\cdots a_{n,m_n}$, where m is the number of inversions of the ordering $m_1, m_2, \ldots, m_n$. Therefore, the proof is complete if $(-1)^k = (-1)^l(-1)^m$.

By 26.2, the factors of the product $a_{i_1,j_1}a_{i_2,j_2}\cdots a_{i_n,j_n}$ can be arranged so that $a_{i_1,j_1}a_{i_2,j_2}\cdots a_{i_n,j_n} = a_{1,m_1}a_{2,m_2}\cdots a_{n,m_n}$ by l successive interchanges of adjacent factors. Each time two adjacent factors $a_{i,j}\,a_{s,t}$ are interchanged, the number of inversions of the column subscripts is either increased by one (if $t > j$) or decreased by one (if $t < j$). Thus, starting with the ordering $j_1, j_2, \ldots, j_n$ which has k inversions, we change the number of inversions by one l times and arrive at the ordering $m_1, m_2, \ldots, m_n$ with m inversions. Therefore, if l is even, then k and m have the same parity (i.e., k and m are both even or both odd), and if l is odd, then k and m have opposite parity (i.e., one is even and the other is odd). In every case, $(-1)^k = (-1)^l(-1)^m$, and, as indicated above, the proof is complete. ▮

EXAMPLE 2 Let $S = \Sigma\,(-1)^k a_{3,j_1}a_{1,j_2}a_{2,j_3}$ be the sum taken over all orderings of the column subscripts j_1, j_2, j_3, where, in each term, k is the number of inversions of j_1, j_2, j_3. The number of inversions of the fixed orderings of the row subscripts 3, 1, 2 is $l = 2$. Then, by Theorem 26.3, $S = (-1)^2|A| = |A|$, where

$$A = \begin{bmatrix} a_{1,1} & a_{1,2} & a_{1,3} \\ a_{2,1} & a_{2,2} & a_{2,3} \\ a_{3,1} & a_{3,2} & a_{3,3} \end{bmatrix}.$$

On the other hand, if S is computed directly, we find

$$\begin{aligned} S &= (-1)^0 a_{3,1}a_{1,2}a_{2,3} + (-1)^1 a_{3,1}a_{1,3}a_{2,2} + (-1)^1 a_{3,2}a_{1,1}a_{2,3} \\ &\quad + (-1)^2 a_{3,2}a_{1,3}a_{2,1} + (-1)^2 a_{3,3}a_{1,1}a_{2,2} + (-1)^3 a_{3,3}a_{1,2}a_{2,1} \\ &= a_{1,1}a_{2,2}a_{3,3} + a_{1,2}a_{2,3}a_{3,1} + a_{1,3}a_{2,1}a_{3,2} - a_{1,3}a_{2,2}a_{3,1} - a_{1,1}a_{2,3}a_{3,2} \\ &\quad - a_{1,2}a_{2,1}a_{3,3} = |A|. \end{aligned}$$

The determinant of a square matrix A can also be expressed as a sum on the row subscripts.

(26.4) theorem Let $T = \Sigma\,(-1)^l a_{i_1,1}\,a_{i_2,2}\cdots a_{i_n,n}$, where in each term the row subscripts $i_1, i_2, \ldots, i_n$ are some ordering of $1, 2, \ldots, n$, and the sum is taken over all possible orderings of the row subscripts. For each term, the exponent l of $(-1)^l$ is the number of inversions of $i_1, i_2, \ldots, i_n$. Then T is the determinant of the n by n matrix $A = [a_{i,j}]$.

The proof of Theorem 26.4 is similar to that of Theorem 26.3 and is left as an exercise (Exercise 26.6).

exercises

26.1 Count the number of inversions in the following orderings of 1, 2, 3, 4, 5, 6, 7: 7, 3, 5, 1, 6, 4, 2; 5, 1, 2, 4, 3, 7, 6; 3, 4, 7, 6, 5, 1, 2.

26.2 Find the signs of the terms $a_{1,5}a_{2,6}a_{3,3}a_{4,1}a_{5,4}a_{6,2}$ and $a_{1,6}a_{2,3}a_{3,2}a_{4,5}a_{5,4}a_{6,1}$ in the expansion of the determinant of the 6 by 6 matrix $A = [a_{i,j}]$.

26.3 Let A be a square triangular matrix; that is, each element below the diagonal is 0. Prove that $|A|$ is equal to the product of the diagonal elements. In particular, the determinant of a square diagonal matrix is equal to the product of the diagonal elements, and $|I_n| = 1$.

26.4 Prove that $|r \cdot A| = r^n|A|$.

26.5 Find the values of the following determinants:

(a) $\begin{vmatrix} x+1 & x^2+x+1 & x \\ x^2-1 & x+2 & 3 \\ x-4 & 0 & x^2 \end{vmatrix}$ (b) $\begin{vmatrix} -3 & 1 & 7 \\ 0 & 2 & 6 \\ -4 & 5 & 1 \end{vmatrix}$

(c) $\begin{vmatrix} 1 & 0 & 0 & 0 & 0 \\ 0 & 1 & 0 & 0 & 0 \\ 0 & 0 & 1 & 0 & 0 \\ 0 & -5 & 0 & 1 & 0 \\ 0 & 0 & 0 & 0 & 1 \end{vmatrix}$

(d) $\begin{vmatrix} \frac{1}{2} & 0 & 0 & 0 \\ 4 & \frac{1}{3} & 0 & 0 \\ -1 & 2 & \frac{1}{5} & 0 \\ 7 & 8 & 10 & \frac{1}{7} \end{vmatrix}$ (e) $\begin{vmatrix} 0 & 0 & 0 & \sqrt{2} \\ 0 & 0 & \sqrt{3} & 0 \\ 0 & \sqrt{5} & 0 & 0 \\ \sqrt{7} & 0 & 0 & 0 \end{vmatrix}$

26.6 Prove Theorem 26.4.

26.7 Prove that if $t > s$, there are $(t-1-s+1)+(t-1-s) = 2t-2s-1$ inversions in the ordering $1, 2, \ldots, s-1, t, s+1, \ldots, t-1, s, t+1, \ldots, n$.

27 properties of determinants

Throughout this section, $A = [a_{i,j}]$ is an n by n matrix.

(27.1) theorem $|A'| = |A|$.

PROOF The matrix A' is the transpose of A (see Exercise 19.22 for the definition of the transpose of a matrix). Thus, $A' = [b_{i,j}]$, where $b_{i,j} = a_{j,i}$ for all i, j. By Definition 26.1,

$$|A'| = \sum (-1)^k b_{1,j_1} b_{2,j_2} \cdots b_{n,j_n} = \sum (-1)^k a_{j_1,1} a_{j_2,2} \cdots a_{j_n,n}$$

where each sum is taken over all possible orderings of $j_1, j_2, \ldots, j_n$ and, in each term, k is the number of inversions of $j_1, j_2, \ldots, j_n$. By Theorem 26.4, the second sum is equal to $|A|$, so that $|A'| = |A|$. ∎

The property given in Theorem 27.1 allows us to replace "row" by "column" in the statement of the remaining properties of determinants.

(27.2) theorem If B is the matrix obtained from A by interchanging two rows (columns), then $|B| = -|A|$.

PROOF Suppose that the rows s and t of A are interchanged to obtain B. Then $B = [b_{i,j}]$, where for all j, $b_{i,j} = a_{i,j}$ if $i \neq s$ or $i \neq t$, and $b_{s,j} = a_{t,j}$, $b_{t,j} = a_{s,j}$. We may assume that $s < t$. Then, by Definition 26.1,

$$|B| = \sum (-1)^k b_{1,j_1} b_{2,j_2} \cdots b_{s,j_s} \cdots b_{t,j_t} \cdots b_{n,j_n}$$

$$= \sum (-1)^k a_{1,j_1} a_{2,j_2} \cdots a_{t,j_s} \cdots a_{s,j_t} \cdots a_{n,j_n}.$$

The row subscripts of the second sum are in the fixed order $1, 2, \ldots, s-1, t, s+1, \ldots, t-1, s, t+1, \ldots, n$, which has $(t-1-s+1) + (t-1-s) = 2t-2s-1$ inversions since $t > s$ (see Exercise 26.7). Therefore, by Theorem 26.3, the second sum is equal to $(-1)^{2t-2s-1}|A| = -|A|$. Hence, $|B| = -|A|$.

If B is the matrix obtained from A by interchanging two columns of A, then B' is obtained from A' by interchanging two rows of A'. Therefore, $|B'| = -|A'|$ by what we have just proved. Hence, by Theorem 27.1,

$$|B| = |B'| = -|A'| = -|A|. \qquad \blacksquare$$

The last part of the proof of Theorem 27.2 illustrates how it follows from Theorem 27.1 that each "row" result for determinants yields a corresponding "column" result. In the remaining theorems, we will omit the proof for columns.

(27.3) theorem If two rows (columns) of A are identical, then $|A| = 0$.

PROOF Let B be the matrix obtained from A by interchanging the identical rows of A. Then $B = A$, and hence, $|B| = |A|$. However, by Theorem 27.2, $|B| = -|A|$. Therefore, $|A| = -|A|$, $2|A| = 0$, and $|A| = 0$. $\blacksquare$

(27.4) theorem If B is the matrix obtained from A by multiplying every element of a row (column) of A by the number r, then $|B| = r|A|$.

PROOF Suppose that every element of row s of A is multiplied by r to obtain the matrix B. Then $B = [b_{i,j}]$, where for all j, $b_{i,j} = a_{i,j}$ if $i \neq s$ and $b_{s,j} = ra_{s,j}$. By Definition 26.1,

$$B = \sum (-1)^k b_{1,j_1} b_{2,j_2} \cdots b_{s,j_s} \cdots b_{n,j_n}$$

$$= \sum (-1)^k a_{1,j_1} a_{2,j_2} \cdots ra_{s,j_s} \cdots a_{n,j_n}$$

$$= r\left(\sum (-1)^k a_{1,j_1} a_{2,j_2} \cdots a_{s,j_s} \cdots a_{n,j_n}\right) = r|A|. \qquad \blacksquare$$

Theorem 27.4 can be used to simplify the expansion of a determinant by removing common factors from rows and columns.

EXAMPLE 1

$$\begin{vmatrix} 18 & -2 & 7 & 5 \\ -2 & 4 & 1 & -1 \\ 6 & -12 & -3 & 3 \\ 0 & 20 & -1 & 10 \end{vmatrix}$$

$$= 2\begin{vmatrix} 9 & -2 & 7 & 5 \\ -1 & 4 & 1 & -1 \\ 3 & -12 & -3 & 3 \\ 0 & 20 & -1 & 10 \end{vmatrix} = 4\begin{vmatrix} 9 & -1 & 7 & 5 \\ -1 & 2 & 1 & -1 \\ 3 & -6 & -3 & 3 \\ 0 & 10 & -1 & 10 \end{vmatrix}$$

$$= 12\begin{vmatrix} 9 & -1 & 7 & 5 \\ -1 & 2 & 1 & -1 \\ 1 & -2 & -1 & 1 \\ 0 & 10 & -1 & 10 \end{vmatrix} = -12\begin{vmatrix} 9 & -1 & 7 & 5 \\ 1 & -2 & -1 & 1 \\ 1 & -2 & -1 & 1 \\ 0 & 10 & -1 & 10 \end{vmatrix}$$

$$= 0$$

by Theorems 27.3 and 27.4.

(27.5) theorem If B is a matrix obtained from A by multiplying each element of a row (column) of A by a number r and adding the product to the corresponding element of another row (column), then $|B| = |A|$.

PROOF Suppose that B is obtained from A by multiplying each element of row s of A by r and adding these elements to corresponding ones in row t ($s \neq t$). Then $B = [b_{i,j}]$, where for all j, $b_{i,j} = a_{i,j}$ if $i \neq t$, and $b_{t,j} = a_{t,j} + ra_{s,j}$. We may assume $s < t$. Then

$$\begin{aligned} |B| &= \sum (-1)^k b_{1,j_1} b_{2,j_2} \cdots b_{s,j_s} \cdots b_{t,j_t} \cdots b_{n,j_n} \\ &= \sum (-1)^k a_{1,j_1} a_{2,j_2} \cdots a_{s,j_s} \cdots (a_{t,j_t} + ra_{s,j_t}) \cdots a_{n,j_n} \\ &= \sum (-1)^k a_{1,j_1} a_{2,j_2} \cdots a_{s,j_s} \cdots a_{t,j_t} \cdots a_{n,j_n} \\ &\qquad + \sum (-1)^k a_{1,j_1} a_{2,j_2} \cdots a_{s,j_s} \cdots ra_{s,j_t} \cdots a_{n,j_n} \\ &= |A| + r \sum (-1)^k a_{1,j_1} a_{2,j_2} \cdots a_{s,j_s} \cdots a_{s,j_t} \cdots a_{n,j_n} \\ &= |A| + 0 = |A| \end{aligned}$$

by Theorem 27.3, since the last sum is the expansion of the determinant of a matrix

$$\begin{matrix} \\ \\ s \\ \\ t \\ \\ \\ \end{matrix}\begin{bmatrix} a_{1,1} & a_{1,2} & \cdots & a_{1,n} \\ \vdots & \vdots & & \vdots \\ a_{s,1} & a_{s,2} & \cdots & a_{s,n} \\ \vdots & \vdots & & \vdots \\ a_{s,1} & a_{s,2} & \cdots & a_{s,n} \\ \vdots & \vdots & & \vdots \\ a_{n,1} & a_{n,2} & \cdots & a_{n,n} \end{bmatrix}.$$

with two identical rows. ■

EXAMPLE 2 Theorems 27.2, 27.4, and 27.5, and the fact that the determinant of a triangular matrix is the product of the diagonal elements (see Exercise 26.3) can be used to find the value of a determinant.

$$\begin{vmatrix} 2 & -\frac{1}{2} & 4 \\ -7 & 3 & 0 \\ 1 & 6 & -5 \end{vmatrix} = -\begin{vmatrix} 1 & 6 & -5 \\ -7 & 3 & 0 \\ 2 & -\frac{1}{2} & 4 \end{vmatrix} = -\begin{vmatrix} 1 & 6 & -5 \\ 0 & 45 & -35 \\ 2 & -\frac{1}{2} & 4 \end{vmatrix}$$

$$= -\begin{vmatrix} 1 & 6 & -5 \\ 0 & 45 & -35 \\ 0 & -\frac{25}{2} & 14 \end{vmatrix} = -45\begin{vmatrix} 1 & 6 & -5 \\ 0 & 1 & -\frac{7}{9} \\ 0 & -\frac{25}{2} & 14 \end{vmatrix}$$

$$= -45\begin{vmatrix} 1 & 6 & -5 \\ 0 & 1 & -\frac{7}{9} \\ 0 & 0 & \frac{77}{18} \end{vmatrix} = (-45)\left(\frac{77}{18}\right) = -\frac{385}{2}.$$

(27.6) theorem Let $A = [a_{i,j}]$, $B = [b_{i,j}]$, and $C = [c_{i,j}]$ be n by n matrices, where for all j, $a_{i,j} = b_{i,j} = c_{i,j}$ if $i \neq s$ and $c_{s,j} = a_{s,j} + b_{s,j}$. Then $|C| = |A| + |B|$.

In Theorem 27.6, the matrices A, B, and C have identical rows except for row s. Row s of A may differ from row s of B, and each element of row s of C is the sum of the corresponding elements in row s of A and B. The proof, which is an easy exercise in the use of Definition 26.1, is left as Exercise 27.7. As in the case of the other properties of determinants, there is a corresponding property for columns.

EXAMPLE 3

$$\begin{vmatrix} x+1 & x^3+3 \\ x^2+2 & x^4+4 \end{vmatrix} = \begin{vmatrix} x & x^3+3 \\ x^2 & x^4+4 \end{vmatrix} + \begin{vmatrix} 1 & x^3+3 \\ 2 & x^4+4 \end{vmatrix}$$

$$= \begin{vmatrix} x & x^3 \\ x^2 & x^4 \end{vmatrix} + \begin{vmatrix} x & 3 \\ x^2 & 4 \end{vmatrix} + \begin{vmatrix} 1 & x^3 \\ 2 & x^4 \end{vmatrix} + \begin{vmatrix} 1 & 3 \\ 2 & 4 \end{vmatrix}$$

$$= (x^5 - x^5) + (4x - 3x^2) + (x^4 - 2x^3) + (4 - 6)$$
$$= x^4 - 2x^3 - 3x^2 + 4x - 2.$$

Theorems 27.2, 27.4, and 27.5 state how the determinant of an n by n matrix A is affected if an elementary transformation is performed on A. By Theorem 27.2, $|I_n^{(i,j)}A| = -|A| = |AI_n^{(i,j)}|$, where $I_n^{(i,j)}$ is an elementary transformation matrix of Type I. If $I_n^{(i,j,c)}$ is an elementary transformation matrix of Type II, then, by Theorem 27.5, $|I_n^{(i,j,c)}A| = |A| = |AI_n^{(i,j,c)}|$. By Theorem 27.4, $|^{(r)}I_n^{(i)}A| = r|A| = |A^{(r)}I_n^{(i)}|$, where $^{(r)}I_n^{(i)}$ is an elementary transformation matrix of Type III. In particular, since $|I_n| = 1$,

$$|I_n^{(i,j)}| = |I_n^{(i,j)}I_n| = -|I_n| = -1$$

$$|I_n^{(i,j,c)}| = |I_n^{(i,j,c)}I_n| = |I_n| = 1,$$

and

$$|^{(r)}I_n^{(i)}| = |^{(r)}I_n^{(i)}I_n| = r|I_n| = r.$$

(27.7) theorem An n by n matrix A is nonsingular if and only if $|A| \neq 0$.

PROOF By the results of §25, A can be carried into canonical form D_k, where k is the rank of A, by a sequence of elementary transformations. We observed above that when an elementary transformation is performed on the matrix A, the determinant of the transformed matrix is $-|A|$, $|A|$, or $r|A|$, where $r \neq 0$, according as the elementary transformation is Type I, Type II, or Type III. Thus, it follows that $|D_k|$ is a nonzero constant multiple of $|A|$. That is, $|D_k| = s|A|$, where $s \neq 0$. We can now show that the following statements are equivalent: (a) $|A| \neq 0$; (b) $|D_k| \neq 0$; (c) $k = n$; (d) A is nonsingular. Since $|D_k| = s|A|$, where $s \neq 0$, it follows that $|A| \neq 0$ if and only if $|D_k| \neq 0$. Thus, (a) and (b) are equivalent statements. The determinant of a diagonal matrix is the product of its diagonal elements (Exercise 26.3). If $k < n$, then D_k has at least one zero on the diagonal, and if $k = n$ each diagonal element of D_k is 1. Therefore, $|D_k| \neq 0$ if and only if $k = n$. That is, statements (b) and (c) are equivalent. Since k is the rank of A, Theorem 20.7 states that A is nonsingular if and only if $k = n$. Hence, statement (c) is equivalent to statement (d). This sequence of equivalent statements yields the result that A is nonsingular if and only if $|A| \neq 0$. ∎

The following theorem is the important "product rule" for determinants.

(27.8) theorem Let $A = [a_{i,j}]$ and $B = [b_{i,j}]$ be n by n matrices. Then $|AB| = |A|\,|B|$.

PROOF First consider the special case where A is an elementary transformation matrix. If $A = I_n^{(i,j)}$, then

$$|I_n^{(i,j)}B| = -|B| \quad \text{and} \quad |I_n^{(i,j)}|\,|B| = (-1)|B| = -|B|.$$

If $A = I_n^{(i,j,c)}$, then

$$|I_n^{(i,j,c)}B| = |B| \quad \text{and} \quad |I_n^{(i,j,c)}|\,|B| = (1)|B| = |B|.$$

Finally, if $A = {}^{(r)}I_n^{(i)}$, then

$$|{}^{(r)}I_n^{(i)}B| = r|B| \quad \text{and} \quad |{}^{(r)}I_n^{(i)}|\,|B| = r|B|.$$

Thus, in every case, $|AB| = |A|\,|B|$. Next, suppose the A is any nonsingular matrix. Then, by Theorem 23.5, $A = E_1E_2 \cdots E_k$ is a product of elementary transformation matrices. By the first part of the proof,

$$\begin{aligned}|A| &= |E_1E_2E_3 \cdots E_k| = |E_1|\,|E_2E_3 \cdots E_k| = |E_1|\,|E_2|\,|E_3 \cdots E_k| \\ &= \cdots = |E_1|\,|E_2|\,|E_3| \cdots |E_k|\end{aligned}$$

and

$$\begin{aligned}|AB| &= |E_1E_2 \cdots E_kB| = |E_1|\,|E_2 \cdots E_kB| \\ &= |E_1|\,|E_2| \cdots |E_k|\,|B| = |A|\,|B|.\end{aligned}$$

If A is singular, then, by Theorem 27.7, $|A| = 0$. Hence, $|A|\,|B| = 0$. Moreover, since rank of $A = k < n$, it follows that rank of $AB \leq k < n$ (see Exercise 24.2). By Theorem 20.7, AB is singular, and, by Theorem 27.7, $|AB| = 0$. Therefore, $|AB| = 0 = |A|\,|B|$, which completes the proof. ▮

(27.9) corollary If P is a nonsingular n by n matrix, then $|P^{-1}| = 1/|P|$.

PROOF Since $P^{-1}P = I_n$, it follows from Theorem 27.8 that $|P^{-1}|\,|P| = |P^{-1}P| = |I_n| = 1$. Hence, $|P^{-1}| = 1/|P|$. ▮

exercises

27.1 Evaluate the following determinants:

(a) $\begin{vmatrix} -2 & 1 & 3 & 7 \\ 4 & -5 & 6 & 0 \\ 2 & 4 & 6 & 8 \\ -1 & 0 & 9 & 2 \end{vmatrix}$ (b) $\begin{vmatrix} -\frac{1}{2} & \frac{3}{2} & 2 & \frac{1}{2} & 3 \\ -\frac{1}{4} & 6 & -3 & 2 & 1 \\ -1 & \frac{5}{2} & 7 & 3 & \frac{2}{3} \\ 0 & 21 & -16 & 7 & -2 \\ 4 & 3 & -6 & 5 & 1 \end{vmatrix}$

(c) $\begin{vmatrix} x+y & y+z & z+w \\ y+z & z+w & x+y \\ z+w & x+y & y+z \end{vmatrix}$ (d) $\begin{vmatrix} 2 & -6 & 4 & 8 \\ 6 & -5 & 4 & 2 \\ -1 & 5 & -2 & -4 \\ 0 & 7 & 14 & 35 \end{vmatrix}$

(e) $\begin{vmatrix} 0 & 1 & 2 & 3 & 4 \\ -1 & 0 & 1 & 2 & 3 \\ -2 & -1 & 0 & 1 & 2 \\ -3 & -2 & -1 & 0 & 1 \\ -4 & -3 & -2 & -1 & 0 \end{vmatrix}$ (f) $\begin{vmatrix} abc & a^2b & ac^2 \\ b & ab & bc \\ bc & ac & c^2 \end{vmatrix}$

(g) $\begin{vmatrix} 1 & a & a^2 & a^3 \\ 1 & b & b^2 & b^3 \\ 1 & c & c^2 & c^3 \\ 1 & d & d^2 & d^3 \end{vmatrix}$

27.2 Which properties of determinants are used to establish each of the following equalities? Do not evaluate the determinants.

(a) $\begin{vmatrix} 6 & 1 & -4 \\ 0 & 3 & 1 \\ \frac{1}{2} & -2 & 5 \end{vmatrix} = \begin{vmatrix} 6 & 0 & \frac{1}{2} \\ 1 & 3 & -2 \\ -4 & 1 & 5 \end{vmatrix} = -\begin{vmatrix} -4 & 1 & 5 \\ 1 & 3 & -2 \\ 6 & 0 & \frac{1}{2} \end{vmatrix}$

(b) $\begin{vmatrix} 3 & -1 & 1 \\ 2 & 8 & 7 \\ -4 & 1 & 5 \end{vmatrix} = \begin{vmatrix} -1 & 1 & 3 \\ 8 & 7 & 2 \\ 1 & 5 & -4 \end{vmatrix} = -\begin{vmatrix} 1 & -1 & -3 \\ 8 & 7 & 2 \\ 1 & 5 & -4 \end{vmatrix}$

(c) $\begin{vmatrix} 3 & 0 & 1 \\ -6 & -4 & -2 \\ 9 & 2 & -1 \end{vmatrix} = 3\begin{vmatrix} 1 & 0 & 1 \\ -2 & -4 & -2 \\ 3 & 2 & -1 \end{vmatrix} = -6\begin{vmatrix} 1 & 0 & 1 \\ 1 & 2 & 1 \\ 3 & 2 & -1 \end{vmatrix}$

(d) $\begin{vmatrix} 8 & 1 & -4 \\ -6 & 0 & 3 \\ 2 & 3 & -1 \end{vmatrix} = 0$

27.3 Show that

$$\begin{vmatrix} x & 0 & 0 & a_3 \\ -1 & x & 0 & a_2 \\ 0 & -1 & x & a_1 \\ 0 & 0 & -1 & a_0 \end{vmatrix} = a_0x^3 + a_1x^2 + a_2x + a_3.$$

27.4 Show that

$$\begin{vmatrix} a+b & b+c & c+a \\ b+c & c+a & a+b \\ c+a & a+b & b+c \end{vmatrix} = 2\begin{vmatrix} a & b & c \\ b & c & a \\ c & a & b \end{vmatrix}.$$

27.5 Show that

$$\begin{vmatrix} au + bv + cw & du + ev + fw \\ ax + by + cz & dx + ey + fz \end{vmatrix}$$

$$= \begin{vmatrix} a & b \\ d & e \end{vmatrix}\begin{vmatrix} u & v \\ x & y \end{vmatrix} + \begin{vmatrix} b & c \\ e & f \end{vmatrix}\begin{vmatrix} v & w \\ y & z \end{vmatrix} + \begin{vmatrix} c & a \\ f & d \end{vmatrix}\begin{vmatrix} w & u \\ z & x \end{vmatrix}.$$

27.6 For

$$A = \begin{bmatrix} 1 & 2 & 4 & 8 \\ 1 & 3 & 9 & 27 \\ 1 & 4 & 16 & 64 \\ 1 & 5 & 25 & 125 \end{bmatrix},$$

evaluate $|A^2|$; $|5 \cdot A|$; $|AA'|$; $|\frac{1}{2} \cdot A^{-1}|$; $|A^3A'A^{-1}|$.

27.7 Prove Theorem 27.6.

27.8 Let A and B be n by n matrices. Prove that $|AB| = |BA|$.

27.9 Prove that if every element of a row (column) of a square matrix A is 0, then $|A| = 0$.

27.10 Let A be a nonsingular matrix such that $A^{-1} = A'$. Prove that $|A| = \pm 1$.

27.11 Let $\mathbf{U} = x_1 \cdot \mathbf{I} + y_1 \cdot \mathbf{J} + z_1 \cdot \mathbf{K}$ and $\mathbf{V} = x_2 \cdot \mathbf{I} + y_2 \cdot \mathbf{J} + z_2 \cdot \mathbf{K}$ be vectors in $\mathcal{E}_3$. Show that $\mathbf{U} \times \mathbf{V}$ can be written in determinant form

$$\begin{vmatrix} \mathbf{I} & \mathbf{J} & \mathbf{K} \\ x_1 & y_1 & z_1 \\ x_2 & y_2 & z_2 \end{vmatrix}$$

28 expansion of a determinant

The labor involved in finding the value of a determinant of an n by n matrix A directly from Definition 26.1 is prohibitive if $n > 3$. For example, if $n = 4$, $\Sigma\,(-1)^k a_{1,j_1} a_{2,j_2} a_{3,j_3} a_{4,j_4}$ has 24 terms. Therefore, methods of evaluating a determinant have been devised. A technique was suggested in Example 2, §27 using the properties of determinants. In this section, we derive another procedure for computing a determinant.

Let $A = [a_{i,j}]$ be an n by n matrix. If the row and column containing an element $a_{i,j}$ is deleted from A, then an $(n-1)$ by $(n-1)$ matrix is obtained. The determinant of this $(n-1)$ by $(n-1)$ matrix is called the *minor of the element* $a_{i,j}$, and is denoted by $\Delta_{i,j}$. By Definition 26.1,

$$\Delta_{i,j} = \sum (-1)^l a_{1,j_1} a_{2,j_2} \cdots a_{i-1,j_{i-1}} a_{i+1,j_{i+1}} \cdots a_{n,j_n}$$

where the sum is taken over all possible orderings $j_1, j_2, \ldots, j_{i-1}, j_{i+1}, \ldots, j_n$ of $1, 2, \ldots, j-1, j+1, \ldots, n$, and, for each term, l is the number of inversions of $j_1, j_2, \ldots, j_{i-1}, j_{i+1}, \ldots, j_n$. The following theorem gives the rule for expanding a determinant according to the elements of any row or column.

(28.1) theorem For any i, $|A| = \Sigma_{j=1}^{n} a_{i,j}(-1)^{i+j}\Delta_{i,j}$; and for any j,

$$|A| = \sum_{i=1}^{n} a_{i,j}(-1)^{i+j}\Delta_{i,j}.$$

PROOF By Theorem 27.1, the second identity, which is the *column expansion,* follows from the first identity, which is the *row expansion*. We first consider the special case, where $i = 1$, and prove

$$|A| = \sum_{j=1}^{n} a_{1,j}(-1)^{1+j}\Delta_{1,j}. \qquad (28\text{–}1)$$

By Definition 26.1,

$$|A| = \sum (-1)^k a_{1,j_1} a_{2,j_2} \cdots a_{n,j_n} = \sum_{j_1=1}^{n} a_{1,j_1} \left[\sum{}' (-1)^k a_{2,j_2} a_{3,j_3} \cdots a_{n,j_n} \right]$$

where, for each value of j_1, the sum $\Sigma' (-1)^k a_{2,j_2} a_{3,j_3} \cdots a_{n,j_n}$ is taken over all orderings $j_2, j_3, \ldots, j_n$ of $1, 2, \ldots, j_1 - 1, j_1 + 1, \ldots, j_n$, and, for each term, k is the number of inversions of $j_1, j_2, j_3, \ldots, j_n$. For a fixed value of j_1, there are $j_1 - 1$ inversions contributed to the total k, because j_1 is in the first position of the ordering $j_1, j_2, j_3, \ldots, j_n$. Hence, there are $k - (j_1 - 1)$ inversions in the ordering $j_2, j_3, \ldots, j_n$. Thus, for each j_1,

$$\Delta_{1,j_1} = \sum{}' (-1)^{k-j_1+1} a_{2,j_2} a_{3,j_3} \cdots a_{n,j_n}$$

and

$$(-1)^{1+j_1} \Delta_{1,j_1} = \sum{}' (-1)^{k+2} a_{2,j_2} a_{3,j_3} \cdots a_{n,j_n}$$
$$= \sum{}' (-1)^{k} a_{2,j_2} a_{3,j_3} \cdots a_{n,j_n}.$$

Therefore,

$$|A| = \sum_{j_1=1}^{n} a_{1,j_1} (-1)^{1+j_1} \Delta_{1,j_1}$$

which is the required identity (28–1).

Consider a matrix B whose first row is the ith row of A and whose remaining rows are the rows of A except for the ith row, in the order $1, 2, 3, \ldots, i-1, i+1, \ldots, n$. The matrix B can be obtained from A by $i-1$ successive interchanges of two rows. By Theorem 27.2, $|B| = (-1)^{i-1}|A|$. The minors of the elements of the first row of B are the minors of the elements of the ith row of A. Therefore, by (28–1),

$$|B| = \sum_{j=1}^{n} a_{i,j} (-1)^{1+j} \Delta_{i,j}$$

and

$$|A| = (-1)^{2(i-1)}|A| = (-1)^{i-1}|B| = \sum_{j=1}^{n} a_{i,j} (-1)^{i+j} \Delta_{i,j}. \quad \blacksquare$$

EXAMPLE 1

$$\begin{vmatrix} 2 & -1 & 3 & 5 \\ 4 & 0 & 1 & 0 \\ 9 & -7 & 4 & 2 \\ 11 & -6 & 2 & 0 \end{vmatrix} = 4(-1)^{2+1} \begin{vmatrix} -1 & 3 & 5 \\ -7 & 4 & 2 \\ -6 & 2 & 0 \end{vmatrix} + 0(-1)^{2+2} \begin{vmatrix} 2 & 3 & 5 \\ 9 & 4 & 2 \\ 11 & 2 & 0 \end{vmatrix}$$
$$+ 1(-1)^{2+3} \begin{vmatrix} 2 & -1 & 5 \\ 9 & -7 & 2 \\ 11 & -6 & 0 \end{vmatrix} + 0(-1)^{2+4} \begin{vmatrix} 2 & -1 & 3 \\ 9 & -7 & 4 \\ 11 & -6 & 2 \end{vmatrix}$$

$$= (-4)\begin{vmatrix} -1 & 3 & 5 \\ -7 & 4 & 2 \\ -6 & 2 & 0 \end{vmatrix} + (-1)\begin{vmatrix} 2 & -1 & 5 \\ 9 & -7 & 2 \\ 11 & -6 & 0 \end{vmatrix}.$$

$$\begin{vmatrix} -1 & 3 & 5 \\ -7 & 4 & 2 \\ -6 & 2 & 0 \end{vmatrix} = (-6)(-1)^{3+1}\begin{vmatrix} 3 & 5 \\ 4 & 2 \end{vmatrix} + 2(-1)^{3+2}\begin{vmatrix} -1 & 5 \\ -7 & 2 \end{vmatrix}$$

$$+ 0(-1)^{3+3}\begin{vmatrix} -1 & 3 \\ -7 & 4 \end{vmatrix}$$

$$= (-6)(6 - 20) + (-2)(-2 + 35) = 18.$$

$$\begin{vmatrix} 2 & -1 & 5 \\ 9 & -7 & 2 \\ 11 & -6 & 0 \end{vmatrix} = 11(-1)^{3+1}\begin{vmatrix} -1 & 5 \\ -7 & 2 \end{vmatrix} + (-6)(-1)^{3+2}\begin{vmatrix} 2 & 5 \\ 9 & 2 \end{vmatrix}$$

$$+ 0(-1)^{3+3}\begin{vmatrix} 2 & -1 \\ 9 & -7 \end{vmatrix}$$

$$= (11)(-2 + 35) + (6)(4 - 45) = 117.$$

Hence, the value of the determinant is $(-4)(18) + (-1)(117) = -189$. The amount of computation can be reduced by using Theorems 27.5 and 27.4. For example,

$$\begin{vmatrix} 2 & -1 & 3 & 5 \\ 4 & 0 & 1 & 0 \\ 9 & -7 & 4 & 2 \\ 11 & -6 & 2 & 0 \end{vmatrix} = \begin{vmatrix} -10 & -1 & 3 & 5 \\ 0 & 0 & 1 & 0 \\ -7 & -7 & 4 & 2 \\ 3 & -6 & 2 & 0 \end{vmatrix} = (-1)^{2+3}\begin{vmatrix} -10 & -1 & 5 \\ -7 & -7 & 2 \\ 3 & -6 & 0 \end{vmatrix}$$

$$= (-3)\begin{vmatrix} -10 & -1 & 5 \\ -7 & -7 & 2 \\ 1 & -2 & 0 \end{vmatrix} = (-3)\begin{vmatrix} -10 & -21 & 5 \\ -7 & -21 & 2 \\ 1 & 0 & 0 \end{vmatrix}$$

$$= (-3)(-21)\begin{vmatrix} -10 & 1 & 5 \\ -7 & 1 & 2 \\ 1 & 0 & 0 \end{vmatrix}$$

$$= (-3)(-21)(-1)^{3+1}\begin{vmatrix} 1 & 5 \\ 1 & 2 \end{vmatrix}$$

$$= (-3)(-21)(2 - 5) = -189.$$

The minor $\Delta_{i,j}$ of an element $a_{i,j}$ of a square matrix A with the sign $(-1)^{i+j}$ attached is called the *cofactor* of $a_{i,j}$, and is denoted by $A_{i,j}$. Thus, $A_{i,j} = (-1)^{i+j}\Delta_{i,j}$. With this notation, the row and column expansions of $|A|$ given in Theorem 26.1 can be written

$$|A| = \sum_{j=1}^{n} a_{i,j}A_{i,j} \quad \text{and} \quad |A| = \sum_{i=1}^{n} a_{i,j}A_{i,j}.$$

A useful identity is obtained if the elements of a given row (column) of A are multiplied by the cofactors of another row (column):

(28.2) theorem If $s \neq k$, then $\Sigma_{j=1}^{n} a_{s,j}A_{k,j} = 0$, and if $t \neq k$, then

$$\sum_{i=1}^{n} a_{i,t}A_{i,k} = 0.$$

PROOF Let $B = [b_{i,j}]$, where, for all j and all $i \neq k$, $b_{i,j} = a_{i,j}$ and $b_{k,j} = a_{s,j}$ for $s \neq k$. Then row k of B is identical with row s of B. Moreover, $B_{k,j} = A_{k,j}$ for all j, since, except for row k, the rows of B are the same as the rows of A. Therefore, by Theorems 28.1 and 27.3,

$$\sum_{j=1}^{n} a_{s,j}A_{k,j} = \sum_{j=1}^{n} b_{k,j}B_{k,j} = |B| = 0.$$

The second identity follows similarly from the second identity of Theorem 28.1. ∎

EXAMPLE 2 Let

$$A = \begin{bmatrix} -1 & 2 & 4 & -6 \\ 3 & 1 & 0 & 2 \\ 4 & -5 & 2 & 1 \\ 0 & 7 & 3 & -2 \end{bmatrix}.$$

The cofactors of the elements in the second row of A are:

$$A_{2,1} = (-1)^{2+1}\begin{vmatrix} 2 & 4 & -6 \\ -5 & 2 & 1 \\ 7 & 3 & -2 \end{vmatrix} = -148;$$

$$A_{2,2} = (-1)^{2+2}\begin{vmatrix} -1 & 4 & -6 \\ 4 & 2 & 1 \\ 0 & 3 & -2 \end{vmatrix} = -33;$$

$$A_{2,3} = (-1)^{2+3}\begin{vmatrix} -1 & 2 & -6 \\ 4 & -5 & 1 \\ 0 & 7 & -2 \end{vmatrix} = 155;$$

$$A_{2,4} = (-1)^{2+4}\begin{vmatrix} -1 & 2 & 4 \\ 4 & -5 & 2 \\ 0 & 7 & 3 \end{vmatrix} = 117.$$

Multiplying the elements of the first row of A by the cofactors of the second row, which are computed above, we obtain

$$\sum_{j=1}^{4} a_{1,j}A_{2,j} = (-1)(-148) + 2(-33) + 4(155) + (-6)(117)$$

$$= 148 - 66 + 620 - 702 = 0$$

as an example of Theorem 28.2.

exercises

28.1 Let

$$A = \begin{bmatrix} -2 & 1 & 0 & 3 \\ 4 & 7 & 6 & 2 \\ -10 & 1 & 1 & 4 \\ 3 & 2 & 1 & -5 \end{bmatrix}$$

Compute the minors of elements 0, 6, and -10. Also compute the cofactors of these elements.

28.2 For the matrix

$$B = \begin{bmatrix} 2 & -1 & 0 & 3 \\ -5 & 6 & 1 & 10 \\ 0 & 2 & -3 & 4 \\ 7 & -5 & 2 & 1 \end{bmatrix}$$

compute the following cofactors: $B_{1,4}$, $B_{3,2}$, $B_{4,1}$, $B_{2,2}$, and $B_{1,3}$.

28.3 Use Theorem 28.1 to find the values of the determinants given in Exercises 26.5 and 27.1.

28.4 Prove, in detail, that the second identity of Theorem 28.1 follows from the first identity and Theorem 27.1.

28.5 For the matrix A in Exercise 28.1, verify Theorem 28.2 by evaluating the following sums:

$$\sum_{j=1}^{4} a_{1,j}A_{3,j}; \quad \sum_{j=1}^{4} a_{3,j}A_{2,j}; \quad \sum_{i=1}^{4} a_{i,2}A_{i,4}; \quad \sum_{i=1}^{4} a_{i,3}A_{i,1}.$$

28.6 For the matrix B in Exercise 28.2, evaluate the following sums:

$$\sum_{i=1}^{4} b_{i,4}B_{i,4}; \quad \sum_{j=1}^{4} b_{2,j}B_{2,j}; \quad \sum_{j=1}^{4} b_{2,j}B_{4,j}; \quad \sum_{i=1}^{4} b_{i,3}B_{i,2}.$$

29 some applications of determinants

In §28, we defined a minor of an element in a square matrix. More generally, if $A = [a_{i,j}]$ is an m by n matrix, then a *minor of A* is the determinant of a square k by k submatrix of A ($k \leq m$, $k \leq n$) obtained from A by choosing any k rows and any k columns of A. Such a minor is called a *k-rowed minor of A*.

EXAMPLE 1 If

$$A = \begin{bmatrix} -\frac{2}{3} & \sqrt{5} & 1 & 0 \\ -\sqrt{3} & \frac{1}{2} & 4 & 7 \\ 3 & 1 & -1 & 2 \end{bmatrix}$$

then the three-rowed minors of A are

$$\begin{vmatrix} -\frac{2}{3} & \sqrt{5} & 1 \\ -\sqrt{3} & \frac{1}{2} & 4 \\ 3 & 1 & -1 \end{vmatrix}, \quad \begin{vmatrix} -\frac{2}{3} & \sqrt{5} & 0 \\ -\sqrt{3} & \frac{1}{2} & 7 \\ 3 & 1 & 2 \end{vmatrix}, \quad \begin{vmatrix} -\frac{2}{3} & 1 & 0 \\ -\sqrt{3} & 4 & 7 \\ 3 & -1 & 2 \end{vmatrix}$$

and

$$\begin{vmatrix} \sqrt{5} & 1 & 0 \\ \frac{1}{2} & 4 & 7 \\ 1 & -1 & 2 \end{vmatrix}$$

There are 18 two-rowed minors of A, and the one-rowed minors are just the elements of A.

If $A \neq \mathbf{O}$ is an m by n matrix, then there is a positive integer t such that A has a nonzero t-rowed minor, and A does not have any nonzero s-rowed minor for $s > t$. The positive integer t is called the *determinantal rank* of A. If A is the zero matrix, then A has no nonzero minors, and the determinantal rank of A is 0.

EXAMPLE 2 The 4 by 3 matrix

$$A = \begin{bmatrix} 2 & -\frac{1}{2} & -1 \\ 1 & 0 & \frac{1}{3} \\ 7 & -1 & -1 \\ 4 & -\frac{1}{2} & -\frac{1}{3} \end{bmatrix}$$

has a nonzero two-rowed minor. For example,

$$\begin{vmatrix} 2 & -\frac{1}{2} \\ 1 & 0 \end{vmatrix} = \frac{1}{2}.$$

The only s-rowed minors of A with $s > 2$ are the three-rowed minors:

$$\begin{vmatrix} 2 & -\frac{1}{2} & -1 \\ 1 & 0 & \frac{1}{3} \\ 7 & -1 & -1 \end{vmatrix} = \begin{vmatrix} 2 & -\frac{1}{2} & -\frac{5}{3} \\ 1 & 0 & 0 \\ 7 & -1 & -\frac{10}{3} \end{vmatrix} = - \begin{vmatrix} -\frac{1}{2} & -\frac{5}{3} \\ -1 & -\frac{10}{3} \end{vmatrix} = 0;$$

$$\begin{vmatrix} 2 & -\frac{1}{2} & -1 \\ 1 & 0 & \frac{1}{3} \\ 4 & -\frac{1}{2} & -\frac{1}{3} \end{vmatrix} = \begin{vmatrix} 2 & -\frac{1}{2} & -\frac{5}{3} \\ 1 & 0 & 0 \\ 4 & -\frac{1}{2} & -\frac{5}{3} \end{vmatrix} = - \begin{vmatrix} -\frac{1}{2} & -\frac{5}{3} \\ -\frac{1}{2} & -\frac{5}{3} \end{vmatrix} = 0;$$

$$\begin{vmatrix} 2 & -\frac{1}{2} & -1 \\ 7 & -1 & -1 \\ 4 & -\frac{1}{2} & -\frac{1}{3} \end{vmatrix} = \begin{vmatrix} -5 & \frac{1}{2} & -1 \\ 0 & 0 & -1 \\ \frac{5}{3} & -\frac{1}{6} & -\frac{1}{3} \end{vmatrix} = \begin{vmatrix} -5 & \frac{1}{2} \\ \frac{5}{3} & -\frac{1}{6} \end{vmatrix} = 0;$$

$$\begin{vmatrix} 1 & 0 & \frac{1}{3} \\ 7 & -1 & -1 \\ 4 & -\frac{1}{2} & -\frac{1}{3} \end{vmatrix} = \begin{vmatrix} 1 & 0 & 0 \\ 7 & -1 & -\frac{10}{3} \\ 4 & -\frac{1}{2} & -\frac{5}{3} \end{vmatrix} = \begin{vmatrix} -1 & -\frac{10}{3} \\ -\frac{1}{2} & -\frac{5}{3} \end{vmatrix} = 0.$$

Thus, A has a nonzero two-rowed minor, and A has no nonzero s-rowed minor for $s > 2$. Therefore, the determinantal rank of A is 2.

An echelon form for A is

$$A^* = \begin{bmatrix} 1 & 0 & \frac{1}{3} \\ 0 & 1 & \frac{10}{3} \\ 0 & 0 & 0 \\ 0 & 0 & 0 \end{bmatrix}$$

Hence, the rank of A is 2, and in this example, we observe that the determinantal rank of A agrees with the rank of A as previously defined.

The result noted in Example 2 is true in general.

(29.1) theorem Let k be the rank of the m by n matrix A, and let t be the determinantal rank of A. Then $t = k$.

PROOF We may suppose that $A \neq \mathbf{O}$, since if $A = \mathbf{O}$, both the rank and determinantal rank of A are zero. Since the determinantal rank of A is t, there is a t by t submatrix B of A such that $|B| \neq 0$. Let C be the t by n submatrix of A that contains the same t rows as B. Then the columns of C form a set of n vectors in $\mathcal{V}_t$ that contains the t vectors in $\mathcal{V}_t$ which are the columns of B. Thus, the column space of C contains the column space of B. Consequently, the dimension of the column space of B is not greater than the dimension of the column space of C. That is,

$$\text{column rank of } B \leq \text{column rank of } C.$$

The rows of C form a set of t vectors in $\mathcal{V}_n$ that is a subset of the set of m vectors in $\mathcal{V}_n$ which are the rows of A. Therefore, the row space of C is contained in the row space of A. Hence, the dimension of the row space of C is not greater than the dimension of the row space of A. Consequently,

$$\text{row rank of } C \leq \text{row rank of } A.$$

Since $|B| \neq 0$, it follows from Theorems 27.7 and 20.7 that B has rank t. We have

$$\begin{aligned} t &= \text{rank of } B = \text{column rank of } B \leq \text{column rank of } C \\ &= \text{row rank of } C \leq \text{row rank of } A = \text{rank of } A = k. \end{aligned}$$

Thus, $t \leq k$.

Since row rank of A = rank of $A = k$, the row space of A has dimension k. Therefore, A has k linearly independent rows (as vectors in $\mathcal{V}_n$). Let D be the k by n submatrix of A with these k linearly independent rows. Then row rank of $D = k$. Since column rank of D = row rank of D, the column space of D has dimension k. Thus, D contains k linearly independent columns (as vectors in $\mathcal{V}_k$). Let E be the k by k submatrix of D with these k linearly independent columns. Then column rank of E = rank of $E = k$. By Theorems 20.7 and 27.7, $|E| \neq 0$. Thus, A contains a nonzero k-rowed minor. Therefore, the determinantal rank of A is greater than or equal to k. That is, $t \geq k$. We have proved that $t \leq k$ and that $t \geq k$. Therefore, $t = k$. ∎

Theorem 29.1 provides a convenient method for determining the rank of a matrix.

EXAMPLE 3 In the matrix

$$A = \begin{bmatrix} 2 & 1 & -3 & 0 & -4 & 0 \\ -5 & 0 & 2 & 1 & 25 & 0 \\ 2/3 & 0 & -1 & 0 & 16 & 1 \end{bmatrix}$$

the three-rowed minor

$$\begin{vmatrix} 1 & 0 & 0 \\ 0 & 1 & 0 \\ 0 & 0 & 1 \end{vmatrix} = 1 \neq 0.$$

Therefore, by Theorem 29.1, A has rank 3.

As a second application of determinants, we derive another method for finding the inverse of a nonsingular matrix that is of more theoretical than practical interest. Let $A = [a_{i,j}]$ be an n by n matrix. The *adjoint* of A is the n by n matrix that has the cofactor $A_{j,i}$ of $a_{j,i}$ in the (i, j)-position for all i, j. Thus, if we denote the adjoint of A by A^{adj}, then A^{adj} is the transpose of the matrix $[A_{i,j}]$. That is, $A^{\mathrm{adj}} = [A_{i,j}]'$.

EXAMPLE 4 In Example 2, §28, we computed the cofactors $A_{2,1} = -148$, $A_{2,2} = -33$, $A_{2,3} = 155$, and $A_{2,4} = 117$ for the matrix

$$A = \begin{bmatrix} -1 & 2 & 4 & -6 \\ 3 & 1 & 0 & 2 \\ 4 & -5 & 2 & 1 \\ 0 & 7 & 3 & -2 \end{bmatrix}$$

The reader can verify that the remaining cofactors of A are: $A_{1,1} = -65$, $A_{1,2} = -3$, $A_{1,3} = 73$, $A_{1,4} = 99$, $A_{3,1} = 34$, $A_{3,2} = 24$, $A_{3,3} = -98$, $A_{3,4} = -63$, $A_{4,1} = 64$, $A_{4,2} = -12$, $A_{4,3} = -113$, $A_{4,4} = -90$. Therefore,

$$A^{\text{adj}} = \begin{bmatrix} -65 & -3 & 73 & 99 \\ -148 & -33 & 155 & 117 \\ 34 & 24 & -98 & -63 \\ 64 & -12 & -113 & -90 \end{bmatrix}' = \begin{bmatrix} -65 & -148 & 34 & 64 \\ -3 & -33 & 24 & -12 \\ 73 & 155 & -98 & -113 \\ 99 & 117 & -63 & -90 \end{bmatrix}$$

We note that

$$AA^{\text{adj}} = \begin{bmatrix} -1 & 2 & 4 & -6 \\ 3 & 1 & 0 & 2 \\ 4 & -5 & 2 & 1 \\ 0 & 7 & 3 & -2 \end{bmatrix} \begin{bmatrix} -65 & -148 & 34 & 64 \\ -3 & -33 & 24 & -12 \\ 73 & 155 & -98 & -113 \\ 99 & 117 & -63 & -90 \end{bmatrix}$$

$$= \begin{bmatrix} -243 & 0 & 0 & 0 \\ 0 & -243 & 0 & 0 \\ 0 & 0 & -243 & 0 \\ 0 & 0 & 0 & -243 \end{bmatrix} = (-243) \cdot I_4.$$

Also, $|A| = -243$, so that $AA^{\text{adj}} = |A| \cdot I_4$. This is an example of the following theorem.

(29.2) theorem For any n by n matrix $A = [a_{i,j}]$, $AA^{\text{adj}} = A^{\text{adj}}A = |A| \cdot I_n$.

PROOF By the definition of matrix multiplication, the element in the ith row, jth column of AA^{adj} is $\sum_{k=1}^{n} a_{i,k}A_{j,k}$ (since $A_{j,k}$ is in the (k, j)-position of A^{adj}). If $j = i$, then $\sum_{k=1}^{n} a_{i,k}A_{j,k} = |A|$ by Theorem 28.1, and if $j \neq i$, then $\sum_{k=1}^{n} a_{i,k}A_{j,k} = 0$ by Theorem 28.2. Hence, each diagonal element of AA^{adj} is $|A|$, and the elements not on the diagonal are zero. Therefore, $AA^{\text{adj}} = |A| \cdot I_n$. The identity $A^{\text{adj}}A = |A| \cdot I_n$ is proved similarly. ∎

(29.3) corollary If A is a nonsingular n by n matrix, then $A^{-1} = (1/|A|) \cdot A^{\text{adj}}$.

PROOF

$$[(1/|A|) \cdot A^{\text{adj}}]A = (1/|A|) \cdot (A^{\text{adj}}A)$$
$$= (1/|A|) \cdot (|A| \cdot I_n) = [(1/|A|)|A|] \cdot I_n = I_n. \quad ∎$$

Computationally, the method 23.4 for finding the inverse of a nonsingular matrix is preferable to the formula given in Corollary 29.3. For example, if $n = 4$, to compute $(1/|A|) \cdot A^{\text{adj}}$ requires the evaluation of $|A|$ and the sixteen 3 by 3 cofactors $A_{i,j}$.

EXAMPLE 5 In Example 4, $A^{-1} = (1/|A|) \cdot A^{\mathrm{adj}}$

$$= \left(-\frac{1}{243}\right) \cdot \begin{bmatrix} -65 & -148 & 34 & 64 \\ -3 & -33 & 24 & -12 \\ 73 & 155 & -98 & -113 \\ 99 & 117 & -63 & -90 \end{bmatrix}$$

$$= \begin{bmatrix} {}^{65}\!/_{243} & {}^{148}\!/_{243} & -{}^{34}\!/_{243} & -{}^{64}\!/_{243} \\ {}^{1}\!/_{81} & {}^{11}\!/_{81} & -{}^{8}\!/_{81} & {}^{4}\!/_{81} \\ -{}^{73}\!/_{243} & -{}^{155}\!/_{243} & {}^{98}\!/_{243} & {}^{113}\!/_{243} \\ -{}^{11}\!/_{27} & -{}^{13}\!/_{27} & {}^{7}\!/_{27} & {}^{10}\!/_{27} \end{bmatrix}$$

If $A = [a_{i,j}]$ is a nonsingular n by n matrix, then the system of n linear equations in n unknowns, which can be written as the single matrix equation $AX = B$, has a unique solution $X = A^{-1}B$ (Theorem 20.8). Using the form of A^{-1} given in Corollary 29.3,

$$\begin{bmatrix} x_1 \\ x_2 \\ \vdots \\ x_n \end{bmatrix} = \frac{1}{|A|} \cdot A^{\mathrm{adj}} \begin{bmatrix} b_1 \\ b_2 \\ \vdots \\ b_n \end{bmatrix} = \frac{1}{|A|} \cdot [A_{i,j}]' \begin{bmatrix} b_1 \\ b_2 \\ \vdots \\ b_n \end{bmatrix}$$

$$= \frac{1}{|A|} \cdot \begin{bmatrix} \sum_{k=1}^{n} A_{k,1} b_k \\ \sum_{k=1}^{n} A_{k,2} b_k \\ \vdots \\ \sum_{k=1}^{n} A_{k,n} b_k \end{bmatrix} = \begin{bmatrix} \sum_{k=1}^{n} A_{k,1} b_k/|A| \\ \sum_{k=1}^{n} A_{k,2} b_k/|A| \\ \vdots \\ \sum_{k=1}^{n} A_{k,n} b_k/|A| \end{bmatrix}$$

Hence, $x_i = \sum_{k=1}^{n} A_{k,i} b_k/|A|$ for $i = 1, 2, \ldots, n$. Note that $\sum_{k=1}^{n} A_{k,i} b_k = \sum_{k=1}^{n} b_k A_{k,i}$ is the expansion of the determinant

$$\begin{vmatrix} a_{1,1} & a_{1,2} & \cdots & a_{1,i-1} & b_1 & a_{1,i+1} & \cdots & a_{1,n} \\ a_{2,1} & a_{2,2} & \cdots & a_{2,i-1} & b_2 & a_{2,i+1} & \cdots & a_{2,n} \\ \vdots & \vdots & & \vdots & \vdots & \vdots & & \vdots \\ a_{n,1} & a_{n,2} & \cdots & a_{n,i-1} & b_n & a_{n,i+1} & \cdots & a_{n,n} \end{vmatrix}$$

This method for solving the system $AX = B$, where A is nonsingular is

Cramer's Rule, which was mentioned in the introductory remarks of this chapter.

EXAMPLE 6 The system of equations

$$\begin{aligned} 2x_1 - 3x_2 + x_3 - x_4 &= 5 \\ x_1 + 4x_2 - 2x_3 + 3x_4 &= 1 \\ 3x_1 - x_2 + 4x_3 + x_4 &= -2 \\ -x_1 + 2x_2 - 5x_3 + 7x_4 &= 0 \end{aligned}$$

has

$$A = \begin{bmatrix} 2 & -3 & 1 & -1 \\ 1 & 4 & -2 & 3 \\ 3 & -1 & 4 & 1 \\ -1 & 2 & -5 & 7 \end{bmatrix}$$

as matrix of coefficients. The matrix A is nonsingular with $|A| = 291$. Using Cramer's Rule, the solution of the given system is

$$x_1 = \frac{\begin{vmatrix} 5 & -3 & 1 & -1 \\ 1 & 4 & -2 & 3 \\ -2 & -1 & 4 & 1 \\ 0 & 2 & -5 & 7 \end{vmatrix}}{291} = \frac{630}{291},$$

$$x_2 = \frac{\begin{vmatrix} 2 & 5 & 1 & -1 \\ 1 & 1 & -2 & 3 \\ 3 & -2 & 4 & 1 \\ -1 & 0 & -5 & 7 \end{vmatrix}}{291} = -\frac{167}{291},$$

$$x_3 = \frac{\begin{vmatrix} 2 & -3 & 5 & -1 \\ 1 & 4 & 1 & 3 \\ 3 & -1 & -2 & 1 \\ -1 & 2 & 0 & 7 \end{vmatrix}}{291} = -\frac{589}{291},$$

$$x_4 = \frac{\begin{vmatrix} 2 & -3 & 1 & 5 \\ 1 & 4 & -2 & 1 \\ 3 & -1 & 4 & -2 \\ -1 & 2 & -5 & 0 \end{vmatrix}}{291} = -\frac{283}{291}.$$

exercises

29.1 List the two-rowed minors of the matrix A of Example 1.

29.2 Evaluate the three-rowed minors of the matrix

$$A = \begin{bmatrix} 1 & 1 & 1 & 1 & 1 \\ 2 & -1 & 2 & -1 & 2 \\ 0 & 1 & 0 & 1 & 0 \\ 0 & 0 & 1 & 0 & 0 \end{bmatrix}.$$

29.3 Let A be an m by n matrix such that every s-rowed minor of A is zero ($s < m$, $s < n$). Prove that if $t > s$, then every t-rowed minor of A is zero. Prove that the rank of A is less than s.

29.4 Let A be an m by n matrix which has a nonzero s-rowed minor. Suppose that every $(s + 1)$-rowed minor of A is zero. Prove that the rank of A is s.

29.5 Find the adjoints of the following matrices:

(a) $\begin{bmatrix} 3 & -1 & 2 \\ 0 & 1 & -5 \\ 6 & 7 & 4 \end{bmatrix}$ (b) $\begin{bmatrix} 0 & 0 & 1 & 0 \\ 0 & 1 & -2 & 0 \\ 0 & 0 & 0 & 1 \\ 1 & 0 & 0 & 0 \end{bmatrix}$ (c) $\begin{bmatrix} 1 & 1 & 1 & 1 \\ 2 & 2 & 2 & 2 \\ 3 & 3 & 3 & 3 \\ 4 & 4 & 4 & 4 \end{bmatrix}$

29.6 Let A be an n by n matrix ($n \geq 2$). Prove that $A^{\text{adj}} = \mathbf{O}$ if and only if rank of $A < n - 1$.

***29.7** Prove that $|A^{\text{adj}}| = |A|^{n-1}$ for any n by n matrix $A (n \geq 2)$.

29.8 Solve the following systems of linear equations by Cramer's Rule:

(a)
$$\begin{aligned} x_1 + x_2 + x_3 + x_4 &= 1 \\ x_1 - 4x_2 + 6x_3 + x_4 &= 3 \\ 4x_1 + 2x_2 + 6x_3 - 2x_4 &= 4 \\ 6x_1 - 5x_2 + 5x_3 + 3x_4 &= 7 \end{aligned}$$

(b)
$$\begin{aligned} 2x_1 - 3x_2 + x_3 &= 0 \\ x_1 - 5x_2 - x_3 &= 1 \\ 4x_1 + 2x_2 + 3x_3 &= 5 \end{aligned}$$

(c)
$$\begin{aligned} x_1 - 3x_2 + 2x_3 &= 0 \\ 2x_1 - x_2 + x_3 &= 0 \\ 6x_1 + 3x_2 - 5x_3 &= 0 \end{aligned}$$

29.9 Find the inverses of the nonsingular matrices given in Exercise 29.5 by the method of Corollary 29.3.

29.10 Prove that a homogeneous system of n linear equations in n unknowns, $AX = \mathbf{O}$, has a nontrivial solution if and only if $|A| = 0$.

7

similarity of matrices

30 characteristic values

In §22 we studied the effect of a change of basis on the matrix of a linear transformation L of a finite dimensional vector space $\mathcal{V}$ into itself. We found that if P is the nonsingular matrix describing the change of basis in $\mathcal{V}$ [see equations (21–1)], then the matrices A and B of L with respect to the two bases of $\mathcal{V}$ were related by the equation $B = Q^{-1}AQ$, where $Q = P'$. In this situation, the matrices A and B were said to be similar (see Definition 22.5). Now we would like to characterize the linear transformations of a vector space $\mathcal{V}$ into itself that have a diagonal matrix with respect to a suitable basis in $\mathcal{V}$. This problem is equivalent to that of characterizing the n by n matrices that are similar to a diagonal matrix. It is particularly easy to describe the properties of a linear transformation of $\mathcal{V}$ into itself that has a diagonal matrix with respect to some basis of $\mathcal{V}$.

Throughout this chapter, $\mathcal{V}$ is an n-dimensional real vector space and L is a linear transformation of $\mathcal{V}$ into itself. Certain vectors in $\mathcal{V}$ play a central role in the discussion of similarity, namely, those which L sends into a scalar multiple of themselves.

(30.1) definition A real number r such that $L(\mathbf{V}) = r \cdot \mathbf{V}$ for some nonzero vector $\mathbf{V} \in \mathcal{V}$ is called a *characteristic value of L*. A nonzero vector $\mathbf{V}$ satisfying $L(\mathbf{V}) = r \cdot \mathbf{V}$ is called a *characteristic vector of L* (belonging to the characteristic value r).

Characteristic values and vectors of a linear transformation are often called *eigenvalues* and *eigenvectors,* respectively.

EXAMPLE 1 Let L be the linear transformation of $\mathcal{V}_3$ that has the matrix

$$A = \begin{bmatrix} 2 & 4 & 1 \\ 1 & -2 & -1 \\ 0 & 0 & 0 \end{bmatrix}$$

with respect to the basis $\{\mathbf{E}_1^{(3)}, \mathbf{E}_2^{(3)}, \mathbf{E}_3^{(3)}\}$. Then r is a characteristic value of L if and only if there is a nonzero vector $(x_1, x_2, x_3) \in \mathcal{V}_3$ such that

$$A \begin{bmatrix} x_1 \\ x_2 \\ x_3 \end{bmatrix} = r \cdot \begin{bmatrix} x_1 \\ x_2 \\ x_3 \end{bmatrix} = (r \cdot I_3) \begin{bmatrix} x_1 \\ x_2 \\ x_3 \end{bmatrix}$$

which can be written in the form

$$(A - r \cdot I_3) \begin{bmatrix} x_1 \\ x_2 \\ x_3 \end{bmatrix} = \begin{bmatrix} 0 \\ 0 \\ 0 \end{bmatrix}. \tag{30–1}$$

Thus, the problem of finding the characteristic values of L is reduced to that of finding the values of r for which the homogeneous system of equations (30–1) has a nontrivial solution. By Corollary 20.6, this system has a nontrivial solution if and only if the rank of $A - r \cdot I_3$ is less than 3, that is, if and only if $|A - r \cdot I_3| = 0$. We have

$$|A - r \cdot I_3| = \begin{vmatrix} 2-r & 4 & 1 \\ 1 & -2-r & -1 \\ 0 & 0 & -r \end{vmatrix} = -r \begin{vmatrix} 2-r & 4 \\ 1 & -2-r \end{vmatrix}$$
$$= -r(-4 + r^2 - 4) = -r(r^2 - 8)$$
$$= -r(r - 2\sqrt{2})(r + 2\sqrt{2}) = 0.$$

Thus, the characteristic values of L are 0, $2\sqrt{2}$, and $-2\sqrt{2}$. Corresponding to the characteristic value $r = 0$, we have $A - r \cdot I_3 = A - 0 \cdot I_3 = A$. The system (30–1) is

$$\begin{bmatrix} 2 & 4 & 1 \\ 1 & -2 & -1 \\ 0 & 0 & 0 \end{bmatrix} \begin{bmatrix} x_1 \\ x_2 \\ x_3 \end{bmatrix} = \begin{bmatrix} 0 \\ 0 \\ 0 \end{bmatrix}.$$

Solving this system by the methods of Chapter 3, we obtain $(x_1, x_2, x_3) = (s/4, -3s/8, s)$, where s is any real number. According to Definition 30.1, characteristic vectors are nonzero vectors. Thus, for every $s \neq 0$, $(s/4, -3s/8, s)$ is a characteristic vector of L that belongs to the characteristic value $r = 0$. That is, $L[(s/4, -3s/8, s)] = 0 \cdot (s/4, -3s/8, s) = (0, 0, 0)$.

Corresponding to the characteristic value $r = 2\sqrt{2}$, we find

$$A - r \cdot I_3 = A - 2\sqrt{2} \cdot I_3$$
$$= \begin{bmatrix} 2 - 2\sqrt{2} & 4 & 1 \\ 1 & -2 - 2\sqrt{2} & -1 \\ 0 & 0 & -2\sqrt{2} \end{bmatrix}.$$

The system (30–1) is

$$\begin{bmatrix} 2 - 2\sqrt{2} & 4 & 1 \\ 1 & -2 - 2\sqrt{2} & -1 \\ 0 & 0 & -2\sqrt{2} \end{bmatrix} \begin{bmatrix} x_1 \\ x_2 \\ x_3 \end{bmatrix} = \begin{bmatrix} 0 \\ 0 \\ 0 \end{bmatrix}.$$

The solution of this latter system of homogeneous equations is $(x_1, x_2, x_3) = ([2 + 2\sqrt{2}]s, s, 0)$, for any real number s. For every $s \neq 0$, $([2 + 2\sqrt{2}]s, s, 0)$ is a characteristic vector of L belonging to the characteristic value $r = 2\sqrt{2}$. Therefore,

$$L[([2 + 2\sqrt{2}]s, s, 0)] = 2\sqrt{2} \cdot ([2 + 2\sqrt{2}]s, s, 0)$$
$$= ([4\sqrt{2} + 8]s, 2\sqrt{2}s, 0).$$

This result can be verified directly. Indeed,

$$\begin{bmatrix} 2 & 4 & 1 \\ 1 & -2 & 1 \\ 0 & 0 & 0 \end{bmatrix} \begin{bmatrix} (2 + 2\sqrt{2})s \\ s \\ 0 \end{bmatrix} = \begin{bmatrix} (4\sqrt{2} + 8)s \\ 2\sqrt{2}s \\ 0 \end{bmatrix}.$$

The task of finding the characteristic vectors belonging to the characteristic value $-2\sqrt{2}$ is left to the reader (Exercise 30.2).

The example above serves to illustrate the following theorem.

(30.2) theorem Let L be a linear transformation of $\mathcal{V}$, and let $\{\mathbf{U}_1, \mathbf{U}_2, \ldots, \mathbf{U}_n\}$ be any basis of $\mathcal{V}$. Let A be the matrix of L with respect to this basis. A real number r is a characteristic value of L if and only if r is a real root of the equation $|A - x \cdot I_n| = 0$. The characteristic vectors of L that belong to a characteristic value r of L are those nonzero vectors $\mathbf{V} \in \mathcal{V}$ with coordinates $(x_1, x_2, \ldots, x_n)$ with respect to the basis $\{\mathbf{U}_1, \mathbf{U}_2, \ldots, \mathbf{U}_n\}$, which satisfy

$$(A - r \cdot I_n) \begin{bmatrix} x_1 \\ x_2 \\ \vdots \\ x_n \end{bmatrix} = \begin{bmatrix} 0 \\ 0 \\ \vdots \\ 0 \end{bmatrix}.$$

The proof of Theorem 30.2 follows step-by-step the argument given in Example 1 and is left as an exercise (Exercise 30.8). In Example 1,

$|A - x \cdot I_n| = 0$ is the equation $x^3 - 8x = x(x^2 - 8) = 0$, which has three distinct real roots. However, this is not generally the case. The equation $|A - x \cdot I_n| = 0$ is always a polynomial equation of degree n with real coefficients. Such an equation has n roots (not necessarily distinct), but some of them may be complex numbers. For example, if

$$A = \begin{bmatrix} 2 & -5 & 1 \\ 1 & -2 & -1 \\ 0 & 0 & 0 \end{bmatrix},$$

then

$$|A - x \cdot I_3| = \begin{vmatrix} 2-x & -5 & 1 \\ 1 & -2-x & -1 \\ 0 & 0 & -x \end{vmatrix}$$
$$= -x(-4 + x^2 + 5) = -x(x^2 + 1).$$

In this case, the roots of the equation $-x(x^2 + 1) = 0$ are 0, i, and $-i$. A linear transformation L of $\mathcal{V}_3$ with matrix A has 0 as its only characteristic value according to Theorem 30.2.

EXAMPLE 2 Let $\mathcal{V}$ be the vector space of polynomials of degree at most 3 (see Example 5, §9). That is,

$$\mathcal{V} = \{a_3x^3 + a_2x^2 + a_1x + a_0 | a_3, a_2, a_1, a_0 \in R\}.$$

Ordinary differentiation of polynomials is a linear transformation of $\mathcal{V}$ into itself, where

$$D(a_3x^3 + a_2x^2 + a_1x + a_0) = 3a_3x^2 + 2a_2x + a_1.$$

The set of polynomials $\{x^3, x^2, x, 1\}$ is a basis of $\mathcal{V}$. Since $D(x^3) = 3x^2$, $D(x^2) = 2x$, $D(x) = 1$, and $D(1) = 0$, it follows that the matrix of D with respect to the basis $\{x^3, x^2, x, 1\}$ is

$$A = \begin{bmatrix} 0 & 0 & 0 & 0 \\ 3 & 0 & 0 & 0 \\ 0 & 2 & 0 & 0 \\ 0 & 0 & 1 & 0 \end{bmatrix}.$$

Therefore,

$$|A - x \cdot I_4| = \begin{vmatrix} -x & 0 & 0 & 0 \\ 3 & -x & 0 & 0 \\ 0 & 2 & -x & 0 \\ 0 & 0 & 1 & -x \end{vmatrix} = x^4.$$

The equation $x^4 = 0$ has 0 as a repeated real root. Thus, 0 is the only characteristic value of D. Corresponding to the characteristic value 0, we solve the homogeneous system of linear equations

$$\begin{bmatrix} 0 & 0 & 0 & 0 \\ 3 & 0 & 0 & 0 \\ 0 & 2 & 0 & 0 \\ 0 & 0 & 1 & 0 \end{bmatrix} \begin{bmatrix} a_3 \\ a_2 \\ a_1 \\ a_0 \end{bmatrix} = \begin{bmatrix} 0 \\ 0 \\ 0 \\ 0 \end{bmatrix}$$

and find that $(0, 0, 0, a_0)$, where a_0 is an arbitrary real number, is the solution. That is, the polynomials $0x^3 + 0x^2 + 0x + a_0 1 = a_0$, where $a_0 \neq 0$, are the characteristic vectors belonging to the characteristic value 0. In other words, the nonzero constant polynomials are the characteristic vectors of the linear transformation D.

(30.3) definition Let A be a real n by n matrix. The determinant $|A - x \cdot I_n|$, which is a polynomial of degree n in x, is called the *characteristic polynomial of A*. The equation $|A - x \cdot I_n| = 0$ is called the *characteristic equation of A*. The n (possibly complex) roots of this equation are called the *characteristic roots of A*.

Thus, the characteristic values of a linear transformation L are the real characteristic roots of any matrix A of L.

(30.4) theorem Similar matrices have the same characteristic polynomial (and hence the same characteristic roots).

PROOF If $B = P^{-1}AP$, then

$$\begin{aligned} B - x \cdot I_n &= P^{-1}AP - x \cdot P^{-1}P = P^{-1}AP - P^{-1}(x \cdot I_n)P \\ &= P^{-1}(A - x \cdot I_n)P. \end{aligned}$$

Thus,

$$|B - x \cdot I_n| = |P^{-1}(A - x \cdot I_n)P| = |P^{-1}|\,|A - x \cdot I_n|\,|P| = |A - x \cdot I_n|,$$

by Theorem 27.8 and Corollary 27.9. ∎

It should be noted that matrices with the same characteristic polynomial are not necessarily similar. The n by n triangular matrix

$$A = \begin{bmatrix} 1 & a_{1,2} & a_{1,3} & \cdots & a_{1,n} \\ 0 & 1 & a_{2,3} & \cdots & a_{2,n} \\ 0 & 0 & 1 & \cdots & a_{3,n} \\ 0 & 0 & 0 & \cdot & \cdot \\ \cdot & \cdot & \cdot & \cdot & \cdot \\ \cdot & \cdot & \cdot & & \cdot \\ \cdot & \cdot & \cdot & \cdot & a_{n-1,n} \\ 0 & 0 & 0 & \cdots & 1 \end{bmatrix}$$

has characteristic polynomial $(1 - x)^n$, which is also the characteristic polynomial of I_n. Suppose that some $a_{i,j} \neq 0$ so that $A \neq I_n$. Then A and I_n are not similar since I_n is similar only to itself. Indeed, $P^{-1}I_nP = P^{-1}P = I_n$ for any nonsingular P.

exercises

30.1 Let L be the linear transformation of $\mathcal{V}_4$ into itself that has matrix

$$A = \begin{bmatrix} 2 & -3 & 4 & 1 \\ 0 & 3 & 1 & 2 \\ 0 & 0 & 1 & 5 \\ 0 & 0 & -2 & 3 \end{bmatrix}$$

with respect to the basis $E^{(4)}$. Find the characteristic values and characteristic vectors of L.

30.2 In Example 1, find the characteristic vectors that belong to the characteristic value $-2\sqrt{2}$.

30.3 Let L be the linear transformation of $\mathcal{E}_3$ into itself defined by $L(\mathbf{V}) = (x_0 \cdot \mathbf{I} + y_0 \cdot \mathbf{J}) \times \mathbf{V}$. Find the matrix of L with respect to the basis $\{\mathbf{I}, \mathbf{J}, \mathbf{K}\}$. Prove that 0 is the only characteristic value of L.

30.4 Let L be the linear transformation of $\mathcal{E}_3$ into itself that has matrix

$$B = \begin{bmatrix} 0 & 1 & 0 \\ 0 & 0 & 1 \\ -8 & 6 & 3 \end{bmatrix}$$

with respect to the basis $\{\mathbf{I}, \mathbf{J}, \mathbf{K}\}$. Find the characteristic values and characteristic vectors of L. Show that $\mathcal{E}_3$ has a basis $\{\mathbf{U}, \mathbf{V}, \mathbf{W}\}$, where $\mathbf{U}$, $\mathbf{V}$, and $\mathbf{W}$ are characteristic vectors of L. Find the matrix of L with respect to the basis $\{\mathbf{U}, \mathbf{V}, \mathbf{W}\}$.

30.5 Let $\mathcal{V}$ be the subspace of the vector space of polynomials $R[x]$ consisting of the polynomials of degree at most k. Let D be the linear transformation of $\mathcal{V}$ into itself which is ordinary differentiation of polynomials. Show that the characteristic vectors of D are the nonzero constant polynomials in $\mathcal{V}$. This is a generalization of Example 2.

30.6 Prove that the set of all characteristic vectors of a linear transformation L of $\mathcal{V}$ that belong to the same characteristic value r form a subspace of $\mathcal{V}$.

30.7 Prove that a linear transformation L of $\mathcal{V}$ is nonsingular if and only if 0 is not a characteristic value of L.

30.8 Prove Theorem 30.2.

30.9 Prove that if the n by n matrices A and B are similar, then $|A| = |B|$.

30.10 Find the characteristic roots of the following matrices:

(a) $\begin{bmatrix} \frac{1}{2} & 2 \\ -\frac{1}{3} & 5 \end{bmatrix}$ (b) $\begin{bmatrix} 0 & 0 & 1 & 0 \\ 0 & 0 & 0 & 1 \\ 0 & 1 & 0 & 0 \\ 1 & 0 & 0 & 0 \end{bmatrix}$ (c) $\begin{bmatrix} 1 & 2 & 3 \\ 0 & 1 & 2 \\ 0 & -2 & 1 \end{bmatrix}$

(d) $\begin{bmatrix} 0 & 1 & 0 \\ 0 & 0 & 1 \\ -4 & 2 & 0 \end{bmatrix}$ (e) $\begin{bmatrix} -4 & 0 & 0 & 0 \\ 6 & 2 & 0 & 0 \\ -5 & 3 & 2 & 0 \\ 1 & -7 & 6 & 3 \end{bmatrix}$ (f) $\begin{bmatrix} 1 & 2 & 3 \\ 2 & 3 & 4 \\ 3 & 4 & 5 \end{bmatrix}$

30.11 Show that the characteristic roots of an n by n triangular matrix are the diagonal elements of the matrix.

30.12 Show that the matrix

$$\begin{bmatrix} a & b \\ b & c \end{bmatrix}$$

has real characteristic roots, for any real numbers a, b, c.

31 matrices similar to diagonal matrices

In this section we find the necessary and sufficient condition for a real n by n matrix A to be similar to a real diagonal matrix.

(31.1) theorem Let A be a real n by n matrix, and let L be the linear transformation of $\mathcal{V}$ that has matrix A with respect to some basis of $\mathcal{V}$. The matrix A is similar to a diagonal matrix

$$D = \begin{bmatrix} r_1 & 0 & \cdots & 0 \\ 0 & r_2 & & \cdot \\ \cdot & & \cdot & \cdot \\ \cdot & & \cdot & \cdot \\ \cdot & & \cdot & 0 \\ 0 & \cdots & 0 & r_n \end{bmatrix} \tag{31-1}$$

where $r_1, r_2, \ldots, r_n$ are real numbers, if and only if there is a basis of $\mathcal{V}$ consisting of characteristic vectors of L.

PROOF By Theorem 22.4, A is similar to D if and only if there is a basis $\{\mathbf{W}_1, \mathbf{W}_2, \ldots, \mathbf{W}_n\}$ of $\mathcal{V}$ such that L has matrix D with respect to this basis. For any basis $\{\mathbf{W}_1, \mathbf{W}_2, \ldots, \mathbf{W}_n\}$, the coordinates of $L(\mathbf{W}_i)$ are the elements in the ith column of the matrix of L with respect to this basis. Thus, L has matrix D if and only if

$$\begin{aligned} L(\mathbf{W}_i) &= 0 \cdot \mathbf{W}_1 + \cdots + 0 \cdot \mathbf{W}_{i-1} + r_i \cdot \mathbf{W}_i + 0 \cdot \mathbf{W}_{i+1} + \cdots + 0 \cdot \mathbf{W}_n \\ &= r_i \cdot \mathbf{W}_i \end{aligned}$$

for $i = 1, 2, \ldots, n$. However, $L(\mathbf{W}_i) = r_i \cdot \mathbf{W}_i$ if and only if the vector $\mathbf{W}_i$ in the basis $\{\mathbf{W}_1, \mathbf{W}_2, \ldots, \mathbf{W}_n\}$ is a characteristic vector of L. ∎

Suppose that the condition of Theorem 31.1 is satisfied. That is, $\mathcal{V}$ has a basis $\{\mathbf{W}_1, \mathbf{W}_2, \ldots, \mathbf{W}_n\}$ consisting of characteristic vectors of L. Let A be the matrix of L with respect to a basis $\{\mathbf{U}_1, \mathbf{U}_2, \ldots, \mathbf{U}_n\}$ of $\mathcal{V}$. By the proof of Theorem 31.1, L has a diagonal matrix D with respect to the basis $\{\mathbf{W}_1, \mathbf{W}_2, \ldots, \mathbf{W}_n\}$. Let $P = [p_{i,j}]$ be the matrix of the change of basis from the **U**-basis to the **W**-basis. Then, for $i = 1, 2, \ldots, n$, the row $(p_{i,1}, p_{i,2}, \ldots, p_{i,n})$ of P gives the coordinates of $\mathbf{W}_i$ with respect to the **U**-basis [see equations (21–1)]. Moreover, by the results of §22, $D = (P')^{-1}AP'$. Setting $Q = P'$, we have $D = Q^{-1}AQ$, where the columns of Q are the coordinates of the characteristic vectors $\mathbf{W}_i$ with respect to the **U**-basis. Thus, given a basis of characteristic vectors of L we have a method for constructing a nonsingular matrix Q such that $Q^{-1}AQ$ is a diagonal matrix. Conversely, it is not difficult to show that if Q is any nonsingular matrix such that $Q^{-1}AQ = D$, where D is a diagonal matrix, then the columns of Q must be the coordinates of n characteristic vectors of L which form a basis of $\mathcal{V}$ (see Exercise 31.1).

EXAMPLE 1 Let

$$A = \begin{bmatrix} 2 & -2 & 2 & 1 \\ -1 & 3 & 0 & 3 \\ 0 & 0 & 4 & -2 \\ 0 & 0 & 2 & -1 \end{bmatrix}.$$

Let L be the linear transformation of $\mathcal{V}_4$ that has matrix A with respect to the basis $E^{(4)}$. The characteristic equation of A is

$$\begin{vmatrix} 2-x & -2 & 2 & 1 \\ -1 & 3-x & 0 & 3 \\ 0 & 0 & 4-x & -2 \\ 0 & 0 & 2 & -1-x \end{vmatrix} = x(x-3)(x-4)(x-1) = 0.$$

Thus, the characteristic roots of A are 0, 3, 4, and 1, and these are the characteristic values of L. Corresponding to these characteristic values, we have the following systems of homogeneous equations, which we solve for the characteristic vectors of L:

$$\begin{bmatrix} 2 & -2 & 2 & 1 \\ -1 & 3 & 0 & 3 \\ 0 & 0 & 4 & -2 \\ 0 & 0 & 2 & -1 \end{bmatrix}\begin{bmatrix} x_1 \\ x_2 \\ x_3 \\ x_4 \end{bmatrix} = \begin{bmatrix} 0 \\ 0 \\ 0 \\ 0 \end{bmatrix}, \quad \begin{bmatrix} -1 & -2 & 2 & 1 \\ -1 & 0 & 0 & 3 \\ 0 & 0 & 1 & -2 \\ 0 & 0 & 2 & -4 \end{bmatrix}\begin{bmatrix} x_1 \\ x_2 \\ x_3 \\ x_4 \end{bmatrix} = \begin{bmatrix} 0 \\ 0 \\ 0 \\ 0 \end{bmatrix};$$

$$\begin{bmatrix} -2 & -2 & 2 & 1 \\ -1 & -1 & 0 & 3 \\ 0 & 0 & 0 & -2 \\ 0 & 0 & 2 & -5 \end{bmatrix}\begin{bmatrix} x_1 \\ x_2 \\ x_3 \\ x_4 \end{bmatrix} = \begin{bmatrix} 0 \\ 0 \\ 0 \\ 0 \end{bmatrix}, \quad \begin{bmatrix} 1 & -2 & 2 & 1 \\ -1 & 2 & 0 & 3 \\ 0 & 0 & 3 & -2 \\ 0 & 0 & 2 & -2 \end{bmatrix}\begin{bmatrix} x_1 \\ x_2 \\ x_3 \\ x_4 \end{bmatrix} = \begin{bmatrix} 0 \\ 0 \\ 0 \\ 0 \end{bmatrix}.$$

The solutions are $(-3s, -2s, s/2, s)$, $(3s, s, 2s, s)$, $(-s, s, 0, 0)$ and $(2s, s, 0, 0)$, where s is any real number. Choosing $s = 1$ in each case, we obtain the vectors $(-3, -2, \frac{1}{2}, 1)$, $(3, 1, 2, 1)$, $(-1, 1, 0, 0)$ and $(2, 1, 0, 0)$. Since

$$\begin{vmatrix} -3 & -2 & \frac{1}{2} & 1 \\ 3 & 1 & 2 & 1 \\ -1 & 1 & 0 & 0 \\ 2 & 1 & 0 & 0 \end{vmatrix} = \frac{9}{2} \neq 0,$$

it follows from Corollary 21.2 that these characteristic vectors of L form a basis of $\mathcal{V}_4$. By Theorem 31.1, A is similar to a diagonal matrix D. Also, by the remarks above, the matrix

$$Q = \begin{bmatrix} -3 & 3 & -1 & 2 \\ -2 & 1 & 1 & 1 \\ \frac{1}{2} & 2 & 0 & 0 \\ 1 & 1 & 0 & 0 \end{bmatrix}$$

is a nonsingular matrix such that $Q^{-1}AQ = D$. Moreover,

$$D = \begin{bmatrix} 0 & 0 & 0 & 0 \\ 0 & 3 & 0 & 0 \\ 0 & 0 & 4 & 0 \\ 0 & 0 & 0 & 1 \end{bmatrix}$$

where the diagonal elements of D are the characteristic roots of A. Since D is the matrix of L with respect to the basis $\{(-3, -2, \frac{1}{2}, 1), (3, 1, 2, 1), (-1, 1, 0, 0), (2, 1, 0, 0)\}$, it is apparent that the null space of L is spanned by the vector $(-3, -2, \frac{1}{2}, 1)$ and that the range space of L is spanned by the vectors $(3, 1, 2, 1)$, $(-1, 1, 0, 0)$, and $(2, 1, 0, 0)$.

Example 1 provides an illustration of the following theorem.

(31.2) theorem If A is similar to a diagonal matrix D, then the diagonal elements of D are the characteristic roots of A.

PROOF If A is similar to a diagonal matrix D [given by (31–1)], then, by Theorem 30.4,

$$|A - x \cdot I_n| = |D - x \cdot I_n| = \begin{vmatrix} r_1 - x & 0 & \cdots & 0 \\ 0 & r_2 - x & & \cdot \\ \cdot & & \cdot & \cdot \\ \cdot & & & \cdot \\ \cdot & & \cdot & 0 \\ 0 & \cdots & 0 & r_n - x \end{vmatrix}$$

$$= (r_1 - x)(r_2 - x) \cdots (r_n - x).$$

Therefore, $r_1, r_2, \ldots, r_n$ are the characteristic roots of A. ∎

In Example 1, the matrix A has four distinct real characteristic roots, and we found that $\mathcal{V}_4$ has a basis consisting of characteristic vectors of the associated linear transformation L. This is an example of the following theorem.

(31.3) theorem If a (real) n by n matrix A has n real distinct characteristic roots, then A is similar to a (real) diagonal matrix.

PROOF Let L be a linear transformation of $\mathcal{V}$ that has matrix A with respect to a basis $\{\mathbf{U}_1, \mathbf{U}_2, \ldots, \mathbf{U}_n\}$ of $\mathcal{V}$, and denote the characteristic roots of A by $r_1, r_2, \ldots, r_n$. Let $\mathbf{W}_i$ be a characteristic vector of L belonging to r_i for $i = 1, 2, \ldots, n$. By Theorem 31.1, it is sufficient to show that $\{\mathbf{W}_1, \mathbf{W}_2, \ldots, \mathbf{W}_n\}$ is a basis of $\mathcal{V}$, which is the case if $\{\mathbf{W}_1, \mathbf{W}_2, \ldots, \mathbf{W}_n\}$ is a linearly independent set. Assume that this set is linearly dependent. Each $\mathbf{W}_i \neq 0$, since $\mathbf{W}_i$ is a characteristic vector. By Theorem 10.2, there is an integer k, where $2 \leq k \leq n$, such that $\{\mathbf{W}_1, \mathbf{W}_2, \ldots, \mathbf{W}_{k-1}\}$ is a linearly independent set and

$$\mathbf{W}_k = c_1 \cdot \mathbf{W}_1 + c_2 \cdot \mathbf{W}_2 + \cdots + c_{k-1} \cdot \mathbf{W}_{k-1} \tag{31–2}$$

where at least one $c_i \neq 0$. Let $(w_{1,i}, w_{2,i}, \ldots, w_{n,i})$ for $i = 1, 2, \ldots, k$ be the coordinates of $\mathbf{W}_i$ with respect to the basis $\{\mathbf{U}_1, \mathbf{U}_2, \ldots, \mathbf{U}_n\}$. Then equation (31–2) can be written as a matrix equation:

$$\begin{bmatrix} w_{1,k} \\ w_{2,k} \\ \vdots \\ w_{n,k} \end{bmatrix} = c_1 \cdot \begin{bmatrix} w_{1,1} \\ w_{2,1} \\ \vdots \\ w_{n,1} \end{bmatrix} + c_2 \cdot \begin{bmatrix} w_{1,2} \\ w_{2,2} \\ \vdots \\ w_{n,2} \end{bmatrix} + \cdots + c_{k-1} \cdot \begin{bmatrix} w_{1,k-1} \\ w_{2,k-1} \\ \vdots \\ w_{n,k-1} \end{bmatrix}. \tag{31–3}$$

Multiply equation (31–3) by the matrix A, recalling that

$$A \begin{bmatrix} w_{1,i} \\ w_{2,i} \\ \vdots \\ w_{n,i} \end{bmatrix} = r_i \cdot \begin{bmatrix} w_{1,i} \\ w_{2,i} \\ \vdots \\ w_{n,i} \end{bmatrix}$$

for $i = 1, 2, \ldots, k$. This is the case since each $\mathbf{W}_i$ is a characteristic vector. We obtain

$$r_k \cdot \mathbf{W}_k = (c_1 r_1) \cdot \mathbf{W}_1 + (c_2 r_2) \cdot \mathbf{W}_2 + \cdots + (c_{k-1} r_{k-1}) \cdot \mathbf{W}_{k-1}. \tag{31–4}$$

From equation (31–2), we have

$$r_k \cdot \mathbf{W}_k = (c_1 r_k) \cdot \mathbf{W}_1 + (c_2 r_k) \cdot \mathbf{W}_2 + \cdots + (c_{k-1} r_k) \cdot \mathbf{W}_{k-1}. \tag{31–5}$$

Subtracting equation (31–5) from (31–4) yields

$$\mathbf{0} = c_1(r_1 - r_k) \cdot \mathbf{W}_1 + c_2(r_2 - r_k) \cdot \mathbf{W}_2 + \cdots + c_{k-1}(r_{k-1} - r_k) \cdot \mathbf{W}_{k-1}. \tag{31–6}$$

Since $r_k \neq r_i$ for $i = 1, 2, \ldots, k - 1$ and at least one $c_i \neq 0$, it follows that one of the coefficients in equation (31–6) is not zero. But this contradicts the fact that $\{\mathbf{W}_1, \mathbf{W}_2, \ldots, \mathbf{W}_{k-1}\}$ is a linearly independent set. Hence, the assumption that the set $\{\mathbf{W}_1, \mathbf{W}_2, \ldots, \mathbf{W}_n\}$ is linearly dependent is false. That is, $\{\mathbf{W}_1, \mathbf{W}_2, \ldots, \mathbf{W}_n\}$ is a linearly independent set, as required. ∎

We have found the condition for a real n by n matrix A to be similar to a diagonal matrix (Theorem 31.1), and we have shown that this condition is satisfied if A has n real distinct characteristic roots (Theorem 31.3). Moreover, by Theorem 31.2, if not all of the characteristic roots of A are real, then A is not similar to a (real) diagonal matrix. In this case, it follows from Theorem 31.1 that the characteristic vectors of a linear transformation defined by A do not span $\mathcal{V}$. If all of the characteristic roots of A are real, but not distinct, then A may or may not be similar to a diagonal matrix, as shown in the following examples.

EXAMPLE 2 Let $A = I_n$. Then

$$|A - x \cdot I_n| = \begin{vmatrix} 1-x & 0 & \cdots & 0 \\ 0 & 1-x & & \cdot \\ \cdot & & \cdot & \cdot \\ \cdot & & & \cdot \\ \cdot & & \cdot & 0 \\ 0 & \cdots & 0 & 1-x \end{vmatrix}$$

$$= (1-x)(1-x)\cdots(1-x).$$

Hence, the characteristic roots are $r_1 = r_2 = \cdots = r_n = 1$. However, A is similar to a diagonal matrix, since A is the diagonal matrix I_n. The linear transformation L of $\mathcal{V}$ defined by $A = I_n$ (with respect to any basis of $\mathcal{V}$) is the identity transformation, and any basis of $\mathcal{V}$ is a basis consisting of characteristic vectors of L.

EXAMPLE 3 Let

$$A = \begin{bmatrix} 1 & -4 & 3 \\ 0 & 1 & 6 \\ 0 & 0 & 2 \end{bmatrix}.$$

Then

$$|A - x \cdot I_3| = \begin{vmatrix} 1-x & -4 & 3 \\ 0 & 1-x & 6 \\ 0 & 0 & 2-x \end{vmatrix} = (1-x)(1-x)(2-x).$$

The characteristic roots of A are $r_1 = r_2 = 1$, $r_3 = 2$. Let $\mathcal{V} = \mathcal{V}_3$, and let L be the linear transformation with matrix A with respect to the basis $E^{(3)}$. The characteristic vectors of L that belong to the characteristic value

$r_1 = r_2 = 1$ are the vectors (x_1, x_2, x_3), where

$$\begin{bmatrix} 0 & -4 & 3 \\ 0 & 0 & 6 \\ 0 & 0 & 1 \end{bmatrix} \begin{bmatrix} x_1 \\ x_2 \\ x_3 \end{bmatrix} = \begin{bmatrix} 0 \\ 0 \\ 0 \end{bmatrix}.$$

Thus, $(x_1, x_2, x_3) = (s, 0, 0) = s \cdot (1, 0, 0)$, where s is any real number. The characteristic vectors (x_1, x_2, x_3) that belong to $r_3 = 2$ satisfy

$$\begin{bmatrix} -1 & -4 & 3 \\ 0 & -1 & 6 \\ 0 & 0 & 0 \end{bmatrix} \begin{bmatrix} x_1 \\ x_2 \\ x_3 \end{bmatrix} = \begin{bmatrix} 0 \\ 0 \\ 0 \end{bmatrix}.$$

Thus, for r_3, $(x_1, x_2, x_3) = (-21s, 6s, s) = s \cdot (-21, 6, 1)$, where s is any real number. Therefore, the subspace of $\mathcal{V}$ spanned by the characteristic vectors of L is spanned by $\{(1, 0, 0), (-21, 6, 1)\}$. This subspace has dimension two. Hence, $\mathcal{V}$ does not have a basis consisting of characteristic vectors, so that, by Theorem 31.1, A is not similar to a diagonal matrix.

exercises

31.1 Let L be a linear transformation of $\mathcal{V}$ that has matrix A with respect to a basis $\{\mathbf{U}_1, \mathbf{U}_2, \ldots, \mathbf{U}_n\}$ of $\mathcal{V}$. Suppose that Q is a nonsingular matrix such that $Q^{-1}AQ$ is a diagonal matrix. Prove that the columns of Q are the coordinates with respect to $\{\mathbf{U}_1, \mathbf{U}_2, \ldots, \mathbf{U}_n\}$ of n characteristic vectors of L that form a basis of $\mathcal{V}$.

31.2 In Example 1, find Q^{-1} and compute the product $Q^{-1}AQ$.

31.3 Which of the following matrices are similar to diagonal matrices?

(a) $\begin{bmatrix} 2 & -1 & 0 & 0 \\ -2 & 3 & 0 & 0 \\ 2 & 0 & 4 & 2 \\ 1 & 3 & -2 & -1 \end{bmatrix}$ (b) $\begin{bmatrix} -2 & 0 & 1 \\ 0 & 1 & -7 \\ 0 & 1 & 0 \end{bmatrix}$

(c) $\begin{bmatrix} 0 & 0 & 0 & 1 \\ 0 & 0 & 1 & 0 \\ 0 & 1 & 0 & 0 \\ 1 & 0 & 0 & 0 \end{bmatrix}$ (d) $\begin{bmatrix} 3 & -1 & -1 & -2 \\ 1 & 1 & -1 & -1 \\ 1 & 0 & 0 & -1 \\ 0 & -1 & 1 & 1 \end{bmatrix}$

31.4 For those matrices A of Exercise 31.3 that are similar to a diagonal matrix D, find a nonsingular matrix Q such that $Q^{-1}AQ = D$.

31.5 Prove that if A is similar to a diagonal matrix, then A' is similar to a diagonal matrix.

***31.6** Prove that if A is similar to a diagonal matrix, then A^n (where n is a positive integer) is similar to a diagonal matrix, and the characteristic roots of A^n are the nth powers of the characteristic roots of A.

32 orthogonal linear transformations and orthogonal matrices

Recall that in Example 1, §15 we discussed the rotations T_φ of $\mathcal{E}_2$ into $\mathcal{E}_2$, where $\mathcal{E}_2$ is the two-dimensional Euclidean vector space consisting of all geometric vectors in the XY plane of a Cartesian coordinate system in space. For any vector $\mathbf{V} = x \cdot \mathbf{I} + y \cdot \mathbf{J} \in \mathcal{E}_2$,

$$\begin{aligned} T_\varphi(\mathbf{V}) &= T_\varphi(x \cdot \mathbf{I} + y \cdot \mathbf{J}) \\ &= (x \cos \varphi - y \sin \varphi) \cdot \mathbf{I} + (x \sin \varphi + y \cos \varphi) \cdot \mathbf{J}. \end{aligned}$$

Let $\mathbf{W} = s \cdot \mathbf{I} + t \cdot \mathbf{J}$ be another vector in $\mathcal{E}_2$, so that

$$\begin{aligned} T_\varphi(\mathbf{W}) &= T_\varphi(s \cdot \mathbf{I} + t \cdot \mathbf{J}) \\ &= (s \cos \varphi - t \sin \varphi) \cdot \mathbf{I} + (s \sin \varphi + t \cos \varphi) \cdot \mathbf{J}. \end{aligned}$$

Then by Theorem 7.2, the inner product of $T_\varphi(\mathbf{V})$ and $T_\varphi(\mathbf{W})$ is

$$\begin{aligned} T_\varphi(\mathbf{V}) \circ T_\varphi(\mathbf{W}) &= (x \cos \varphi - y \sin \varphi)(s \cos \varphi - t \sin \varphi) \\ &\quad + (x \sin \varphi + y \cos \varphi)(s \sin \varphi + t \cos \varphi) \\ &= xs \cos^2 \varphi - xt \cos \varphi \sin \varphi - ys \sin \varphi \cos \varphi + yt \sin^2 \varphi \\ &\quad + xs \sin^2 \varphi + xt \sin \varphi \cos \varphi + ys \cos \varphi \sin \varphi + yt \cos^2 \varphi \\ &= xs(\cos^2 \varphi + \sin^2 \varphi) + yt(\sin^2 \varphi + \cos^2 \varphi) \\ &= xs + yt = \mathbf{V} \circ \mathbf{W}. \end{aligned}$$

Thus, the inner product of two vectors in $\mathcal{E}_2$ is left unchanged when each vector is mapped by the linear transformation T_φ. This fact implies that the length of a vector, $(\mathbf{V} \circ \mathbf{V})^{1/2}$, is unchanged. Moreover, since $\cos \theta = (\mathbf{V} \circ \mathbf{W})/(\mathbf{V} \circ \mathbf{V})^{1/2}(\mathbf{W} \circ \mathbf{W})^{1/2}$, where θ is the angle between $\mathbf{V}$ and $\mathbf{W}$, the angle between two vectors is preserved by T_φ.

We now define a class of linear transformations of a Euclidean vector space $\mathcal{E}$ into itself that has the properties exhibited by the rotations T_φ of $\mathcal{E}_2$.

(32.1) definition Let $\mathcal{E}$ be a Euclidean vector space. A linear transformation L of $\mathcal{E}$ into itself such that $\mathcal{I}(\mathbf{V}, \mathbf{V}) = \mathcal{I}[L(\mathbf{V}), L(\mathbf{V})]$ for all $\mathbf{V} \in \mathcal{E}$ is called an *orthogonal transformation* of $\mathcal{E}$.

Thus, an orthogonal transformation of $\mathcal{E}$ preserves the length $[\mathcal{I}(\mathbf{V}, \mathbf{V})]^{1/2}$ of each vector $\mathbf{V} \in \mathcal{E}$. It is not difficult to show that an orthogonal transformation leaves the inner product $\mathcal{I}(\mathbf{V}, \mathbf{W})$ of two vectors unchanged (see Exercise 32.1). Therefore, if the angle θ between two vectors in $\mathcal{E}$ is defined by $\cos \theta = \mathcal{I}(\mathbf{V}, \mathbf{W})/[\mathcal{I}(\mathbf{V}, \mathbf{V})]^{1/2}[\mathcal{I}(\mathbf{W}, \mathbf{W})]^{1/2}$, an orthogonal transformation also preserves angles.

EXAMPLE 1 The rotation T_φ of $\mathcal{E}_2$ is an orthogonal transformation of $\mathcal{E}_2$ with respect to the ordinary inner product of geometric vectors, since $\mathbf{V} \circ \mathbf{V} = T_\varphi(\mathbf{V}) \circ T_\varphi(\mathbf{V})$ for all $\mathbf{V} \in \mathcal{E}_2$. Since $T_\varphi(\mathbf{I}) = \cos \varphi \cdot \mathbf{I} + \sin \varphi \cdot \mathbf{J}$ and $T_\varphi(\mathbf{J}) = -\sin \varphi \cdot \mathbf{I} + \cos \varphi \cdot \mathbf{J}$, it follows that the matrix of T_φ with respect to the orthonormal basis $\{\mathbf{I}, \mathbf{J}\}$ of $\mathcal{E}_2$ is

$$A = \begin{bmatrix} \cos\varphi & -\sin\varphi \\ \sin\varphi & \cos\varphi \end{bmatrix}.$$

The matrix A is nonsingular since $|A| = \cos^2\varphi + \sin^2\varphi = 1$. Moreover,

$$AA' = \begin{bmatrix} \cos\varphi & -\sin\varphi \\ \sin\varphi & \cos\varphi \end{bmatrix} \begin{bmatrix} \cos\varphi & \sin\varphi \\ -\sin\varphi & \cos\varphi \end{bmatrix}$$

$$= \begin{bmatrix} \cos^2\varphi + \sin^2\varphi & \cos\varphi\sin\varphi - \sin\varphi\cos\varphi \\ \sin\varphi\cos\varphi - \cos\varphi\sin\varphi & \sin^2\varphi + \cos^2\varphi \end{bmatrix}$$

$$= \begin{bmatrix} 1 & 0 \\ 0 & 1 \end{bmatrix} = I_2.$$

Therefore, $A' = A^{-1}$. Thus, the orthogonal transformation T_φ is nonsingular and the matrix A of T_φ with respect to the orthonormal basis $\{\mathbf{I}, \mathbf{J}\}$ has the property that the inverse of A is equal to the transpose of A.

Example 1 suggests the following theorem.

(32.2) theorem Let L be an orthogonal transformation of a Euclidean vector space $\mathcal{E}$ of dimension n. Let $A = [a_{i,j}]$ be the matrix of L with respect to an orthonormal basis $\{\mathbf{U}_1, \mathbf{U}_2, \ldots, \mathbf{U}_n\}$ of $\mathcal{E}$. Then A is nonsingular, and $A^{-1} = A'$.

PROOF For $j = 1, 2, \ldots, n$, the jth column of A gives the coordinates $(a_{1,j}, a_{2,j}, \ldots, a_{n,j})$ of $L(\mathbf{U}_j)$ with respect to the given basis. Therefore, for $i = 1, 2, \ldots, n$, the ith row of A' gives the coordinates $(a_{1,i}, a_{2,i}, \ldots, a_{n,i})$ of $L(\mathbf{U}_i)$. Since $\{\mathbf{U}_1, \mathbf{U}_2, \ldots, \mathbf{U}_n\}$ is an orthonormal basis of $\mathcal{E}$, it follows from Theorem 11.7 that

$$\mathcal{I}[L(\mathbf{U}_i), L(\mathbf{U}_j)] = \sum_{k=1}^{n} a_{k,i}a_{k,j}.$$

By the definition of matrix multiplication, this is the element in the (i, j)-position of $A'A$ for all i, j. Since L is an orthogonal transformation, $\mathcal{I}(\mathbf{U}_i, \mathbf{U}_j) = \mathcal{I}[L(\mathbf{U}_i), L(\mathbf{U}_j)]$. Moreover, $\mathcal{I}(\mathbf{U}_i, \mathbf{U}_j) = 0$ if $i \neq j$, and $\mathcal{I}(\mathbf{U}_i, \mathbf{U}_i) = 1$, again using the fact that $\{\mathbf{U}_1, \mathbf{U}_2, \ldots, \mathbf{U}_n\}$ is an orthonormal basis. Hence, the matrix $A'A$ has 1 in each diagonal position and 0 elsewhere. That is, $A'A = I_n$. This proves that A is nonsingular, and $A^{-1} = A'$. ∎

(32.3) definition A (real) nonsingular n by n matrix A such that $A^{-1} = A'$ is called an *orthogonal matrix*.

By Theorem 32.2, the matrix of an orthogonal transformation L of a Euclidean vector space $\mathcal{E}$ with respect to an orthonormal basis is an orthogonal matrix. In particular, L is a nonsingular linear transformation.

EXAMPLE 2 Let L be an orthogonal transformation of $\mathcal{E}_2$, and let

$$A = \begin{bmatrix} a_{1,1} & a_{1,2} \\ a_{2,1} & a_{2,2} \end{bmatrix}$$

be the matrix of L with respect to the orthonormal basis $\{\mathbf{I}, \mathbf{J}\}$ of $\mathcal{E}_2$. Then A is an orthogonal matrix. Since $A^{-1} = A'$, it follows that $AA' = I_2$. Using the properties of determinants given by Theorems 27.8 and 27.1, this matrix equation yields

$$1 = |I_2| = |AA'| = |A|\,|A'| = |A|\,|A| = |A|^2.$$

Therefore, $|A| = \pm 1$. Thus, we have

$$\begin{bmatrix} a_{1,1} & a_{1,2} \\ a_{2,1} & a_{2,2} \end{bmatrix}\begin{bmatrix} a_{1,1} & a_{2,1} \\ a_{1,2} & a_{2,2} \end{bmatrix} = \begin{bmatrix} 1 & 0 \\ 0 & 1 \end{bmatrix}, \quad \text{and} \quad \begin{vmatrix} a_{1,1} & a_{1,2} \\ a_{2,1} & a_{2,2} \end{vmatrix} = \pm 1.$$

Consequently, the elements of the matrix A are subject to the following conditions:

$$a_{1,1}^2 + a_{1,2}^2 = 1, \qquad a_{1,1}a_{2,1} + a_{1,2}a_{2,2} = 0,$$
$$a_{2,1}^2 + a_{2,2}^2 = 1, \quad \text{and} \quad a_{1,1}a_{2,2} - a_{2,1}a_{1,2} = \pm 1.$$

Suppose first that $|A| = a_{1,1}a_{2,2} - a_{2,1}a_{1,2} = 1$. Then

$$a_{1,1}a_{2,1}^2 + a_{1,2}a_{2,2}a_{2,1} = 0$$
$$a_{1,1}a_{2,2}^2 - a_{2,1}a_{1,2}a_{2,2} = a_{2,2}$$

and adding, we obtain $a_{1,1}(a_{2,1}^2 + a_{2,2}^2) = a_{1,1} = a_{2,2}$. The equation $a_{1,1}a_{2,1} + a_{1,2}a_{2,2} = 0$, now becomes $a_{1,1}(a_{2,1} + a_{1,2}) = 0$. Therefore, either $a_{1,1} = 0$ or $a_{1,2} = -a_{2,1}$. If $a_{1,1} = 0$, then the original equations yield $a_{1,2} = -a_{2,1} = \pm 1$. Thus, in every case,

$$A = \begin{bmatrix} u & -v \\ v & u \end{bmatrix}$$

where $u^2 + v^2 = 1$. We now observe that there is an angle φ such that $u = \cos\varphi$ and $v = \sin\varphi$. Therefore, when $|A| = 1$, the matrix of the orthogonal transformation L is the matrix of the rotation T_φ. That is, L is a rotation of $\mathcal{E}_2$.

If $|A| = -1$, an analysis similar to the one above shows that

$$A = \begin{bmatrix} u & v \\ v & -u \end{bmatrix}$$

where $u^2 + v^2 = 1$. Again, there is an angle φ such that $u = \cos\varphi$ and

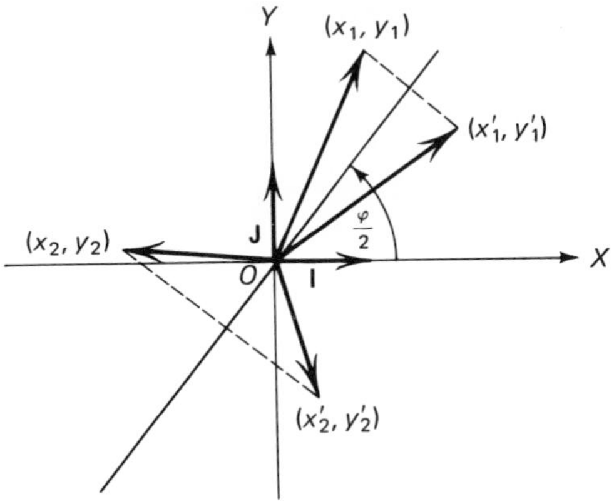

FIGURE 52

$v = \sin\varphi$. Thus, a vector $\mathbf{V} \in \mathcal{E}_2$ with coordinates (x, y) is sent into $L(\mathbf{V})$ with coordinates (x', y'), where

$$\begin{bmatrix} \cos\varphi & \sin\varphi \\ \sin\varphi & -\cos\varphi \end{bmatrix} \begin{bmatrix} x \\ y \end{bmatrix} = \begin{bmatrix} x' \\ y' \end{bmatrix}.$$

It is left as an exercise (Exercise 32.9) to show that in this case L is a *reflection* of $\mathcal{E}_2$ through the line that makes an angle $\varphi/2$ with the X axis at the origin of the coordinate system (see Figure 52). This example shows that the *only* orthogonal transformations of $\mathcal{E}_2$ are rotations and reflections.

Suppose that $\{\mathbf{U}_1, \mathbf{U}_2, \ldots, \mathbf{U}_n\}$ is an orthonormal basis of $\mathcal{E}$ and that $\{\mathbf{W}_1, \mathbf{W}_2, \ldots, \mathbf{W}_n\}$ is any set of n vectors in $\mathcal{E}$. Then

$$\begin{aligned} \mathbf{W}_1 &= p_{1,1} \cdot \mathbf{U}_1 + p_{1,2} \cdot \mathbf{U}_2 + \cdots + p_{1,n} \cdot \mathbf{U}_n \\ \mathbf{W}_2 &= p_{2,1} \cdot \mathbf{U}_1 + p_{2,2} \cdot \mathbf{U}_2 + \cdots + p_{2,n} \cdot \mathbf{U}_n \\ &\vdots \\ \mathbf{W}_n &= p_{n,1} \cdot \mathbf{U}_1 + p_{n,2} \cdot \mathbf{U}_2 + \cdots + p_{n,n} \cdot \mathbf{U}_n. \end{aligned} \tag{32–1}$$

Since $\{\mathbf{U}_1, \mathbf{U}_2, \ldots, \mathbf{U}_n\}$ is an orthonormal basis, it follows from Theorem 11.7 that

$$\mathcal{G}(\mathbf{W}_i, \mathbf{W}_j) = \sum_{k=1}^{n} p_{i,k}p_{j,k} \tag{32–2}$$

for all i, j. The equations (32–2) imply that $\{\mathbf{W}_1, \mathbf{W}_2, \ldots, \mathbf{W}_n\}$ is an orthonormal basis of $\mathcal{E}$ if and only if the matrix $P = [p_{i,j}]$ is an orthogonal matrix. The details of this argument are left to the reader (Exercise 32.6). Thus, the matrix of a change of basis from one orthonormal basis to another orthonormal basis in a Euclidean vector space is an orthogonal matrix. If the discussion of §22 on the relation between two matrices that are the matrices of the same linear transformation of a vector space $\mathcal{V}$ into itself, is applied to the present situation, we obtain the following theorem.

(32.3) theorem Let A and B be n by n matrices, and let $\mathcal{E}$ be an n-dimensional Euclidean vector space. Then A and B are matrices of the same linear transformation L of $\mathcal{E}$ into itself (with respect to suitable orthonormal bases in $\mathcal{E}$) if and only if there exists an orthogonal matrix Q such that $B = Q^{-1}AQ$.

Theorem 32.3 suggests the next definition.

(32.4) definition Let A and B be n by n matrices. The matrix B is *orthogonally equivalent* to the matrix A if there exists an orthogonal matrix Q such that $B = Q^{-1}AQ$.

Orthogonal equivalence is an equivalence relation on the set $\mathfrak{A}_n$ of all real n by n matrices. From Definition 32.4 it is immediately obvious that orthogonal equivalence is a special case of similarity.

If all of the characteristic roots $r_1, r_2, \ldots, r_n$ of an n by n matrix A are real, then A is orthogonally equivalent to a triangular matrix with the characteristic roots of A on the diagonal. This is the content of the next theorem. The proof of Theorem 32.5, which proceeds by induction, describes an iterative process for constructing an orthogonal matrix Q such that $Q^{-1}AQ$ is in triangular form. The result of this theorem is necessary in the simplification of quadratic forms, which is the topic discussed in Chapter 8. Since our later work does not depend upon the method of proof used here, some readers may well accept this result on faith and avoid the rather lengthy, albeit fascinating, derivation.

(32.5) theorem Let A be an n by n matrix such that all of the characteristic roots $r_1, r_2, \ldots, r_n$ of A are real. There exists an orthogonal matrix Q such that

$$Q^{-1}AQ = \begin{bmatrix} r_1 & b^*_{1,2} & b^*_{1,3} & \cdots & & b^*_{1,n} \\ 0 & r_2 & b^*_{2,3} & \cdots & & b^*_{2,n} \\ \cdot & 0 & r_3 & & & \cdot \\ \cdot & \cdot & 0 & & & \cdot \\ & \cdot & \cdot & \cdot & & \\ \cdot & \cdot & \cdot & & \cdot & \cdot \\ & & \cdot & & \cdot & \\ & & & & & b^*_{n-1,n} \\ 0 & 0 & 0 & \cdots & 0 & r_n \end{bmatrix}. \tag{32–3}$$

PROOF The proof is by induction on n. If $n = 1$, then $A = [a_{1,1}]$, $r_1 = a_{1,1}$ is the characteristic root of A, and $I_1AI_1 = I_1^{-1}AI_1 = [r_1]$, where $I_1 = [1]$ is an orthogonal matrix. Therefore, let $n > 1$, and assume that the theorem is true for $(n - 1)$ by $(n - 1)$ matrices. Let L be the linear transformation of an n-dimensional Euclidean vector space $\mathcal{E}$ into itself that has matrix A with respect to an orthonormal basis $\{\mathbf{U}_1, \mathbf{U}_2, \ldots, \mathbf{U}_n\}$ of $\mathcal{E}$. Then r_1, $r_2, \ldots, r_n$ are characteristic values of L. Let $\mathbf{W}_1$ be a characteristic vector of L that belongs to the characteristic value r_1. Since $\mathbf{W}_1 \neq \mathbf{0}$, there is a basis of $\mathcal{E}$ that contains $\mathbf{W}_1$. By the Gram-Schmidt orthogonalization process (Theorem 11.6), there is an orthonormal basis $\{\bar{\mathbf{W}}_1, \bar{\mathbf{W}}_2, \ldots, \bar{\mathbf{W}}_n\}$ of $\mathcal{E}$ such that $\bar{\mathbf{W}}_1 = (1/|\mathbf{W}_1|) \cdot \mathbf{W}_1$. We next observe that $\bar{\mathbf{W}}_1$ is a characteristic vector of L that belongs to the characteristic value r_1. Indeed, since $\mathbf{W}_1$ is a characteristic vector of L belonging to r_1, we have $L(\mathbf{W}_1) = r_1 \cdot \mathbf{W}_1$, and hence

$$L(\bar{\mathbf{W}}_1) = L\left(\frac{1}{|\mathbf{W}_1|} \cdot \mathbf{W}_1\right) = \frac{1}{|\mathbf{W}_1|} \cdot L(\mathbf{W}_1) = \frac{1}{|\mathbf{W}_1|} \cdot (r_1 \cdot \mathbf{W}_1)$$
$$= r_1 \cdot \left(\frac{1}{|\mathbf{W}_1|} \cdot \mathbf{W}_1\right) = r_1 \cdot \bar{\mathbf{W}}_1.$$

Let $P = [p_{i,j}]$ be the matrix of the change of basis from the $\mathbf{U}$-basis to the $\bar{\mathbf{W}}$-basis. Then

$$\bar{\mathbf{W}}_1 = p_{1,1} \cdot \mathbf{U}_1 + p_{1,2} \cdot \mathbf{U}_2 + \cdots + p_{1,n} \cdot \mathbf{U}_n$$

so that the coordinates of $\bar{\mathbf{W}}_1$ with respect to the basis $\{\mathbf{U}_1, \mathbf{U}_2, \ldots, \mathbf{U}_n\}$ are $(p_{1,1}, p_{1,2}, \ldots, p_{1,n})$. Since A is the matrix of the linear transformation L with respect to the basis $\{\mathbf{U}_1, \mathbf{U}_2, \ldots, \mathbf{U}_n\}$, and $\bar{\mathbf{W}}_1$ is a characteristic vector of L that belongs to the characteristic value r_1, we have

$$A\begin{bmatrix} p_{1,1} \\ p_{1,2} \\ \vdots \\ p_{1,n} \end{bmatrix} = r_1 \cdot \begin{bmatrix} p_{1,1} \\ p_{1,2} \\ \vdots \\ p_{1,n} \end{bmatrix}. \tag{32–4}$$

The matrix $P = [p_{i,j}]$ is the matrix of the change of basis from one orthonormal basis to another, hence, it is orthogonal. Therefore,

$$P\begin{bmatrix} p_{1,1} \\ p_{1,2} \\ \vdots \\ p_{1,n} \end{bmatrix} = \begin{bmatrix} 1 \\ 0 \\ \vdots \\ 0 \end{bmatrix}$$

which can be written

$$\begin{bmatrix} p_{1,1} \\ p_{1,2} \\ \vdots \\ p_{1,n} \end{bmatrix} = P^{-1}\begin{bmatrix} 1 \\ 0 \\ \vdots \\ 0 \end{bmatrix}. \tag{32–5}$$

Using equations (32–4) and (32–5), we obtain

$$A\left(P^{-1}\begin{bmatrix} 1 \\ 0 \\ \vdots \\ 0 \end{bmatrix}\right) = A\begin{bmatrix} p_{1,1} \\ p_{1,2} \\ \vdots \\ p_{1,n} \end{bmatrix} = r_1 \cdot \begin{bmatrix} p_{1,1} \\ p_{1,2} \\ \vdots \\ p_{1,n} \end{bmatrix} = r_1 \cdot \left(P^{-1}\begin{bmatrix} 1 \\ 0 \\ \vdots \\ 0 \end{bmatrix}\right)$$

$$= P^{-1}\left(r_1 \cdot \begin{bmatrix} 1 \\ 0 \\ \vdots \\ 0 \end{bmatrix}\right) = P^{-1}\begin{bmatrix} r_1 \\ 0 \\ \vdots \\ 0 \end{bmatrix}.$$

Multiplying the above equation on the left by P, we have

$$PAP^{-1}\begin{bmatrix} 1 \\ 0 \\ \vdots \\ 0 \end{bmatrix} = \begin{bmatrix} r_1 \\ 0 \\ \vdots \\ 0 \end{bmatrix}$$

which implies that the first column of PAP^{-1} is

$$\begin{bmatrix} r_1 \\ 0 \\ \vdots \\ 0 \end{bmatrix}.$$

Let $Q_1 = P^{-1}$. Then Q_1 is an orthogonal matrix (see Exercise 32.5) such that

$$Q_1^{-1}AQ_1 = \begin{bmatrix} r_1 & b_{1,2} & \cdots & b_{1,n} \\ 0 & b_{2,2} & \cdots & b_{2,n} \\ \vdots & \vdots & & \vdots \\ 0 & b_{n,2} & \cdots & b_{n,n} \end{bmatrix}.$$

Let B be the $(n - 1)$ by $(n - 1)$ matrix

$$\begin{bmatrix} b_{2,2} & \cdots & b_{2,n} \\ \vdots & & \vdots \\ b_{n,2} & \cdots & b_{n,n} \end{bmatrix}.$$

Then $|A - x \cdot I_n| = |Q_1^{-1}AQ_1 - x \cdot I_n| = (r_1 - x)|B - x \cdot I_{n-1}|$. Thus, the characteristic roots of B are $r_2, r_3, \ldots, r_n$. By the induction assumption, there is an $(n - 1)$ by $(n - 1)$ orthogonal matrix Q_2 such that

$$Q_2^{-1}BQ_2 = \begin{bmatrix} r_2 & b^*_{2,3} & \cdots & b^*_{2,n} \\ 0 & r_3 & \ddots & \vdots \\ \vdots & 0 & \ddots & \\ & \vdots & \ddots & b^*_{n-1,n} \\ 0 & 0 & \cdots & r_n \end{bmatrix}.$$

Then the matrix

$$Q_3 = \begin{bmatrix} 1 & 0 & \cdots & 0 \\ 0 & & & \\ \vdots & & \begin{bmatrix} & Q_2 & \end{bmatrix} & \\ 0 & & & \end{bmatrix}$$

is an n by n orthogonal matrix such that

$$Q_3^{-1}Q_1^{-1}AQ_1Q_3 = \begin{bmatrix} r_1 & b_{1,2}^* & b_{1,3}^* & \cdots & & b_{1,n}^* \\ 0 & r_2 & b_{2,3}^* & \cdots & & b_{2,n}^* \\ \cdot & 0 & r_3 & & & \cdot \\ \cdot & \cdot & 0 & & & \cdot \\ \cdot & & \cdot & \cdot & & \cdot \\ & \cdot & \cdot & & \cdot & \cdot \\ & \cdot & \cdot & & \cdot & \\ & & & & & b_{n-1,n}^* \\ 0 & 0 & 0 & \cdots & 0 & r_n \end{bmatrix}.$$

The matrix $Q = Q_1Q_3$, being the product of orthogonal matrices, is orthogonal. Hence, $Q^{-1}AQ$ has the triangular form (32–3). ∎

EXAMPLE 2 Let

$$A = \begin{bmatrix} 0 & 1 & 0 \\ 0 & 0 & 1 \\ -8 & 6 & 3 \end{bmatrix}$$

be the matrix considered in Exercise 30.4. We will follow the method given in the proof of Theorem 32.5 to construct an orthogonal matrix Q such that $Q^{-1}AQ$ is in triangular form. Let L be the linear transformation of the Euclidean vector $\mathcal{V}_3$ (with the $E^{(3)}$-inner product) that has matrix A with respect to the orthonormal basis $E^{(3)} = \{(1, 0, 0), (0, 1, 0), (0, 0, 1)\}$. The characteristic roots of A are easily calculated; they are $r_1 = 1$, $r_2 = -2$, and $r_3 = 4$. These characteristic roots of A are characteristic values of the linear transformation L. A characteristic vector of L belonging to the characteristic value $r_1 = 1$ is $\mathbf{W}_1 = (1, 1, 1)$. Indeed,

$$A\begin{bmatrix} 1 \\ 1 \\ 1 \end{bmatrix} = \begin{bmatrix} 0 & 1 & 0 \\ 0 & 0 & 1 \\ -8 & 6 & 3 \end{bmatrix}\begin{bmatrix} 1 \\ 1 \\ 1 \end{bmatrix} = \begin{bmatrix} 1 \\ 1 \\ 1 \end{bmatrix} = 1 \cdot \begin{bmatrix} 1 \\ 1 \\ 1 \end{bmatrix}.$$

Then $\bar{\mathbf{W}}_1 = 1/\sqrt{3} \cdot (1, 1, 1)$ is a unit characteristic vector belonging to $r_1 = 1$. By inspection, we find that $(1, -1, 0)$ is a vector orthogonal to $\bar{\mathbf{W}}_1$, and $\bar{\mathbf{W}}_2 = 1/\sqrt{2} \cdot (1, -1, 0)$ is a unit vector orthogonal to $\bar{\mathbf{W}}_1$. Again by inspection, we see that $(1, 1, -2)$ is a vector orthogonal to both $\bar{\mathbf{W}}_1$ and $\bar{\mathbf{W}}_2$, and $\bar{\mathbf{W}}_3 = 1/\sqrt{6}\cdot(1, 1, -2)$ is a unit vector orthogonal to $\bar{\mathbf{W}}_1$ and $\bar{\mathbf{W}}_2$. Therefore, $\{\bar{\mathbf{W}}_1, \bar{\mathbf{W}}_2, \bar{\mathbf{W}}_3\}$ is an orthonormal basis of $\mathcal{V}_3$ with $\bar{\mathbf{W}}_1$ a characteristic vector of L belonging to $r_1 = 1$. The matrix

$$P_1 = \begin{bmatrix} \dfrac{1}{\sqrt{3}} & \dfrac{1}{\sqrt{3}} & \dfrac{1}{\sqrt{3}} \\ \dfrac{1}{\sqrt{2}} & -\dfrac{1}{\sqrt{2}} & 0 \\ \dfrac{1}{\sqrt{6}} & \dfrac{1}{\sqrt{6}} & -\dfrac{2}{\sqrt{6}} \end{bmatrix}$$

is the orthogonal matrix that is the matrix of the change of basis from the basis $E^{(3)}$ to the basis $\{\bar{\mathbf{W}}_1, \bar{\mathbf{W}}_2, \bar{\mathbf{W}}_3\}$. Then

$$Q_1 = P_1^{-1} = \begin{bmatrix} \dfrac{1}{\sqrt{3}} & \dfrac{1}{\sqrt{2}} & \dfrac{1}{\sqrt{6}} \\ \dfrac{1}{\sqrt{3}} & -\dfrac{1}{\sqrt{2}} & \dfrac{1}{\sqrt{6}} \\ \dfrac{1}{\sqrt{3}} & 0 & -\dfrac{2}{\sqrt{6}} \end{bmatrix}$$

is an orthogonal matrix such that

$$Q_1^{-1}AQ_1 = \begin{bmatrix} \dfrac{1}{\sqrt{3}} & \dfrac{1}{\sqrt{3}} & \dfrac{1}{\sqrt{3}} \\ \dfrac{1}{\sqrt{2}} & -\dfrac{1}{\sqrt{2}} & 0 \\ \dfrac{1}{\sqrt{6}} & \dfrac{1}{\sqrt{6}} & -\dfrac{2}{\sqrt{6}} \end{bmatrix} \begin{bmatrix} 0 & 1 & 0 \\ 0 & 0 & 1 \\ -8 & 6 & 3 \end{bmatrix} \begin{bmatrix} \dfrac{1}{\sqrt{3}} & \dfrac{1}{\sqrt{2}} & \dfrac{1}{\sqrt{6}} \\ \dfrac{1}{\sqrt{3}} & -\dfrac{1}{\sqrt{2}} & \dfrac{1}{\sqrt{6}} \\ \dfrac{1}{\sqrt{3}} & 0 & -\dfrac{2}{\sqrt{6}} \end{bmatrix}$$

$$= \begin{bmatrix} 1 & -\dfrac{15}{\sqrt{6}} & -\dfrac{3}{\sqrt{2}} \\ 0 & -\dfrac{1}{2} & \dfrac{\sqrt{3}}{2} \\ 0 & \dfrac{9\sqrt{3}}{2} & \dfrac{5}{2} \end{bmatrix}.$$

The 2 by 2 matrix

$$B = \begin{bmatrix} -\dfrac{1}{2} & \dfrac{\sqrt{3}}{2} \\ \dfrac{9\sqrt{3}}{2} & \dfrac{5}{2} \end{bmatrix}$$

has characteristic equation $x^2 - 2x - 8 = (x + 2)(x - 4) = 0$. Thus, the characteristic roots of B are $r_2 = -2$ and $r_3 = 4$. Now r_2 and r_3 are regarded as characteristic values of a linear transformation M of $\mathcal{V}_2$ with matrix B with respect to the orthonormal basis $E^{(2)} = \{(1, 0), (0, 1)\}$. A characteristic vector (x_1, x_2) of M belonging to $r_2 = -2$ satisfies

$$\begin{bmatrix} \frac{3}{2} & \frac{\sqrt{3}}{2} \\ \frac{9\sqrt{3}}{2} & \frac{9}{2} \end{bmatrix} \begin{bmatrix} x_1 \\ x_2 \end{bmatrix} = \begin{bmatrix} 0 \\ 0 \end{bmatrix}.$$

Thus, $(1, -\sqrt{3})$ is such a characteristic vector, and $\bar{\mathbf{U}}_1 = (\frac{1}{2}, -\sqrt{3}/2)$ is a unit characteristic vector belonging to $r_2 = -2$. By inspection, we find that $\bar{\mathbf{U}}_2 = (\sqrt{3}/2, \frac{1}{2})$ is a unit vector orthogonal to $\bar{\mathbf{U}}_1$. Therefore, $\{\bar{\mathbf{U}}_1, \bar{\mathbf{U}}_2\}$ is an orthonormal basis of $\mathcal{V}_2$ with $\bar{\mathbf{U}}_1$ a characteristic vector of M that belongs to $r_2 = -2$. The matrix

$$P_2 = \begin{bmatrix} \frac{1}{2} & -\frac{\sqrt{3}}{2} \\ \frac{\sqrt{3}}{2} & \frac{1}{2} \end{bmatrix}$$

is the orthogonal matrix that is the matrix of the change of basis from the basis $E^{(2)}$ to the basis $\{\bar{\mathbf{U}}_1, \bar{\mathbf{U}}_2\}$. Then

$$Q_2 = P_1^{-1} = \begin{bmatrix} \frac{1}{2} & \frac{\sqrt{3}}{2} \\ -\frac{\sqrt{3}}{2} & \frac{1}{2} \end{bmatrix}$$

is an orthogonal matrix such that

$$Q_2^{-1}BQ_2 = \begin{bmatrix} \frac{1}{2} & -\frac{\sqrt{3}}{2} \\ \frac{\sqrt{3}}{2} & \frac{1}{2} \end{bmatrix} \begin{bmatrix} -\frac{1}{2} & \frac{\sqrt{3}}{2} \\ \frac{9\sqrt{3}}{2} & \frac{5}{2} \end{bmatrix} \begin{bmatrix} \frac{1}{2} & \frac{\sqrt{3}}{2} \\ -\frac{\sqrt{3}}{2} & \frac{1}{2} \end{bmatrix}$$

$$= \begin{bmatrix} -2 & -\frac{4}{\sqrt{3}} \\ 0 & 4 \end{bmatrix}.$$

The matrix

$$Q_3 = \begin{bmatrix} 1 & 0 & 0 \\ 0 & \frac{1}{2} & \frac{\sqrt{3}}{2} \\ 0 & -\frac{\sqrt{3}}{2} & \frac{1}{2} \end{bmatrix}$$

is a 3 by 3 orthogonal matrix such that

$$Q_3^{-1}(Q_1^{-1}AQ_1)Q_3$$

$$= \begin{bmatrix} 1 & 0 & 0 \\ 0 & \frac{1}{2} & -\frac{\sqrt{3}}{2} \\ 0 & \frac{\sqrt{3}}{2} & \frac{1}{2} \end{bmatrix} \begin{bmatrix} 1 & -\frac{15}{\sqrt{6}} & -\frac{3}{\sqrt{2}} \\ 0 & -\frac{1}{2} & \frac{\sqrt{3}}{2} \\ 0 & \frac{9\sqrt{3}}{2} & \frac{5}{2} \end{bmatrix} \begin{bmatrix} 1 & 0 & 0 \\ 0 & \frac{1}{2} & \frac{\sqrt{3}}{2} \\ 0 & -\frac{\sqrt{3}}{2} & \frac{1}{2} \end{bmatrix}$$

$$= \begin{bmatrix} 1 & -6\sqrt{6} & -24\sqrt{2} \\ 0 & -2 & -4\sqrt{3} \\ 0 & 0 & 4 \end{bmatrix}.$$

The matrix

$$Q = Q_1Q_3 = \begin{bmatrix} \frac{1}{\sqrt{3}} & 0 & \frac{2}{\sqrt{6}} \\ \frac{1}{\sqrt{3}} & -\frac{1}{\sqrt{2}} & -\frac{1}{\sqrt{6}} \\ \frac{1}{\sqrt{3}} & \frac{1}{\sqrt{2}} & -\frac{1}{\sqrt{6}} \end{bmatrix}$$

is an orthogonal matrix such that

$$Q^{-1}AQ = \begin{bmatrix} 1 & -6\sqrt{6} & -24\sqrt{2} \\ 0 & -2 & -4\sqrt{3} \\ 0 & 0 & 4 \end{bmatrix}.$$

The reader should note that in Example 2, we regarded the matrix A as the matrix of a linear transformation of the Euclidean vector space $\mathcal{V}_3$ with the $E^{(3)}$-inner product and orthonormal basis $E^{(3)}$. Then, in the second step of the reduction of A to triangular form, we considered the matrix B as the matrix of a linear transformationn of $\mathcal{V}_2$ with the $E^{(2)}$-inner product and orthonormal basis $E^{(2)}$. This method can be followed in general. That is, the k by k matrices, for $n \geq k \geq 2$, involved in the reduction of the given n by n matrix with real characteristic roots to triangular form, may always be considered to be the matrices of linear transformations of the Euclidean vector spaces $\mathcal{V}_k$ with the $E^{(k)}$-inner product and orthonormal basis $E^{(k)}$.

exercises

32.1 Prove that if L is an orthogonal transformation of a Euclidean vector space $\mathcal{E}$, then $\mathcal{I}(\mathbf{V}, \mathbf{W}) = \mathcal{I}[L(\mathbf{V}), L(\mathbf{W})]$ for all vectors $\mathbf{V}, \mathbf{W} \in \mathcal{E}$. Use the fact that $\mathcal{I}(\mathbf{V} - \mathbf{W}, \mathbf{V} - \mathbf{W}) = \mathcal{I}[L(\mathbf{V} - \mathbf{W}), L(\mathbf{V} - \mathbf{W})]$ and expand this identity.

32.2 Let L be the linear transformation of $\mathcal{E}_3$ defined by $L(x \cdot \mathbf{I} + y \cdot \mathbf{J} + z \cdot \mathbf{K}) = x' \cdot \mathbf{I} + y' \cdot \mathbf{J} + z' \cdot \mathbf{K}$, where

$$\begin{aligned} x' &= x \cos \varphi - y \sin \varphi \\ y' &= x \sin \varphi + y \cos \varphi \\ z' &= z \end{aligned}$$

with respect to a rectangular Cartesian coordinate system. Prove that L is an orthogonal transformation of $\mathcal{E}_3$.

32.3 Let L and M be orthogonal transformations of a Euclidean vector space $\mathcal{E}$. Prove that LM is an orthogonal transformation of $\mathcal{E}$.

***32.4** Let A be an orthogonal matrix, and let L be a linear transformation of a Euclidean vector space $\mathcal{E}$ that has matrix A with respect to an orthonormal basis of $\mathcal{E}$. Prove that L is an orthogonal transformation of $\mathcal{E}$. *Hint:* Express the inner product of two vectors in $\mathcal{E}$ as a matrix product.

***32.5** Let A be an orthogonal matrix. Prove (a) $|A| = \pm 1$; (b) A' is orthogonal; (c) A^{-1} is orthogonal; and (d) $Q^{-1}AQ$ is orthogonal, for any orthogonal matrix Q.

32.6 Let $\{\mathbf{U}_1, \mathbf{U}_2, \ldots, \mathbf{U}_n\}$ be an orthonormal basis of a Euclidean vector space $\mathcal{E}$. Let $\{\mathbf{W}_1, \mathbf{W}_2, \ldots, \mathbf{W}_n\}$ be a set of n vectors in $\mathcal{E}$ given by equations (32–1). Prove that $\{\mathbf{W}_1, \mathbf{W}_2, \ldots, \mathbf{W}_n\}$ is an orthonormal basis of $\mathcal{E}$ if and only if the matrix $P = [p_{i,j}]$ is an orthogonal matrix.

32.7 Prove that orthogonal equivalence is an equivalence relation on the set $\mathfrak{A}_n$ of all real n by n matrices.

32.8 For each of the following matrices A, find an orthogonal matrix Q such that $Q^{-1}AQ$ is in triangular form (32–3).

(a) $\begin{bmatrix} 1 & 0 & 1 \\ -1 & -\frac{1}{2} & -\frac{1}{2} \\ -1 & -\frac{3}{2} & -\frac{3}{2} \end{bmatrix}$ (b) $\begin{bmatrix} 1 & \frac{2}{3} \\ 2 & \frac{4}{3} \end{bmatrix}$ (c) $\begin{bmatrix} 0 & 0 & -1 \\ 0 & 4 & -6 \\ -2 & 0 & 1 \end{bmatrix}$

32.9 For the problem stated in Example 1, give the details of the computation which shows that

$$A = \begin{bmatrix} \cos \varphi & \sin \varphi \\ \sin \varphi & -\cos \varphi \end{bmatrix}$$

in the case where $|A| = -1$. Show that the orthogonal transformation L with matrix A is a reflection of $\mathcal{E}_2$.

8

quadratic forms

33 bilinear and quadratic mappings

We recall that if $\mathcal{E}$ is a Euclidean vector space, the inner product $\mathcal{I}(\mathbf{U}, \mathbf{W})$, for $\mathbf{U}, \mathbf{W} \in \mathcal{E}$, is a mapping of ordered pairs of vectors $(\mathbf{U}, \mathbf{W})$ into the real numbers that is bilinear, symmetric, and positive definite [see condition (11.3) and the discussion that follows]. We now generalize this concept by retaining only the bilinearity condition (11.3).

(33.1) definition Let $\mathcal{V}$ be an n-dimensional vector space. A mapping $\mathcal{B}$ from ordered pairs of vectors in $\mathcal{V}$ to the real numbers is called a *bilinear mapping* if $\mathcal{B}$ satisfies

$$\mathcal{B}(r_1 \cdot \mathbf{U}_1 + r_2 \cdot \mathbf{U}_2, s_1 \cdot \mathbf{W}_1 + s_2 \cdot \mathbf{W}_2) = r_1 s_1 \mathcal{B}(\mathbf{U}_1, \mathbf{W}_1) + r_1 s_2 \mathcal{B}(\mathbf{U}_1, \mathbf{W}_2) + r_2 s_1 \mathcal{B}(\mathbf{U}_2, \mathbf{W}_1) + r_2 s_2 \mathcal{B}(\mathbf{U}_2, \mathbf{W}_2),$$

for all vectors $\mathbf{U}_1, \mathbf{U}_2, \mathbf{W}_1, \mathbf{W}_2$ in $\mathcal{V}$ and all real numbers r_1, r_2, s_1, s_2.

Let $\{\mathbf{U}_1, \mathbf{U}_2, \ldots, \mathbf{U}_n\}$ be a basis of $\mathcal{V}$, and let $\mathbf{U} = r_1 \cdot \mathbf{U}_1 + r_2 \cdot \mathbf{U}_2 + \cdots + r_n \cdot \mathbf{U}_n$, $\mathbf{W} = s_1 \cdot \mathbf{U}_1 + s_2 \cdot \mathbf{U}_2 + \cdots + s_n \cdot \mathbf{U}_n$. Then, if $\mathcal{B}$ is a bilinear mapping, it follows from Definition 33.1 that

$$\mathcal{B}(\mathbf{U}, \mathbf{W}) = \mathcal{B}(r_1 \cdot \mathbf{U}_1 + r_2 \cdot \mathbf{U}_2 + \cdots + r_n \cdot \mathbf{U}_n, s_1 \cdot \mathbf{U}_1 + s_2 \cdot \mathbf{U}_2 + \cdots + s_n \cdot \mathbf{U}_n) \qquad (33\text{–}1)$$

$$= \sum_{i,j=1}^{n} r_i s_j \mathcal{B}(\mathbf{U}_i, \mathbf{U}_j) = \sum_{i,j=1}^{n} r_i \mathcal{B}(\mathbf{U}_i, \mathbf{U}_j) s_j.$$

Let $A = [a_{i,j}]$ be an n by n matrix such that $a_{i,j} = \mathcal{B}(\mathbf{U}_i, \mathbf{U}_j)$ for all i, j. Then by the definition of matrix multiplication, equation (33–1) can be written

$$\mathcal{B}(\mathbf{U}, \mathbf{W}) = \sum_{i,j=1}^{n} r_i a_{i,j} s_j = [r_1 \quad r_2 \quad \cdots \quad r_n] A \begin{bmatrix} s_1 \\ s_2 \\ \vdots \\ s_n \end{bmatrix}. \tag{33–2}$$

The matrix A is called the *matrix of* $\mathcal{B}$ with respect to the basis $\{\mathbf{U}_1, \mathbf{U}_2, \ldots, \mathbf{U}_n\}$. Strictly speaking, the right-hand side of equation (33–2) is the 1 by 1 matrix $[\mathcal{B}(\mathbf{U}, \mathbf{W})]$, which we have identified with the real number $\mathcal{B}(\mathbf{U}, \mathbf{W})$, its only element. In the expression for a bilinear mapping, this identification is convenient and causes no confusion.

EXAMPLE 1 As indicated in the discussion preceding Definition 33.1, any inner product $\mathcal{I}$ defined on a Euclidean vector space $\mathcal{E}$ is a bilinear mapping from ordered pairs of vectors in $\mathcal{E}$ to the real numbers. Let $\mathcal{E}$ be a Euclidean vector space of dimension n with inner product $\mathcal{I}$, and let $\{\mathbf{U}_1, \mathbf{U}_2, \ldots, \mathbf{U}_n\}$ be an orthonormal basis of $\mathcal{E}$. Then $\mathcal{I}(\mathbf{U}_i, \mathbf{U}_j) = 0$ if $i \neq j$, and $\mathcal{I}(\mathbf{U}_i, \mathbf{U}_i) = 1$, for $i, j = 1, 2, \ldots, n$. Thus, the n by n matrix $A = [a_{i,j}]$ with $a_{i,j} = \mathcal{I}(\mathbf{U}_i, \mathbf{U}_j)$ for all i, j is the identity matrix I_n. If $\mathbf{U} = r_1 \cdot \mathbf{U}_1 + r_2 \cdot \mathbf{U}_2 + \cdots + r_n \cdot \mathbf{U}_n$ and $\mathbf{W} = s_1 \cdot \mathbf{U}_1 + s_2 \cdot \mathbf{U}_2 + \cdots + s_n \cdot \mathbf{U}_n$, then

$$\mathcal{I}(\mathbf{U}, \mathbf{W}) = \sum_{i,j=1}^{n} r_i \mathcal{I}(\mathbf{U}_i, \mathbf{U}_j) s_j = \sum_{i,j=1}^{n} r_i a_{i,j} s_j$$

$$= [r_1 \quad r_2 \quad \cdots \quad r_n] I_n \begin{bmatrix} s_1 \\ s_2 \\ \vdots \\ s_n \end{bmatrix} = r_1 s_1 + r_2 s_2 + \cdots + r_n s_n.$$

Of course, this is precisely the result given in Theorem 11.7.

If, throughout the discussion above, we take $\mathbf{W} = \mathbf{U}$, then the bilinear mapping $\mathcal{B}$ is restricted to pairs $(\mathbf{U}, \mathbf{U})$ of vectors $\mathbf{U} \in \mathcal{V}$. By (33–2) we have

$$\mathcal{Q}(\mathbf{U}) = \mathcal{B}(\mathbf{U}, \mathbf{U}) = \sum_{i,j=1}^{n} r_i a_{i,j} r_j = [r_1 \quad r_2 \quad \cdots \quad r_n] A \begin{bmatrix} r_1 \\ r_2 \\ \vdots \\ r_n \end{bmatrix}. \tag{33–3}$$

Such a restricted bilinear mapping is called a *quadratic mapping of* $\mathcal{V}$. Although each bilinear mapping $\mathcal{B}$ determines a unique quadratic mapping $\mathcal{Q}$, different bilinear mappings may yield the same quadratic mapping. It follows from this observation that the matrix of a quadratic mapping is not unique (see Example 2).

Conversely, if $A = [a_{i,j}]$ is any n by n matrix, then equation (33–2) can be used to define a bilinear mapping $\mathcal{B}$, where $(r_1, r_2, \ldots, r_n)$ and $(s_1, s_2, \ldots, s_n)$ are the coordinates of **U** and **W**, respectively, relative to a given basis $\{\mathbf{U}_1, \mathbf{U}_2, \ldots, \mathbf{U}_n\}$ of $\mathcal{V}$. Similarly, equation (33–3) can be used to define a quadratic mapping of $\mathcal{V}$.

EXAMPLE 2 Let

$$A = [a_{i,j}] = \begin{bmatrix} 7 & 5 & -1 \\ 2 & 0 & 3 \\ 1 & -2 & 4 \end{bmatrix}$$

define a bilinear mapping $\mathcal{B}$ of ordered pairs of vectors in $\mathcal{V}_3$ with respect to the basis $E^{(3)}$. Then, for $\mathbf{U} = (r_1, r_2, r_3)$ and $\mathbf{W} = (s_1, s_2, s_3)$ in $\mathcal{V}_3$, we have

$$\mathcal{B}(\mathbf{U}, \mathbf{W}) = [r_1 \quad r_2 \quad r_3] \begin{bmatrix} 7 & 5 & -1 \\ 2 & 0 & 3 \\ 1 & -2 & 4 \end{bmatrix} \begin{bmatrix} s_1 \\ s_2 \\ s_3 \end{bmatrix}$$

$$= [7r_1 + 2r_2 + r_3 \quad 5r_1 - 2r_3 \quad -r_1 + 3r_2 + 4r_3] \begin{bmatrix} s_1 \\ s_2 \\ s_3 \end{bmatrix}$$

$$= 7r_1s_1 + 2r_2s_1 + r_3s_1 + 5r_1s_2 - 2r_3s_2 - r_1s_3 + 3r_2s_3 + 4r_3s_3.$$

If $\mathbf{U} = (1, -6, 2)$ and $\mathbf{W} = (-3, 5, 7)$, then $\mathcal{B}(\mathbf{U}, \mathbf{W}) = -63$ and $\mathcal{B}(\mathbf{W}, \mathbf{U}) = 262$. Therefore, $\mathcal{B}$ is not symmetric. Also $\mathcal{B}(\mathbf{U}, \mathbf{U}) = -31$, so that $\mathcal{B}$ is not positive definite.

The corresponding quadratic mapping of $\mathcal{V}_3$ is given by

$$\mathcal{Q}(\mathbf{U}) = [r_1 \quad r_2 \quad r_3] \begin{bmatrix} 7 & 5 & -1 \\ 2 & 0 & 3 \\ 1 & -2 & 4 \end{bmatrix} \begin{bmatrix} r_1 \\ r_2 \\ r_3 \end{bmatrix}$$

$$= 7r_1^2 + 2r_2r_1 + r_3r_1 + 5r_1r_2 - 2r_3r_2 - r_1r_3 + 3r_2r_3 + 4r_3^2$$

$$= 7r_1^2 + 7r_1r_2 + r_2r_3 + 4r_3^2.$$

Note that this quadratic mapping can also be expressed by the matrix

$$\begin{bmatrix} 7 & 7 & 0 \\ 0 & 0 & 1 \\ 0 & 0 & 4 \end{bmatrix}$$

which defines a bilinear mapping $\mathcal{C}$ different from $\mathcal{B}$.

There is a second interpretation of bilinear and quadratic mappings that is important in applications. Let

$$\sum_{i,j=1}^{n} x_i a_{i,j} y_j = x_1 a_{1,1} y_1 + x_1 a_{1,2} y_2 + \cdots + x_1 a_{1,n} y_n + x_2 a_{2,1} y_1 + \cdots + x_n a_{n,n} y_n$$

be a quadratic polynomial in $2n$ variables $x_1, x_2, \ldots, x_n$ and $y_1, y_2, \ldots, y_n$ with real coefficients $a_{i,j}$. Such a polynomial is called a *bilinear form*. If real numbers $r_1, r_2, \ldots, r_n$ and $s_1, s_2, \ldots, s_n$ are substituted for the variables $x_1, x_2, \ldots, x_n$ and $y_1, y_2, \ldots, y_n$, respectively, then $\sum_{i,j=1}^{n} r_i a_{i,j} s_j$ is a *value of the bilinear form*. Note that the value of a bilinear form is given by the same expression as $\mathcal{B}(\mathbf{U}, \mathbf{W})$ in equation (33–2). Therefore, each bilinear form defines a bilinear mapping. If $y_1 = x_1$, $y_2 = x_2, \ldots, y_n = x_n$, then $\sum_{i,j=1}^{n} x_i a_{i,j} y_j = \sum_{i,j=1}^{n} x_i a_{i,j} x_j$ is a quadratic polynomial in the n variables $x_1, x_2, \ldots, x_n$, which is called a *quadratic form*. A *value of the quadratic form* is obtained by substituting real numbers $r_1, r_2, \ldots, r_n$ for the variables $x_1, x_2, \ldots, x_n$. A value of a quadratic form $\sum_{i,j=1}^{n} r_i a_{i,j} r_j$ is given by the same expression as $\mathcal{Q}(\mathbf{U})$ in equation (33–3), so that each quadratic form defines a quadratic mapping of $\mathcal{V}$.

EXAMPLE 3 The bilinear and quadratic forms related to the bilinear mapping of Example 1 have the expressions

$$\sum_{i,j=1}^{3} x_i a_{i,j} y_j = 7x_1y_1 + 2x_2y_1 + x_3y_1 + 5x_1y_2 - 2x_3y_2 - x_1y_3 + 3x_2y_3 + 4x_3y_3$$

$$\sum_{i,j=1}^{3} x_i a_{i,j} x_j = 7x_1^2 + 7x_1x_2 + x_2x_3 + 4x_3^2.$$

Quadratic forms occur in analytic geometry in the equations of central conics in a plane and central quadric surfaces in three-dimensional space. For example, $ax^2 + bxy + cy^2 = k$ is the equation of a conic with center at the origin of a rectangular Cartesian coordinate system in a plane. This equation can be regarded as the equation of all vectors $\mathbf{U} \in \mathcal{E}_2$ that have coordinates (r_1, r_2) for which the value of the quadratic form $ax^2 + bxy + cy^2$ is the number k; such vectors have their terminal point P on the conic (Figure 53). Similarly, $ax^2 + bxy + cy^2 + dxz + eyz + fz^2 = k$ is the equation of a quadric surface with its center at the origin in three-dimensional space. This equation can be thought of as the equation of all vectors $\mathbf{U} = r_1 \cdot \mathbf{I} + r_2 \cdot \mathbf{J} + r_3 \cdot \mathbf{K}$ in $\mathcal{E}_3$ for which $ar_1^2 + br_1r_2 + cr_2^2 + dr_1r_3 + er_2r_3 + fr_3^2 = k$. Such a vector $\mathbf{U}$ has its terminal point on the quadric surface. Quadratic forms also have important applications in other fields, including statistics, differential geometry, and mechanics.

Suppose that $\{\mathbf{W}_1, \mathbf{W}_2, \ldots, \mathbf{W}_n\}$ is a second basis of $\mathcal{V}$ and that $P = [p_{i,j}]$ is the matrix of the change of basis from the basis

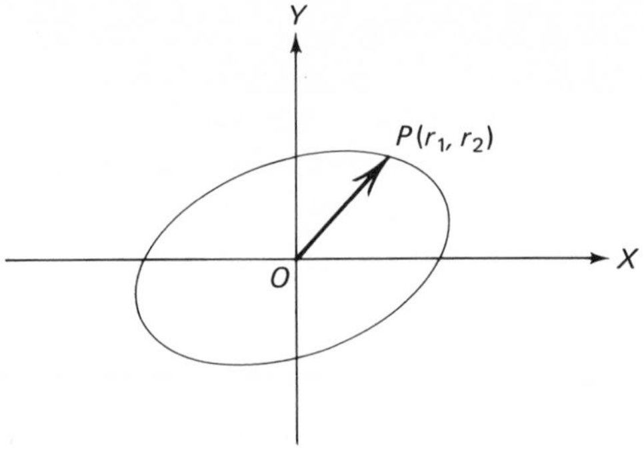

FIGURE 53

$\{\mathbf{U}_1, \mathbf{U}_2, \ldots, \mathbf{U}_n\}$ to this new basis. Then

$$[r_1 \quad r_2 \quad \cdots \quad r_n] = [\bar{r}_1 \quad \bar{r}_2 \quad \cdots \quad \bar{r}_n]P, \quad \text{and} \quad \begin{bmatrix} s_1 \\ s_2 \\ \vdots \\ s_n \end{bmatrix} = P' \begin{bmatrix} \bar{s}_1 \\ \bar{s}_2 \\ \vdots \\ \bar{s}_n \end{bmatrix},$$

where $(\bar{r}_1, \bar{r}_2, \ldots, \bar{r}_n)$ and $(\bar{s}_1, \bar{s}_2, \ldots, \bar{s}_n)$ are the coordinates of **U** and **W**, respectively, with respect to the second basis [see equations (21–3) and (21–4)]. Substituting in equation (33–2), we obtain

$$\mathcal{B}(\mathbf{U}, \mathbf{W}) = [\bar{r}_1 \quad \bar{r}_2 \quad \cdots \quad \bar{r}_n]PAP' \begin{bmatrix} \bar{s}_1 \\ \bar{s}_2 \\ \vdots \\ \bar{s}_n \end{bmatrix}. \tag{33–4}$$

Thus, the matrix of $\mathcal{B}$ with respect to the basis $\{\mathbf{W}_1, \mathbf{W}_2, \ldots, \mathbf{W}_n\}$ is PAP'. If, in particular, $\mathcal{V}$ is a Euclidean vector space and the two given bases are orthonormal, then P is an orthogonal matrix. In this case, $PAP' = Q^{-1}AQ$, where $Q = P' = P^{-1}$. That is, the two matrices of $\mathcal{B}$ are orthogonally equivalent.

(33.2) definition Let A and B be n by n matrices. The matrix B is *congruent* to the matrix A if there exists a nonsingular n by n matrix P such that $B = PAP'$.

Congruence is an equivalence relation on the set $\mathfrak{A}_n$ of all real n by n matrices. Orthogonal equivalence, which is a special case of similarity, is also a special case of congruence. For if $B = Q^{-1}AQ$, where Q is orthogonal, then $B = PAP'$, where $P = Q^{-1} = Q'$.

These results show that the simplification of the expressions for bilinear and quadratic mappings by a change of basis in $\mathcal{V}$ will entail the reduction of matrices under the relations of congruence and orthogonal equivalence. In the sequel, we will be principally concerned with the simplification of quadratic mappings and forms. In this case, we find it sufficient to study the effect of the relations of congruence and orthogonal equivalence on a special class of n by n matrices.

In the expression (33–3) for a quadratic mapping, the terms $r_i a_{i,j} r_j$ and $r_j a_{j,i} r_i$ with $i \neq j$ occur in pairs. The value of the mapping for each vector $\mathbf{U} \in \mathcal{V}$ is unchanged if, for each such pair, we write

$$r_i a_{i,j} r_j + r_j a_{j,i} r_i = r_i \left(\frac{a_{i,j} + a_{j,i}}{2} \right) r_j + r_j \left(\frac{a_{i,j} + a_{j,i}}{2} \right) r_i.$$

Consequently, if $B = [b_{i,j}]$ is an n by n matrix such that $b_{i,j} = b_{j,i} = (a_{i,j} + a_{j,i})/2$ for all pairs i, j, then

$$\mathcal{Q}(\mathbf{U}) = \sum_{i,j=1}^{n} r_i b_{i,j} r_j = [r_1 \quad r_2 \quad \cdots \quad r_n] B \begin{bmatrix} r_1 \\ r_2 \\ \vdots \\ r_n \end{bmatrix}.$$

An n by n matrix $B = [b_{i,j}]$ such that $b_{i,j} = b_{j,i}$, for $i = 1, 2, \ldots, n$ and $j = 1, 2, \ldots, n$, is called a *symmetric matrix*. Thus, each quadratic mapping and the related quadratic form can be expressed in terms of a symmetric matrix.

EXAMPLE 4 The quadratic form of Example 3 can be written

$$\sum_{i,j=1}^{3} x_i a_{i,j} x_j = 7x_1^2 + 7x_1x_2 + x_2x_3 + 4x_3^2$$

$$= 7x_1^2 + \tfrac{7}{2}x_1x_2 + \tfrac{7}{2}x_2x_1 + \tfrac{1}{2}x_2x_3 + \tfrac{1}{2}x_3x_2 + 4x_3^2$$

$$= [x_1 \quad x_2 \quad x_3] \begin{bmatrix} 7 & \tfrac{7}{2} & 0 \\ \tfrac{7}{2} & 0 & \tfrac{1}{2} \\ 0 & \tfrac{1}{2} & 4 \end{bmatrix} \begin{bmatrix} x_1 \\ x_2 \\ x_3 \end{bmatrix} = [x_1 \quad x_2 \quad x_3] B \begin{bmatrix} x_1 \\ x_2 \\ x_3 \end{bmatrix},$$

where B is a symmetric matrix.

Clearly, an n by n matrix B is symmetric if and only if $B' = B$. If B is symmetric and P is any n by n matrix, then PBP' is again symmetric. Indeed, $(PBP')' = (P')'B'P' = PBP'$. Each quadratic mapping of $\mathcal{V}$ can be written with a symmetric matrix B. A change of basis in $\mathcal{V}$ replaces B with the symmetric matrix PBP', which is congruent to B. Thus, our discussion of the simplification of the expression for a quadratic mapping amounts to a study of the relation of congruence on the set of real symmetric matrices.

exercises

33.1 Let $A = [a_{i,j}]$ be an n by n matrix. Prove that a mapping $\mathcal{B}$ of ordered pairs of vectors in a vector space $\mathcal{V}$ of dimension n, defined by equation (33–2), is a bilinear mapping.

33.2 Let $\mathcal{B}$ be a bilinear mapping with matrix A with respect to some basis of $\mathcal{V}$. Prove that $\mathcal{B}$ is a symmetric mapping, that is, $\mathcal{B}(\mathbf{U}, \mathbf{W}) = \mathcal{B}(\mathbf{W}, \mathbf{U})$ for all $\mathbf{U}, \mathbf{W} \in \mathcal{V}$, if and only if A is a symmetric matrix.

33.3 Let $\mathcal{B}$ be the mapping from ordered pairs of vectors in $\mathcal{V}_4$ to the real numbers defined by $\mathcal{B}(\mathbf{U}, \mathbf{W}) = r_1s_2 - r_4s_3$ for $\mathbf{U} = (r_1, r_2, r_3, r_4)$ and $\mathbf{W} = (s_1, s_2, s_3, s_4)$ in $\mathcal{V}_4$. Prove that $\mathcal{B}$ is a bilinear mapping. Show that $\mathcal{B}$ is neither symmetric nor positive definite. Find the matrix of $\mathcal{B}$ with respect to the basis $E^{(4)}$.

33.4 Let $\mathcal{B}$ be a bilinear mapping which has matrix A with respect to some basis of $\mathcal{V}$. Let L be the linear transformation of $\mathcal{V}$ which has matrix A with respect to this basis. If $\mathbf{U}$ has coordinates $(r_1, r_2, \ldots, r_n)$ and $L(\mathbf{W})$ has coordinates $(\bar{s}_1, \bar{s}_2, \ldots, \bar{s}_n)$, show that $\mathcal{B}(\mathbf{U}, \mathbf{W}) = r_1\bar{s}_1 + r_2\bar{s}_2 + \cdots + r_n\bar{s}_n$.

33.5 Write out the quadratic forms with the following matrices:

(a) $\begin{bmatrix} -3 & 2 & 1 \\ 6 & 0 & 2 \\ -1 & 2 & 4 \end{bmatrix}$ (b) $\begin{bmatrix} -3 & 8 & 0 \\ 0 & 0 & 4 \\ 0 & 0 & 4 \end{bmatrix}$ (c) $\begin{bmatrix} -3 & 4 & 0 \\ 4 & 0 & 2 \\ 0 & 2 & 4 \end{bmatrix}$

For (a) and (b), express the quadratic form with a symmetric matrix.

33.6 Express the quadratic form $x_1^2 - 2x_1x_3 - 6x_2^2 + 3x_2x_3 + x_2x_4 + 5x_3^2 - x_3x_4 + x_4^2$ with at least three different matrices.

33.7 Let $\Sigma_{i,j=1}^n x_i a_{i,j} x_j$ and $\Sigma_{i,j=1}^n x_i b_{i,j} x_j$ be quadratic forms such that both of the matrices $A = [a_{i,j}]$ and $B = [b_{i,j}]$ are symmetric. Suppose that the given forms have the same value for every n-tuple $(r_1, r_2, \ldots, r_n)$ of real numbers. Prove that $A = B$.

33.8 Prove that congruence is an equivalence relation on the set $\mathfrak{A}_n$ of all real n by n matrices.

33.9 Express the following quadratic forms with symmetric matrices: $x_1^2 + \sqrt{2}x_1x_2 - 5x_2^2$; $3x_1^2 + 5x_1x_2 - 7x_1x_3 + x_2x_3 - 4x_2^2 - x_3^2$; $x_1x_2 + x_1x_3 + x_1x_4 + x_2x_3 + x_2x_4 + x_3x_4$.

33.10 (a) Let A be an m by n matrix. Show that AA' and $A'A$ are symmetric matrices. (b) Let A be an n by n matrix. Show that $A + A'$ is symmetric.

33.11 Let

$$B = \begin{bmatrix} 7 & \frac{7}{2} & 0 \\ \frac{7}{2} & 0 & \frac{1}{2} \\ 0 & \frac{1}{2} & 4 \end{bmatrix} \quad \text{and} \quad P = \begin{bmatrix} \frac{1}{\sqrt{7}} & 0 & 0 \\ 0 & 0 & \frac{1}{2} \\ \frac{-2}{\sqrt{29}} & \frac{4}{\sqrt{29}} & \frac{-1}{2\sqrt{29}} \end{bmatrix}.$$

Show that

$$PBP' = \begin{bmatrix} 1 & 0 & 0 \\ 0 & 1 & 0 \\ 0 & 0 & -1 \end{bmatrix}.$$

Hence, show that $7x_1^2 + 7x_1x_2 + x_2x_3 + 4x_3^2 = y_1^2 + y_2^2 - y_3^2$, where $[x_1 \ \ x_2 \ \ x_3] = [y_1 \ \ y_2 \ \ y_3]P$.

34 orthogonal reduction of real symmetric matrices

The key result in the theory of the reduction of real symmetric matrices by orthogonal equivalence is the fact that the characteristic roots of a real symmetric matrix are real. In order to prove this result, we utilize the following properties of complex numbers, which are probably familiar to the reader. Recall that if $c = x + yi$ is a complex number, then the *conjugate* of c is the complex number $\bar{c} = x - yi$. For example, the conjugate of $2 + 3i$ is $2 - 3i$, the conjugate of $4 - \frac{1}{2}i$ is $4 + \frac{1}{2}i$, the conjugate of $-3i = 0 - 3i$ is $0 + 3i = 3i$, and the conjugate of $\sqrt{2} = \sqrt{2} + 0i$ is $\sqrt{2} - 0i = \sqrt{2}$.

(34.1) Let c and d be complex numbers. Then (a) $\overline{c + d} = \bar{c} + \bar{d}$; (b) $\overline{cd} = \bar{c}\bar{d}$; (c) $\bar{\bar{c}} = c$; (d) c is a real number if and only if $c = \bar{c}$; (e) $c\bar{c}$ is a nonnegative real number, and $c\bar{c} = 0$ if and only if $c = 0$.

PROOF Let $c = x + yi$ and $d = u + vi$. Then $\bar{c} = x - yi$ and $\bar{d} = u - vi$. Moreover, $c + d = (x + yi) + (u + vi) = (x + u) + (y + v)i$, and $cd = (x + yi)(u + vi) = (xu - yv) + (xv + yu)i$. Thus,

$$\overline{c + d} = (x + u) - (y + v)i = (x - yi) + (u - vi) = \bar{c} + \bar{d}$$

which proves (a). Part (b) follows from $\overline{cd} = (xu - yv) - (xv + yu)i = (x - yi)(u - vi) = \bar{c}\bar{d}$. To prove part (c), we note that $\bar{\bar{c}} = \overline{x - yi} = x + yi = c$. We have $c = \bar{c}$ if and only if $x + yi = x - yi$. The latter equality holds if and only if $y = -y$, or $y = 0$. Thus, $c = \bar{c}$ if and only if $c = x + 0i = x$ is a real number. This proves (d). Finally, $c\bar{c} = (x + yi)(x - yi) = x^2 + y^2$, which is a nonnegative real number. Also, $x^2 + y^2 = 0$ if and only if $x = y = 0$, proving part (e) of (34.1). ∎

(34.2) theorem The characteristic roots of a real symmetric matrix are real.

PROOF Let $A = [a_{i,j}]$ be a real symmetric n by n matrix, and let r be a characteristic root of A. Then $|A - r \cdot I_n| = 0$. Regard the matrix $A - r \cdot I_n$ as a matrix whose elements are complex numbers. (In fact, if r is not real, the diagonal elements of $A - r \cdot I_n$ are not real.) The results of Chapter 3 on the solutions of systems of homogeneous equations still hold if the matrix of coefficients of the system has complex numbers as elements. In particular, a homogeneous system of n equations in n unknowns has a nontrivial solution if and only if the determinant of the matrix of coefficients of the system is zero (see Exercise 29.10). Therefore, since $|A - r \cdot I_n| = 0$, the system of equations $(A - r \cdot I_n)X = \mathbf{O}$ has a nontrivial solution

$$X = C = \begin{bmatrix} c_1 \\ c_2 \\ \vdots \\ c_n \end{bmatrix}$$

where the c_i are complex numbers. The equation $(A - r \cdot I_n)C = \mathbf{O}$ implies $AC = r \cdot C$. Now consider the complex number

$$m = \sum_{i,j=1}^{n} \bar{c}_i a_{i,j} c_j = [\bar{c}_1 \quad \bar{c}_2 \quad \cdots \quad \bar{c}_n]A \begin{bmatrix} c_1 \\ c_2 \\ \vdots \\ c_n \end{bmatrix}$$

where each $\bar{c}_i$ is the conjugate of c_i. Since $m = [\bar{c}_1\ \bar{c}_2 \cdots \bar{c}_n]AC$, and $AC = r \cdot C$, we have

$$\begin{aligned} m &= [\bar{c}_1\ \bar{c}_2 \cdots \bar{c}_n](r \cdot C) = r([\bar{c}_1\ \bar{c}_2 \cdots \bar{c}_n]C) \\ &= r(\bar{c}_1 c_1 + \bar{c}_2 c_2 + \cdots + \bar{c}_n c_n). \end{aligned}$$

Since C is a nontrivial solution of a homogeneous system of equations, it follows that at least one $c_i \neq 0$. Therefore, by 34.1 (e), the expression

$$\bar{c}_1 c_1 + \bar{c}_2 c_2 + \cdots + \bar{c}_n c_n = s$$

is a positive real number. Hence, $m = rs$, and $r = m/s$, where $s \neq 0$ is a real number. To prove that r is a real number, it is sufficient to show that m is a real number. By 34.1 (d), this latter result can be obtained by showing that $m = \bar{m}$. We have

$$\bar{m} = \overline{\sum_{i,j=1}^{n} \bar{c}_i a_{i,j} c_j} = \sum_{i,j=1}^{n} \overline{\bar{c}_i a_{i,j} c_j} = \sum_{i,j=1}^{n} \bar{\bar{c}}_i \bar{a}_{i,j} \bar{c}_j = \sum_{i,j=1}^{n} c_i a_{i,j} \bar{c}_j$$

using (a), (b), (c), and (d) of 34.1. Indeed, by 34.1 (a) the conjugate of a sum of complex numbers is the sum of the conjugates. Therefore,

$$\overline{\sum_{i,j=1}^{n} \bar{c}_i a_{i,j} c_j} = \sum_{i,j=1}^{n} \overline{\bar{c}_i a_{i,j} c_j}.$$

Each term of the latter sum is a conjugate of a product of complex numbers, and hence,

$$\sum_{i,j=1}^{n} \overline{\bar{c}_i a_{i,j} c_j} = \sum_{i,j=1}^{n} \bar{\bar{c}}_i \bar{a}_{i,j} \bar{c}_j$$

by 34.1 (b). By 34.1 (c), $\bar{\bar{c}}_i = c_i$, for $i = 1, 2, \ldots, n$. Since the matrix $A = [a_{i,j}]$ is real, each $a_{i,j}$ is a real number, so that by 34.1 (d), $\bar{a}_{i,j} = a_{i,j}$ for $i, j = 1, 2, \ldots, n$. Therefore,

$$\sum_{i,j=1}^{n} \bar{\bar{c}}_i \bar{a}_{i,j} \bar{c}_j = \sum_{i,j=1}^{n} c_i a_{i,j} \bar{c}_j$$

establishing the result $\bar{m} = \Sigma_{i,j=1}^{n} c_i a_{i,j} \bar{c}_j$. Since it is immaterial which index, i or j, is summed on first in the sum $\Sigma_{i,j=1}^{n} c_i a_{i,j} \bar{c}_j$ (only the order of the summands is affected), we may write

$$\bar{m} = \sum_{j,i=1}^{n} c_i a_{i,j} \bar{c}_j.$$

Each term $c_i a_{i,j} \bar{c}_j$ of the above sum may be written $\bar{c}_j a_{i,j} c_i$, since the multiplication of complex numbers is commutative. Thus,

$$\bar{m} = \sum_{j,i=1}^{n} \bar{c}_j a_{i,j} c_i.$$

As we explained in the discussion of the summation convention in §14, any symbols may be used for the indices of summation. Thus, i and j may be interchanged throughout the latter sum to obtain

$$\bar{m} = \sum_{j,i=1}^{n} \bar{c}_j a_{i,j} c_i = \sum_{i,j=1}^{n} \bar{c}_i a_{j,i} c_j.$$

Finally, since $A = [a_{i,j}]$ is a symmetric matrix, $a_{j,i} = a_{i,j}$ for all i, j. Therefore,

$$\bar{m} = \sum_{i,j=1}^{n} \bar{c}_i a_{j,i} c_j = \sum_{i,j=1}^{n} \bar{c}_i a_{i,j} c_j = m.$$

Since $\bar{m} = m$, m is a real number, and the characteristic root $r = m/s$ of A is real. ■

EXAMPLE 1 Consider the real symmetric matrix

$$A = \begin{bmatrix} 1 & -2 & 3 \\ -2 & 0 & 1 \\ 3 & 1 & 4 \end{bmatrix}.$$

The characteristic equation of A is

$$|A - x \cdot I_3| = \begin{vmatrix} 1-x & -2 & 3 \\ -2 & -x & 1 \\ 3 & 1 & 4-x \end{vmatrix} = -x^3 + 5x^2 + 10x - 29 = 0.$$

Let $f(x) = -x^3 + 5x^2 + 10x - 29$. Then $f(-3) = 15$, $f(-2) = -21$, $f(2) = 3$, and $f(6) = -5$. Therefore, the graph of $y = f(x)$ crosses the X axis three times (see Figure 54). Thus, $f(r_1) = 0$ for $-3 < r_1 < -2$, $f(r_2) = 0$ for $-2 < r_2 < 2$, and $f(r_3) = 0$ for $2 < r_3 < 6$. That is, the equation $-x^3 + 5x^2 + 10x - 29 = 0$ has three real roots, r_1, r_2, r_3, and these are the characteristic roots of the matrix A.

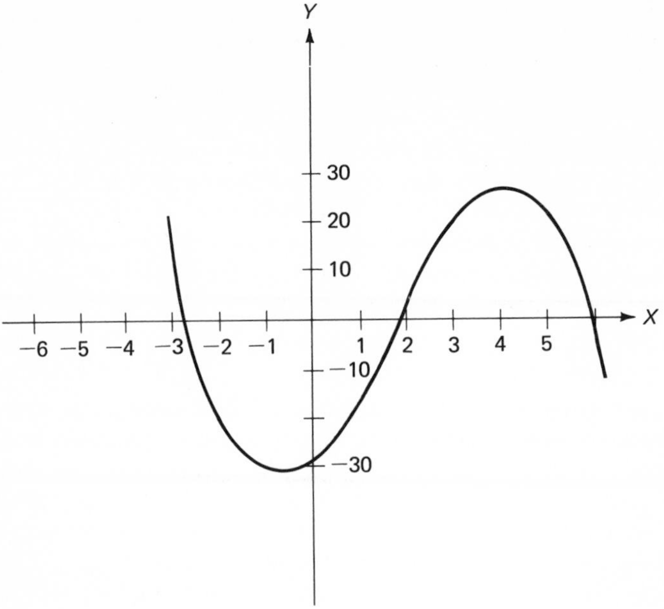

FIGURE 54

As we mentioned at the beginning of this section, Theorem 34.2 is the key in the reduction of a real symmetric matrix to diagonal form by orthogonal equivalence. The next result describes this diagonal matrix exactly.

(34.3) theorem Let A be a real symmetric n by n matrix. There exists an orthogonal matrix P such that

$$P^{-1}AP = \begin{bmatrix} r_1 & 0 & \cdots & 0 \\ 0 & r_2 & & \cdot \\ \cdot & & \cdot & \cdot \\ \cdot & & \cdot & \cdot \\ \cdot & & \cdot & 0 \\ 0 & \cdots & 0 & r_n \end{bmatrix} = D$$

where $r_1, r_2, \ldots, r_n$ are the characteristic roots of A.

PROOF By Theorems 32.5 and 34.2, there exists an orthogonal matrix P such that

$$P^{-1}AP = \begin{bmatrix} r_1 & b^*_{1,2} & \cdots & b^*_{1,n} \\ 0 & r_2 & & b^*_{2,n} \\ \cdot & & \cdot & \cdot \\ \cdot & & \cdot & \cdot \\ \cdot & & & b^*_{n-1,n} \\ 0 & \cdots & 0 & r_n \end{bmatrix}$$

where $r_1, r_2, \ldots, r_n$ are the characteristic roots of A. Since A is symmetric and $P^{-1} = P'$, it follows that $P^{-1}AP$ is symmetric. Therefore, the elements above the diagonal of $P^{-1}AP$ must be 0, which proves the theorem. ∎

(34.4) corollary Two real symmetric matrices are orthogonally equivalent if and only if they have the same characteristic roots.

PROOF Let A and B be n by n real symmetric matrices. If A and B are orthogonally equivalent, then they are similar. By Theorem 30.4, they have the same characteristic roots. Conversely, if A and B have the same characteristic roots, they are orthogonally equivalent to the same diagonal matrix D, by Theorem 34.3. Thus, $P^{-1}AP = Q^{-1}BQ$ for orthogonal matrices P and Q. Therefore, $B = (QP^{-1})A(PQ^{-1}) = (PQ^{-1})^{-1}A(PQ^{-1})$, where PQ^{-1} is an orthogonal matrix. That is, B is orthogonally equivalent to A. ∎

The method for constructing the orthogonal matrix P of Theorem 34.3 is a modification of the method discussed in §31 for finding a nonsingular matrix Q such that $Q^{-1}AQ$ is a diagonal matrix.

Let A be an n by n symmetric matrix, and suppose that A is the matrix of a linear transformation L of $\mathcal{V}_n$ into itself with respect to the basis $E^{(n)}$. By Theorem 34.3, there exists an orthogonal matrix $P = [p_{i,j}]$ such that $AP = PD$, where D is a diagonal matrix with the characteristic values of L on the diagonal. Therefore, for $j = 1, 2, \ldots, n$,

$$A\begin{bmatrix} p_{1,j} \\ p_{2,j} \\ \vdots \\ p_{n,j} \end{bmatrix} = P\begin{bmatrix} 0 \\ \vdots \\ 0 \\ r_j \\ 0 \\ \vdots \\ 0 \end{bmatrix} = r_j \cdot \begin{bmatrix} p_{1,j} \\ p_{2,j} \\ \vdots \\ p_{n,j} \end{bmatrix}. \qquad (34\text{–}1)$$

The columns of P can be regarded as vectors in $\mathcal{V}_n$. Hence, by (34–1), the jth column of P is a characteristic vector of L, which belongs to the characteristic value r_j, for $j = 1, 2, \ldots, n$. Since P is an orthogonal matrix, these characteristic vectors form an orthonormal basis of $\mathcal{V}_n$. (By Example 1, §11, the n-tuple space $\mathcal{V}_n$ is a Euclidean vector space.) Conversely, suppose that $P = [p_{i,j}]$ is an n by n matrix whose columns form an orthonormal basis of $\mathcal{V}_n$ such that the jth column of P is a characteristic vector of L belonging to the characteristic value r_j, for $j = 1, 2, \ldots, n$. Then equation (34–1) is satisfied, which implies that $P^{-1}AP = D$.

EXAMPLE 2 Let

$$A = \begin{bmatrix} 53/18 & 2/9 & 1/18 \\ 2/9 & 19/9 & -2/9 \\ 1/18 & -2/9 & 53/18 \end{bmatrix}.$$

Then $|A - x \cdot I_3| = -(x-3)^2(x-2)$. Thus, the characteristic roots of A are 3, 3, 2. According to the preceding discussion, there is an orthonormal basis of $\mathcal{V}_3$ consisting of characteristic vectors of the linear transformation determined by A, two of which belong to the repeated root 3 and one that belongs to 2. For the repeated root 3, the system of equations for the characteristic vectors is

$$\begin{bmatrix} -1/18 & 2/9 & 1/18 \\ 2/9 & -8/9 & -2/9 \\ 1/18 & -2/9 & -1/18 \end{bmatrix} \begin{bmatrix} x_1 \\ x_2 \\ x_3 \end{bmatrix} = \begin{bmatrix} 0 \\ 0 \\ 0 \end{bmatrix}$$

which is equivalent to the system

$$\begin{bmatrix} 1 & -4 & -1 \\ 0 & 0 & 0 \\ 0 & 0 & 0 \end{bmatrix} \begin{bmatrix} x_1 \\ x_2 \\ x_3 \end{bmatrix} = \begin{bmatrix} 0 \\ 0 \\ 0 \end{bmatrix}.$$

This system has the solution $(x_1, x_2, x_3) = (4r + s, r, s)$ for any real numbers r and s. Selecting $r = 0$, $s = 1$ and $r = -1/2$, $s = 1$, we obtain the orthogonal vectors $(1, 0, 1)$ and $(-1, -1/2, 1)$. Then $(1/\sqrt{2}) \cdot (1, 0, 1)$ and $(2/3) \cdot (-1, -1/2, 1)$ are orthogonal unit vectors. For the characteristic root 2, the system of equations for the characteristic vectors is

$$\begin{bmatrix} 17/18 & 2/9 & 1/18 \\ 2/9 & 1/9 & -2/9 \\ 1/18 & -2/9 & 17/18 \end{bmatrix} \begin{bmatrix} x_1 \\ x_2 \\ x_3 \end{bmatrix} = \begin{bmatrix} 0 \\ 0 \\ 0 \end{bmatrix}.$$

The solution of this system is $(x_1, x_2, x_3) = (-r, 4r, r)$, for any real number r. Choosing $r = 1/3\sqrt{2}$, we obtain a characteristic unit vector $(1/3\sqrt{2}) \cdot (-1, 4, 1)$ belonging to 2. This vector is orthogonal to the characteristic vectors that belong to 3 (see Exercise 34.1). Thus, the set of vectors $\{(1/\sqrt{2}) \cdot (1, 0, 1),\ (2/3) \cdot (-1, -1/2, 1),\ (1/3\sqrt{2}) \cdot (-1, 4, 1)\}$ is an orthonormal basis of $\mathcal{V}_3$ of the type described in the preceding paragraph. Therefore, the orthogonal matrix

$$P = \begin{bmatrix} \dfrac{1}{\sqrt{2}} & -\dfrac{2}{3} & -\dfrac{1}{3\sqrt{2}} \\ 0 & -\dfrac{1}{3} & \dfrac{2\sqrt{2}}{3} \\ \dfrac{1}{\sqrt{2}} & \dfrac{2}{3} & \dfrac{1}{3\sqrt{2}} \end{bmatrix}$$

is a matrix such that

$$P^{-1}AP = \begin{bmatrix} 3 & 0 & 0 \\ 0 & 3 & 0 \\ 0 & 0 & 2 \end{bmatrix}.$$

Example 2 illustrates the general method for finding an orthogonal matrix P that reduces a symmetric matrix A to diagonal form. If $r = r_{i_1} = r_{i_2} = \cdots = r_{i_s}$ is a repeated characteristic root of A, then the matrix of coefficients of the homogeneous system of equations whose solutions are the characteristic vectors belonging to r has rank $n - s$ (see Exercise 34.3). Then the dimension of the subspace of $\mathcal{V}_n$ consisting of the solutions of this system of equations is $n - (n - s) = s$ (see Theorem 14.4). By Theorem 11.7, this subspace has an orthonormal basis. Thus, there are s mutually orthogonal characteristic vectors that belong to r. Since characteristic vectors belonging to distinct characteristic roots are orthogonal (see Exercise 34.1), we can find an orthonormal basis of $\mathcal{V}_n$ of the type required to construct the orthogonal matrix P.

We now apply Theorem 34.3 to quadratic mappings and forms.

(34.5) theorem Let $\mathcal{Q}$ be a quadratic mapping of a Euclidean vector space $\mathcal{E}$ that has a symmetric matrix A with respect to an orthonormal basis $\{\mathbf{U}_1, \mathbf{U}_2, \ldots, \mathbf{U}_n\}$ of $\mathcal{E}$. There exists an orthonormal basis $\{\mathbf{W}_1, \mathbf{W}_2, \ldots, \mathbf{W}_n\}$ of $\mathcal{E}$ such that for $\mathbf{U} = a_1 \cdot \mathbf{W}_1 + a_2 \cdot \mathbf{W}_2 + \cdots + a_n \cdot \mathbf{W}_n \in \mathcal{E}$,

$$\mathcal{Q}(\mathbf{U}) = r_1a_1^2 + r_2a_2^2 + \cdots + r_na_n^2$$

where $r_1, r_2, \ldots, r_n$ are the characteristic roots of A.

PROOF By Theorem 34.3, there is an orthogonal matrix P such that

$$P^{-1}AP = D = \begin{bmatrix} r_1 & 0 & \cdots & 0 \\ 0 & r_2 & & \cdot \\ \cdot & & \cdot & \cdot \\ \cdot & & \cdot & \cdot \\ \cdot & & \cdot & 0 \\ 0 & \cdots & 0 & r_n \end{bmatrix}$$

where $r_1, r_2, \ldots, r_n$ are the characteristic roots of A. Let the orthogonal matrix P^{-1} be the matrix of a change of basis from $\{\mathbf{U}_1, \mathbf{U}_2, \ldots, \mathbf{U}_n\}$ to a new basis $\{\mathbf{W}_1, \mathbf{W}_2, \ldots, \mathbf{W}_n\}$. Then the matrix of $\mathcal{Q}$ with respect to this new basis is $P^{-1}A(P^{-1})' = P^{-1}AP = D$. Therefore,

$$\mathcal{Q}(\mathbf{U}) = [a_1 \quad a_2 \quad \cdots \quad a_n]D \begin{bmatrix} a_1 \\ a_2 \\ \vdots \\ a_n \end{bmatrix} = r_1a_1^2 + r_2a_2^2 + \cdots + r_na_n^2. \qquad \blacksquare$$

(34.6) theorem Let $\Sigma_{i,j=1}^{n} x_i a_{i,j} x_j$ be a real quadratic form, where $A = [a_{i,j}]$ is a symmetric matrix with characteristic roots $r_1, r_2, \ldots, r_n$. There exists an orthogonal matrix Q such that

$$\sum_{i,j=1}^{n} x_i a_{i,j} x_j = r_1 y_1^2 + r_2 y_2^2 + \cdots + r_n y_n^2$$

where $[x_1 \quad x_2 \quad \cdots \quad x_n] = [y_1 \quad y_2 \quad \cdots \quad y_n]Q$.

The proof of Theorem 34.6 is a direct application of Theorem 34.3 and is left as an exercise (Exercise 34.8).

EXAMPLE 3 Let $\sum_{i,j=1}^{3} x_i a_{i,j} x_j$ be the quadratic form with the matrix $A = [a_{i,j}]$ of Example 1. Let P be the orthogonal matrix obtained in Example 1 such that

$$P^{-1}AP = \begin{bmatrix} 3 & 0 & 0 \\ 0 & 3 & 0 \\ 0 & 0 & 2 \end{bmatrix}.$$

Let $Q = P^{-1} = P'$, and let $[x_1 \quad x_2 \quad x_3] = [y_1 \quad y_2 \quad y_3]Q$. Then

$$\sum_{i,j=1}^{3} x_i a_{i,j} x_j = [x_1 \quad x_2 \quad x_3]\, A \begin{bmatrix} x_1 \\ x_2 \\ x_3 \end{bmatrix} = [y_1 \quad y_2 \quad y_3]\, Q\, A\, Q' \begin{bmatrix} y_1 \\ y_2 \\ y_3 \end{bmatrix}$$

$$= [y_1 \quad y_2 \quad y_3] P^{-1}AP \begin{bmatrix} y_1 \\ y_2 \\ y_3 \end{bmatrix} = [y_1 \quad y_2 \quad y_3] \begin{bmatrix} 3 & 0 & 0 \\ 0 & 3 & 0 \\ 0 & 0 & 2 \end{bmatrix} \begin{bmatrix} y_1 \\ y_2 \\ y_3 \end{bmatrix}$$

$$= 3y_1^2 + 3y_2^2 + 2y_3^2.$$

EXAMPLE 4 A central quadric surface, $ax^2 + bxy + cy^2 + dxz + eyz + fz^2 = k$, is said to be *referred to principal axes* if a new rectangular Cartesian coordinate system is chosen so that the equation of the quadric surface is $r_1\bar{x}^2 + r_2\bar{y}^2 + r_3\bar{z}^2 = k$. The lengths of the *semiaxes* of the quadric surface are $\sqrt{|k/r_i|}$, $i = 1, 2, 3$. The orthogonal matrix P that reduces the symmetric matrix

$$A = \begin{bmatrix} a & \frac{b}{2} & \frac{d}{2} \\ \frac{b}{2} & c & \frac{e}{2} \\ \frac{d}{2} & \frac{e}{2} & f \end{bmatrix}$$

of the quadratic form $ax^2 + bxy + cy^2 + dxz + eyz + fz^2$ to diagonal form defines a change of coordinates $[\bar{x} \quad \bar{y} \quad \bar{z}] = [x \quad y \quad z]P$ to a new rectangular Cartesian coordinate system. Thus, to find the equation of the quadric surface when referred to principal axes, it is sufficient to find the characteristic roots r_1, r_2, r_3 of the matrix A. Similar remarks apply to central conics in a plane.

The quadric surface $\frac{53}{18}x^2 + \frac{4}{9}xy + \frac{19}{9}y^2 + \frac{1}{9}xz + (-\frac{4}{9})yz + \frac{53}{18}z^2 = k$ has as the matrix of its quadratic form, the matrix A of Example 1. Hence, there is a new rectangular Cartesian coordinate system defined by $[\bar{x} \quad \bar{y} \quad \bar{z}] = [x \quad y \quad z]P$, where P is the orthogonal matrix obtained in Example 1, such that the equation of the quadric surface is $3\bar{x}^2 + 3\bar{y}^2 + 2\bar{z}^2 = k$. The lengths of the semiaxes are $\sqrt{|k/3|}$, $\sqrt{|k/3|}$, and $\sqrt{|k/2|}$. In this case, if $k > 0$, the surface is an ellipsoid, if $k = 0$, the surface consists of a single point, and if $k < 0$, the surface is imaginary.

EXAMPLE 5 The central conic $5x^2 - 6xy - 3y^2 = 4$ has

$$A = \begin{bmatrix} 5 & -3 \\ -3 & -3 \end{bmatrix}$$

as the symmetric matrix of its quadratic form. Then

$$|A - x \cdot I_2| = \begin{vmatrix} 5 - x & -3 \\ -3 & -3 - x \end{vmatrix} = x^2 - 2x - 24 = (x - 6)(x + 4).$$

Thus, the characteristic roots of A are 6 and -4. Therefore, there is an orthogonal matrix P that determines a new rectangular Cartesian coordinate system in the plane by the change of coordinates $[\bar{x}\ \bar{y}] = [x\ y]P$ such that the equation of the given conic is $6\bar{x}^2 - 4\bar{y}^2 = 4$, or $3/2\bar{x}^2 - \bar{y}^2 = 1$, in the new coordinate system. The lengths of the semiaxes of the conic, which is a hyperbola, are $\sqrt{3}/\sqrt{2}$ and 1.

To find the matrix P, we solve for unit characteristic vectors belonging to the characteristic values 6 and -4. The characteristic vectors belonging to 6 are solutions of the system of equations

$$\begin{bmatrix} -1 & -3 \\ -3 & -9 \end{bmatrix} \begin{bmatrix} x_1 \\ x_2 \end{bmatrix} = \begin{bmatrix} 0 \\ 0 \end{bmatrix}.$$

This system has the solution $(x_1, x_2) = (3r, -r)$, for any real number r. The vector $(3r, -r)$ has length 1 if $9r^2 + r^2 = 10r^2 = 1$. Thus, choosing $r = 1/\sqrt{10}$, it follows that $(3/\sqrt{10}, -1/\sqrt{10})$ is a unit characteristic vector belonging to the characteristic value 6. Similarly, solving the system of equations

$$\begin{bmatrix} 9 & -3 \\ -3 & 1 \end{bmatrix} \begin{bmatrix} x_1 \\ x_2 \end{bmatrix} = \begin{bmatrix} 0 \\ 0 \end{bmatrix}.$$

we find that a unit characteristic vector belonging to -4 is

$(1/\sqrt{10}, 3/\sqrt{10})$. Thus, we obtain the orthogonal matrix

$$P = \begin{bmatrix} \dfrac{3}{\sqrt{10}} & \dfrac{1}{\sqrt{10}} \\ -\dfrac{1}{\sqrt{10}} & \dfrac{3}{\sqrt{10}} \end{bmatrix}.$$

The change of coordinates determined by P is a rotation in the plane through an angle θ, where $\cos\theta = 3/\sqrt{10}$, $\sin\theta = -1/\sqrt{10}$.

exercises

34.1 Prove that characteristic vectors belonging to distinct characteristic roots of a symmetric matrix A are orthogonal.

34.2 In Example 1, use matrix multiplication to check that

$$P^{-1}AP = \begin{bmatrix} 3 & 0 & 0 \\ 0 & 3 & 0 \\ 0 & 0 & 2 \end{bmatrix}.$$

***34.3** Prove that if r is a characteristic root of an n by n symmetric matrix A that is repeated s times, then the matrix $A - r \cdot I_n$ has rank $n - s$.

34.4 Find an orthogonal matrix P such that $P^{-1}AP$ is a diagonal matrix for the following symmetric matrices A:

(a) $\begin{bmatrix} 4 & 0 & 0 \\ 0 & -3 & 1 \\ 0 & 1 & 2 \end{bmatrix}$ (b) $\begin{bmatrix} 0 & 0 & 0 & 1 \\ 0 & 0 & 2 & 0 \\ 0 & 2 & 0 & 0 \\ 1 & 0 & 0 & 0 \end{bmatrix}$

(c) $\begin{bmatrix} 1 & \dfrac{3}{\sqrt{2}} & \dfrac{-3}{\sqrt{2}} \\ \dfrac{3}{\sqrt{2}} & \dfrac{-1}{2} & \dfrac{-3}{2} \\ \dfrac{-3}{\sqrt{2}} & \dfrac{-3}{2} & \dfrac{-1}{2} \end{bmatrix}$ (d) $\begin{bmatrix} 2 & -2 \\ -2 & 5 \end{bmatrix}$

(e) $\begin{bmatrix} 4 & 2 & 0 & 0 \\ 2 & 1 & 0 & 0 \\ 0 & 0 & -3 & -2 \\ 0 & 0 & -2 & -6 \end{bmatrix}$ (f) $\begin{bmatrix} 12 & 3 \\ 3 & 4 \end{bmatrix}$

34.5 Let $\mathcal{V}_4$ be a Euclidean vector space with the $E^{(4)}$-inner product. Write each of the following quadratic mappings $\mathcal{Q}$ of $\mathcal{V}_4$ with a symmetric matrix with respect to the orthonormal basis $E^{(4)}$:
(a) $\mathcal{Q}[(w, x, y, z)] = wx + wy + wz + xy + xz + yz$
(b) $\mathcal{Q}[(w, x, y, z)] = (w + x + y + z)^2$
Find an orthonormal basis $\{\mathbf{W}_1, \mathbf{W}_2, \mathbf{W}_3, \mathbf{W}_4\}$ of $\mathcal{V}_4$ such that for $\mathbf{U} =$

$a_1 \cdot \mathbf{W}_1 + a_2 \cdot \mathbf{W}_2 + a_3 \cdot \mathbf{W}_3 + a_4 \cdot \mathbf{W}_4 \in \mathcal{V}_4$, $\quad \mathcal{Q}(\mathbf{U}) = r_1a_1^2 + r_2a_2^2 + r_3a_3^2 + r_4a_4^2$.

34.6 Write the quadratic form $5x_1x_4 + 5x_2x_3$ with a symmetric matrix. Find an orthogonal matrix Q such that $5x_1x_4 + 5x_2x_3 = \frac{5}{2}(y_1^2 + y_2^2 - y_3^2 - y_4^2)$, where $[x_1 \;\; x_2 \;\; x_3 \;\; x_4] = [y_1 \;\; y_2 \;\; y_3 \;\; y_4]Q$.

34.7 Refer the following conics and quadric surfaces to principal axes and find their semiaxes:
(a) $10x^2 + 4xy + 7y^2 = 100$ (b) $x^2 + 5xy - 11y^2 = 4$
(c) $-x^2 - 2\sqrt{3}\,xz - 4yz + 3z^2 = 25$ (d) $x^2 - 2y^2 + 6yz + 4z^2 = 1$

34.8 Prove Theorem 34.6.

35 congruent reduction of real symmetric matrices

In this section we consider the reduction of real symmetric matrices under the relation of congruence. As indicated earlier, the results we obtain can be applied to the problem of simplifying the expressions for quadratic mappings and quadratic forms.

Denote the set of all real symmetric n by n matrices by $\mathfrak{A}_n^s$. By Definition 33.2, an n by n matrix B is congruent to a matrix $A \in \mathfrak{A}_n^s$ if there is a nonsingular n by n matrix P such that $B = PAP'$. Moreover, B is again a symmetric matrix, that is, $B \in \mathfrak{A}_n^s$. Thus, congruence is a equivalence relation on the set of real symmetric matrices, $\mathfrak{A}_n^s$ (see Exercise 33.8). We will show that each matrix $A \in \mathfrak{A}_n^s$ is congruent to exactly one matrix in a set of canonical matrices that we now describe. Let $D_t^{(s)}$ be an n by n diagonal matrix, with $0 \le s \le t \le n$, such that the first s diagonal elements are 1, the next $t - s$ diagonal elements are -1, and the remaining $n - t$ diagonal elements are 0. For example, if $n = 2$,

$$D_0^{(0)} = \begin{bmatrix} 0 & 0 \\ 0 & 0 \end{bmatrix}, \quad D_1^{(0)} = \begin{bmatrix} -1 & 0 \\ 0 & 0 \end{bmatrix}, \quad D_1^{(1)} = \begin{bmatrix} 1 & 0 \\ 0 & 0 \end{bmatrix},$$

$$D_2^{(0)} = \begin{bmatrix} -1 & 0 \\ 0 & -1 \end{bmatrix}, \quad D_2^{(1)} = \begin{bmatrix} 1 & 0 \\ 0 & -1 \end{bmatrix}, \quad \text{and} \quad D_2^{(2)} = \begin{bmatrix} 1 & 0 \\ 0 & 1 \end{bmatrix}.$$

If $n = 3$, there are ten $D_t^{(s)}$ matrices.

(35.1) theorem Let A be a real symmetric n by n matrix. There exists a nonsingular matrix P such that $PAP' = D_t^{(s)}$, where t is the rank of A and s is the number of positive characteristic roots of A.

PROOF By Theorem 34.3, there exists an orthogonal matrix Q_1 such that $Q_1^{-1}AQ_1 = D$, where D is the diagonal matrix with the characteristic roots $r_1, r_2, \ldots, r_n$ of A on the diagonal. Clearly, the rank of D is equal to the number of nonzero characteristic roots. Moreover, A is equivalent to D.

Therefore, if A has rank t, then D has rank t and there are exactly t nonzero characteristic roots. We may assume that $r_1, r_2, \ldots, r_s$ are positive, $r_{s+1}, r_{s+2}, \ldots, r_t$ are negative, and $r_{t+1}, r_{t+2}, \ldots, r_n$ are zero. Let Q_2 be the n by n diagonal matrix that has $1/\sqrt{r_i}$ for $i = 1, 2, \ldots, s$ in the first s positions on the diagonal, $1/\sqrt{-r_i}$ for $i = s+1, s+2, \ldots, t$ in the next $t - s$ diagonal positions, and 1 in the remaining places on the diagonal. Then Q_2 is nonsingular and symmetric. We have $Q_2DQ_2' = Q_2DQ_2 = D_t^{(s)}$, where t is the rank of A and s is the number of positive characteristic roots of A. Let $P = Q_2Q_1^{-1}$. Then $P' = (Q_1^{-1})'Q_2' = Q_1Q_2'$, since Q_1 is orthogonal. Therefore,

$$PAP' = Q_2Q_1^{-1}AQ_1Q_2' = Q_2DQ_2' = D_t^{(s)}. \qquad \blacksquare$$

EXAMPLE 1 Let A be the matrix of Example 1, §34. Then the matrix Q_1 of Theorem 35.1 is

$$\begin{bmatrix} \frac{1}{\sqrt{2}} & \frac{-2}{3} & \frac{-1}{3\sqrt{2}} \\ 0 & \frac{-1}{3} & \frac{2\sqrt{2}}{3} \\ \frac{1}{\sqrt{2}} & \frac{2}{3} & \frac{1}{3\sqrt{2}} \end{bmatrix}, \quad \text{and} \quad Q_1^{-1}AQ_1 = \begin{bmatrix} 3 & 0 & 0 \\ 0 & 3 & 0 \\ 0 & 0 & 2 \end{bmatrix}.$$

The matrix Q_2 of Theorem 35.1 is the diagonal matrix

$$\begin{bmatrix} \frac{1}{\sqrt{3}} & 0 & 0 \\ 0 & \frac{1}{\sqrt{3}} & 0 \\ 0 & 0 & \frac{1}{\sqrt{2}} \end{bmatrix}.$$

Therefore,

$$P = Q_2Q_1^{-1} = \begin{bmatrix} \frac{1}{\sqrt{3}} & 0 & 0 \\ 0 & \frac{1}{\sqrt{3}} & 0 \\ 0 & 0 & \frac{1}{\sqrt{2}} \end{bmatrix} \begin{bmatrix} \frac{1}{\sqrt{2}} & 0 & \frac{1}{\sqrt{2}} \\ \frac{-2}{3} & \frac{-1}{3} & \frac{2}{3} \\ \frac{-1}{3\sqrt{2}} & \frac{2\sqrt{2}}{3} & \frac{1}{3\sqrt{2}} \end{bmatrix}$$

$$= \begin{bmatrix} \frac{1}{\sqrt{6}} & 0 & \frac{1}{\sqrt{6}} \\ \frac{-2}{3\sqrt{3}} & \frac{-1}{3\sqrt{3}} & \frac{2}{3\sqrt{3}} \\ \frac{-1}{6} & \frac{2}{3} & \frac{1}{6} \end{bmatrix}.$$

Then for this matrix P,

$$PAP' = \begin{bmatrix} 1 & 0 & 0 \\ 0 & 1 & 0 \\ 0 & 0 & 1 \end{bmatrix} = D_3^{(3)}.$$

The number of positive characteristic roots of a real symmetric matrix A is called the *index* of A. Theorem 35.1 yields the following results for quadratic mappings and forms.

(35.2) theorem Let $\mathcal{Q}$ be a quadratic mapping of a vector space $\mathcal{V}$ that has a symmetric matrix A with respect to a basis $\{\mathbf{U}_1, \mathbf{U}_2, \ldots, \mathbf{U}_n\}$ of $\mathcal{V}$. There exists a basis $\{\mathbf{W}_1, \mathbf{W}_2, \ldots, \mathbf{W}_n\}$ of $\mathcal{V}$ such that for $\mathbf{U} = a_1 \cdot \mathbf{W}_1 + a_2 \cdot \mathbf{W}_2 + \cdots + a_n \cdot \mathbf{W}_n \in \mathcal{V}$,

$$\mathcal{Q}(\mathbf{U}) = a_1^2 + \cdots + a_s^2 - a_{s+1}^2 - \cdots - a_t^2$$

where t is the rank of A and s is the index of A.

PROOF By Theorem 35.1, there is a nonsingular matrix P such that $PAP' = D_t^{(s)}$, where t is the rank of A and s is the index of A. Let P be the matrix of a change of basis from $\{\mathbf{U}_1, \mathbf{U}_2, \ldots, \mathbf{U}_n\}$ to a new basis $\{\mathbf{W}_1, \mathbf{W}_2, \ldots, \mathbf{W}_n\}$. Then the matrix of $\mathcal{Q}$ with respect to the new basis is $PAP' = D_t^{(s)}$. Therefore,

$$\begin{aligned} \mathcal{Q}(\mathbf{U}) &= [a_1 \quad a_2 \quad \cdots \quad a_n] D_t^{(s)} \begin{bmatrix} a_1 \\ a_2 \\ \vdots \\ a_n \end{bmatrix} \\ &= a_1^2 + \cdots + a_s^2 - a_{s+1}^2 - \cdots - a_t^2. \end{aligned}$$ ∎

(35.3) theorem Let $\Sigma_{i,j=1}^n x_i a_{i,j} x_j$ be a real quadratic form, where $A = [a_{i,j}]$ is a symmetric matrix with rank t and index s. There exists a nonsingular matrix P such that

$$\sum_{i,j=1}^{n} x_i a_{i,j} x_j = y_1^2 + \cdots + y_s^2 - y_{s+1}^2 - \cdots - y_t^2$$

where $[x_1 \quad x_2 \quad \cdots \quad x_n] = [y_1 \quad y_2 \quad \cdots \quad y_n]P$.

The proof of Theorem 35.3 is left as an exercise (Exercise 35.4).

EXAMPLE 2 Let $\Sigma_{i,j=1}^3 x_i a_{i,j} x_j$ be the quadratic form with matrix $A = [a_{i,j}]$, where A is the matrix of Example 1, §34. Then A has rank 3 and index 3. Let P be the matrix of Example 1 of this section, so that $PAP' = D_3^{(3)}$. Then

$$\sum_{i,j=1}^{3} x_i a_{i,j} x_j = y_1^2 + y_2^2 + y_3^2$$

where $[x_1 \quad x_2 \quad x_3] = [y_1 \quad y_2 \quad y_3]P$.

The complete set of values of a quadratic form $\Sigma_{i,j=1}^{n} x_i a_{i,j} x_j$ for all choices of n real numbers $x_1 = r_1, x_2 = r_2, \ldots, x_n = r_n$ is called the *range of values* of the form. It is important to note in Theorem 35.3 that the two forms $\Sigma_{i,j=1}^{n} x_i a_{i,j} x_j$ and $y_1^2 + \cdots + y_s^2 - y_s^2 - \cdots - y_t^2$ have the same range of values. This is because the change of variables $[x_1 \quad x_2 \quad \cdots \quad x_n] = [y_1 \quad y_2 \quad \cdots \quad y_n]P$ is given by a nonsingular matrix P. Indeed, the value of $y_1^2 + \cdots + y_s^2 - y_{s+1}^2 - \cdots - y_t^2$ when $(y_1, y_2, \ldots, y_n) = (\bar{r}_1, \bar{r}_2, \ldots, \bar{r}_n)$ is the same as the value of $\Sigma_{i,j=1}^{n} x_i a_{i,j} x_j$ when $(x_1, x_2, \ldots, x_n) = (r_1, r_2, \ldots, r_n) = (\bar{r}_1, \bar{r}_2, \ldots, \bar{r}_n)P$, and conversely, the value of $\Sigma_{i,j=1}^{n} x_i a_{i,j} x_j$ when $(x_1, x_2, \ldots, x_n) = (r_1, r_2, \ldots, r_n)$ is the same as the value of $y_1^2 + \cdots + y_s^2 - y_{s+1}^2 - \cdots - y_t^2$ when $(y_1, y_2, \ldots, y_n) = (\bar{r}_1, \bar{r}_2, \ldots, \bar{r}_n) = (r_1, r_2, \ldots, r_n)P^{-1}$. Thus, in Example 2, the value of the form $\Sigma_{i,j=1}^{3} x_i a_{i,j} x_j$ is always nonnegative, and is zero only when $x_1 = x_2 = x_3 = 0$.

Theorem 35.1 states that each real symmetric n by n matrix is congruent to an n by n diagonal $D_t^{(s)}$ matrix. To prove that for a given n, the set of n by n $D_t^{(s)}$ matrices is a set of canonical matrices for real symmetric matrices under the relation of congruence, we must show that a symmetric matrix A is congruent to exactly one $D_t^{(s)}$ matrix. This result is a consequence of the following theorem.

(35.4) theorem If the n by n matrices $D_t^{(s)}$ and $D_v^{(u)}$ are congruent, then $t = v$ and $s = u$.

PROOF If $D_t^{(s)}$ and $D_v^{(u)}$ are congruent, then they are equivalent. Hence, $t =$ rank of $D_t^{(s)} =$ rank of $D_v^{(u)} = v$. If $t = 0$, then $s = u = 0$, since $0 \leq s \leq t$ and $0 \leq u \leq v = t$. Therefore, we may suppose that $t > 0$. Assume that $s \neq u$, say $s < u$. We will find that this assumption leads to a contradiction, which will prove the theorem. Consider the quadratic form

$$x_1^2 + \cdots + x_s^2 - x_{s+1}^2 - \cdots - x_t^2$$

with matrix $D_t^{(s)}$. By hypothesis, there is a nonsingular matrix $P = [p_{i,j}]$ such that $PD_t^{(s)}P' = D_v^{(u)} = D_t^{(u)}$. Hence,

$$x_1^2 + \cdots + x_s^2 - x_{s+1}^2 - \cdots - x_t^2 = y_1^2 + \cdots + y_u^2 - y_{u+1}^2 - \cdots - y_t^2 \qquad (35\text{–}1)$$

where $[x_1 \quad x_2 \quad \cdots \quad x_n] = [y_1 \quad y_2 \quad \cdots \quad y_n]P$. We will now determine values for the x's and y's that show that (35–1) is impossible. Select $y_{u+1} = \cdots = y_n = 0$. Then the equations connecting the x's and the y's are

$$\begin{aligned} x_1 &= p_{1,1}y_1 + \cdots + p_{u,1}y_u \\ &\vdots \\ x_s &= p_{1,s}y_1 + \cdots + p_{u,s}y_u \\ x_{s+1} &= p_{1,s+1}y_1 + \cdots + p_{u,s+1}y_u \\ &\vdots \\ x_n &= p_{1,n}y_1 + \cdots + p_{u,n}y_u. \end{aligned}$$

If we set $x_1 = \cdots = x_s = 0$, then the first s equations become a homogeneous system to solve for $y_1, y_2, \ldots, y_u$. Since $s < u$, it follows from Corollary 14.3 that there is a nontrivial solution $y_1 = a_1$, $y_2 = a_2, \ldots, y_u = a_u$. Then

$$\begin{aligned} x_{s+1} &= p_{1,s+1}a_1 + \cdots + p_{u,s+1}a_u = b_{s+1} \\ &\vdots \\ x_n &= p_{1,n}a_1 + \cdots + p_{u,n}a_u = b_n \end{aligned}$$

are determined by the last $n - s$ equations. Substituting in (35–1), we obtain

$$- b_{s+1}^2 - \cdots - b_t^2 = a_1^2 + \cdots + a_u^2. \qquad (35\text{–}2)$$

Since not every a_i is zero, the right-hand side of equation (35–2) is positive. However, the left-hand side of this equation is less than or equal to zero. This contradiction proves the theorem. ▮

Now, if a symmetric matrix A is congruent to $D_t^{(s)}$ and $D_v^{(u)}$, then since congruence is an equivalence relation, $D_t^{(s)}$ is congruent to $D_v^{(u)}$. By Theorem 35.4, $t = v$ and $s = u$. That is, $D_t^{(s)} = D_v^{(u)}$. Hence, each symmetric matrix is congruent to exactly one $D_t^{(s)}$ matrix. Theorem 35.4 leads to the result given in Corollary 35.5.

(35.5) corollary Let A and B be real symmetric n by n matrices. Then A is congruent to B if and only if A and B have the same rank and index.

PROOF Suppose rank of A = rank of $B = t$ and index of A = index of $B = s$. By Theorem 35.1, A is congruent to $D_t^{(s)}$ and B is congruent to $D_t^{(s)}$. Therefore, A is congruent to B. Conversely, assume that A is congruent to B. By Theorem 35.1, A is congruent to $D_t^{(s)}$, where t is the rank of A and s is the index of A. Similarly, B is congruent to $D_v^{(u)}$, where v is the rank of B and u is the index of B. We have $D_t^{(s)}$ is congruent to A, A is congruent to B, and B is congruent to $D_v^{(u)}$. Therefore, $D_t^{(s)}$ is congruent to $D_v^{(u)}$. By Theorem 35.4, $t = v$ and $s = u$. Hence, A and B have the same rank and index. ▮

The theory of the reduction of a real symmetric matrix A to canonical form and the application of this theory to the simplification of quadratic mappings and forms is now complete. However, the construction of a nonsingular matrix P such that $PAP' = D_t^{(s)}$ by the method of the proof of Thoerem 35.1 is lengthy, since it involves first reducing A to diagonal form by orthogonal equivalence (see Example 1, §34). Fortunately, there is a simpler way to construct P.

Let E denote an elementary transformation matrix of any type, and let A be a symmetric matrix. The matrix EAE' is congruent to A. Moreover, E' is an elementary transformation matrix that performs the same elementary transformation on the columns of A as E performs on the rows. For example, if

$$A = \begin{bmatrix} 1 & -\tfrac{1}{2} & -3 \\ -\tfrac{1}{2} & \tfrac{5}{2} & 6 \\ -3 & 6 & 5 \end{bmatrix} \quad \text{and} \quad E = I_3^{(1,3,3)},$$

then

$$EAE' = \begin{bmatrix} 1 & -\tfrac{1}{2} & 0 \\ -\tfrac{1}{2} & \tfrac{5}{2} & \tfrac{9}{2} \\ 0 & \tfrac{9}{2} & -4 \end{bmatrix}.$$

That is, the first fow of A is multiplied by 3 and added to the third row, and then the first column of A is multiplied by 3 and added to the third column. Since A is symmetric, a sequence of these *elementary congruence transformations* that reduces A to a $D_t^{(s)}$ matrix can be found. Continuing this example, where we now denote the matrix E by E_1, we have

$$E_2E_1AE_1'E_2' = \begin{bmatrix} 1 & 0 & 0 \\ 0 & \tfrac{9}{4} & \tfrac{9}{2} \\ 0 & \tfrac{9}{2} & -4 \end{bmatrix}, \quad \text{where} \quad E_2 = I_3^{(1,2,1/2)}$$

$$E_3E_2E_1AE_1'E_2'E_3' = \begin{bmatrix} 1 & 0 & 0 \\ 0 & 1 & 3 \\ 0 & 3 & -4 \end{bmatrix}, \quad \text{where} \quad E_3 = {}^{(2/3)}I_3^{(2)}$$

$$E_4E_3E_2E_1AE_1'E_2'E_3'E_4' = \begin{bmatrix} 1 & 0 & 0 \\ 0 & 1 & 0 \\ 0 & 0 & -13 \end{bmatrix}, \quad \text{where} \quad E_4 = I_3^{(2,3,-3)}$$

$$E_5E_4E_3E_2E_1AE_1'E_2'E_3'E_4'E_5' = \begin{bmatrix} 1 & 0 & 0 \\ 0 & 1 & 0 \\ 0 & 0 & -1 \end{bmatrix} = D_3^{(2)}, \quad \text{where } E_5 = {}^{(1/\sqrt{13})}I_3^{(3)}.$$

Thus, $PAP' = D_3^{(2)}$, where $P = E_5E_4E_3E_2E_1$. Moreover, the matrix P can be computed by performing the sequence of elementary transformations on the identity matrix I_n. In this example,

$$P = \begin{bmatrix} 1 & 0 & 0 \\ \dfrac{1}{3} & \dfrac{2}{3} & 0 \\ \dfrac{2}{\sqrt{13}} & \dfrac{-2}{\sqrt{13}} & \dfrac{1}{\sqrt{13}} \end{bmatrix}.$$

The $D_t^{(s)}$ matrix obtained by this reduction must be the same as the one obtained by the more complicated procedure of Theorem 35.1, since we have proved that each symmetric matrix is congruent to exactly one $D_t^{(s)}$ matrix.

EXAMPLE 3 Let

$$A = \begin{bmatrix} 1 & -6 & -3 \\ -6 & 40 & 19 \\ -3 & 19 & 10 \end{bmatrix}.$$

Then A is carried to a $D_t^{(s)}$ matrix by the following sequence of elementary congruence transformations:

$$A \xrightarrow{\text{Type II}} \begin{bmatrix} 1 & 0 & -3 \\ 0 & 4 & 1 \\ -3 & 1 & 10 \end{bmatrix} \xrightarrow{\text{Type II}} \begin{bmatrix} 1 & 0 & 0 \\ 0 & 4 & 1 \\ 0 & 1 & 1 \end{bmatrix}$$

$$\xrightarrow{\text{Type II}} \begin{bmatrix} 1 & 0 & 0 \\ 0 & 4 & 0 \\ 0 & 0 & 3/4 \end{bmatrix} \xrightarrow{\text{Type III}} \begin{bmatrix} 1 & 0 & 0 \\ 0 & 1 & 0 \\ 0 & 0 & 3/4 \end{bmatrix} \xrightarrow{\text{Type III}} \begin{bmatrix} 1 & 0 & 0 \\ 0 & 1 & 0 \\ 0 & 0 & 1 \end{bmatrix} = D_3^{(3)}$$

The matrix P such that $PAP' = D_3^{(3)}$ is

$$P = \begin{bmatrix} 1 & 0 & 0 \\ 3 & \dfrac{1}{2} & 0 \\ \sqrt{3} & \dfrac{-1}{2\sqrt{3}} & \dfrac{2}{\sqrt{3}} \end{bmatrix}.$$

A quadratic form $\Sigma_{i,j=1}^{n} x_i a_{i,j} x_j$ is *positive definite* if all of its values are positive except for $x_1 = x_2 = \cdots = x_n = 0$. The form is called *positive semidefinite* if all of its values are greater than or equal to zero. As a consequence of Theorem 35.3, a form is positive definite if and only if $s = t = n$ and is positive semidefinite if and only if $s = t$. For example, the quadratic form of Example 2 is positive definite.

exercises

35.1 Show that the set of real symmetric matrices, $\mathfrak{A}_n^s$, is a subspace of the vector space $\mathfrak{A}_n$ of all real n by n matrices. What is the dimension of the subspace $\mathfrak{A}_n^s$?

35.2 List the ten $D_t^{(s)}$ matrices for $n = 3$.

35.3 How many $D_t^{(s)}$ matrices are there for $n = 4$? Show that for any n, there are $(n^2 + 3n + 2)/2$ $D_t^{(s)}$ matrices.

35.4 Prove Theorem 35.3.

35.5 Find a nonsingular matrix P such that $PAP' = D_t^{(s)}$ for the following matrices A. Also, determine the rank and index of each matrix.

(a) $\begin{bmatrix} -5 & -\frac{1}{2} & -\frac{2}{3} \\ -\frac{1}{2} & 0 & 6 \\ -\frac{2}{3} & 6 & 1 \end{bmatrix}$ (b) $\begin{bmatrix} 4 & -1 & 6 & 2 \\ -1 & 3 & 0 & 5 \\ 6 & 0 & -1 & 1 \\ 2 & 5 & 1 & -7 \end{bmatrix}$

(c) $\begin{bmatrix} 1 & 1 & 1 & 1 \\ 1 & 1 & 1 & 1 \\ 1 & 1 & 1 & 1 \\ 1 & 1 & 1 & 1 \end{bmatrix}$ (d) $\begin{bmatrix} 5 & 3 & -1 \\ 3 & -2 & 0 \\ -1 & 0 & 4 \end{bmatrix}$

(e) $\begin{bmatrix} 12 & 16 & -8 & 6 \\ 16 & -2 & 0 & 4 \\ -8 & 0 & -24 & 20 \\ 6 & 4 & 20 & 0 \end{bmatrix}$ (f) $\begin{bmatrix} 1 & \frac{1}{2} & -3 \\ \frac{1}{2} & 2 & 1 \\ -3 & 1 & 15 \end{bmatrix}$

***35.6** A real symmetric matrix A is *positive definite* (*positive semidefinite*) if the quadratic form with matrix A is positive definite (positive semidefinite). Which of the matrices in Exercise 35.5 are positive definite? Which are positive semidefinite?

35.7 Prove that a real symmetric n by n matrix A is positive definite if and only if A is congruent to I_n. Use this result to show that the determinant of a positive definite matrix is positive.

35.8 Prove that a symmetric matrix A is positive definite if and only if there exists a nonsingular matrix B such that $A = B'B$.

35.9 Find a nonsingular matrix P such that each of the following quadratic forms has the expression $y_1^2 + \cdots + y_s^2 - y_{s+1}^2 - \cdots - y_t^2$, where

$$[x_1 \quad x_2 \quad \cdots \quad x_n] = [y_1 \quad y_2 \quad \cdots \quad y_n]P.$$

(a) $x_1x_2 + 2x_1x_3 + 4x_1x_4 + x_2x_3 + x_4^2$
(b) $x_1^2 + 2x_1x_2 + x_2^2 + 2x_1x_3 + 2x_2x_3 + x_3^2$
(c) $2x_1^2 + 8x_1x_2 - 12x_1x_3 + 7x_2^2 - 24x_2x_3 + 15x_3^2$
(d) $x_1^2 - 2x_1x_2 + \frac{3}{2}x_1x_3 + 3x_2^2 + 3x_2x_3 + 4x_3^2$
(e) $x_1^2 - 12x_1x_2 - 6x_1x_3 + 37x_2^2 + 38x_2x_3 + 10x_3^2$
(f) $x_1x_2 + x_1x_3 + x_1x_4 + x_2x_3 + x_2x_4 + x_3x_4$

Which of these forms are positive definite? Which are positive semidefinite?

answers to selected exercises

Section 1

1.4 $\sqrt{109}$; $\sqrt{454}/2$; $\sqrt{3}$.

Section 2

2.3 (a) $\overrightarrow{OP} - (\overrightarrow{OQ} + \overrightarrow{OR}) = \overrightarrow{OP} + [-(\overrightarrow{OQ} + \overrightarrow{OR})]$ by the definition of subtraction. $-(\overrightarrow{OQ} + \overrightarrow{OR}) = -\overrightarrow{OQ} + (-\overrightarrow{OR})$ by the result of Example 2. Substituting, we have

$$\begin{aligned}\overrightarrow{OP} &- (\overrightarrow{OQ} + \overrightarrow{OR})\\ &= \overrightarrow{OP} + [-\overrightarrow{OQ} + (-\overrightarrow{OR})]\\ &= [\overrightarrow{OP} + (-\overrightarrow{OQ})] + (-\overrightarrow{OR}) \quad \text{by the associative law of addition,}\\ &= (\overrightarrow{OP} - \overrightarrow{OQ}) - \overrightarrow{OR} \quad \text{by the definition of subtraction.}\end{aligned}$$

Section 3

3.3 (a) First show that the three vectors have the same length. Then show that they have the same direction (or are the zero vector) by considering the following cases: $\overrightarrow{OP} = \vec{0}$; $\overrightarrow{OP} \neq \vec{0}$ and either $r = 0$ or $s = 0$; $\overrightarrow{OP} \neq \vec{0}$ and $r > 0$, $s > 0$; $\overrightarrow{OP} \neq \vec{0}$ and $r > 0$, $s < 0$; $\overrightarrow{OP} \neq \vec{0}$ and $r < 0$, $s > 0$; $\overrightarrow{OP} \neq \vec{0}$ and $r < 0$, $s < 0$.

3.6 If $r \neq 0$ and $\overrightarrow{OP} \neq \vec{0}$, then the length of $r \cdot \overrightarrow{OP}$ is $|r|\ |OP| \neq 0$.

3.8 $r = -\frac{1}{5}$; $r = 5k$ and $s = k$, for any real number k.

3.9 $r = k$, $s = -2\sqrt{2}k$, and $t = -3k$, for any real number k.

Section 4

4.1 (a) $\sqrt{2}$; (b) 9; (c) $-15\sqrt{3}/8$; (d) 0; (e) $-\frac{2}{3}$; (f) $-\frac{1}{2}$.

4.2 120°.

4.3 $-6 \leq \overrightarrow{OP} \circ \overrightarrow{OQ} \leq 6$.

Section 5

5.3 The given vectors $\overrightarrow{OP}$, $\overrightarrow{OQ}$, and $\overrightarrow{OR}$ are mutually orthogonal, so that each one is perpendicular to the plane determined by the other two. Thus, $\overrightarrow{OP}$ is perpendicular to the plane determined by $\overrightarrow{OQ}$ and $\overrightarrow{OR}$. By Definition 5.1, $\overrightarrow{OQ} \times \overrightarrow{OR}$ is perpendicular to the plane determined by $\overrightarrow{OQ}$ and $\overrightarrow{OR}$. Hence, $\overrightarrow{OP}$ and $\overrightarrow{OQ} \times \overrightarrow{OR}$ are collinear. Therefore, the angle between $\overrightarrow{OP}$ and $\overrightarrow{OQ} \times \overrightarrow{OR}$ is 0° or 180°. By Definition 5.1, the length of $\overrightarrow{OP} \times (\overrightarrow{OQ} \times \overrightarrow{OR})$ is zero. That is, $\overrightarrow{OP} \times (\overrightarrow{OQ} \times \overrightarrow{OR}) = \vec{0}$. A similar argument shows that $(\overrightarrow{OP} \times \overrightarrow{OQ}) \times \overrightarrow{OR} = \vec{0}$.

5.6 The last three products are defined. The first two of these are real numbers, and the last is a vector.

Section 6

6.2 $2 \cdot \mathbf{I} - 1 \cdot \mathbf{J} + 3 \cdot \mathbf{K}$; $-5 \cdot \mathbf{I} + 1 \cdot \mathbf{K}$; $-3 \cdot \mathbf{K}$; $1 \cdot \mathbf{I} + 1 \cdot \mathbf{J} + 1 \cdot \mathbf{K}$; $7 \cdot \mathbf{I} + 6 \cdot \mathbf{J} + 4 \cdot \mathbf{K}$, $-1 \cdot \mathbf{I} - 2 \cdot \mathbf{J}$.

6.3 $\overrightarrow{OP} = \frac{5}{2} \cdot (\mathbf{I} + \mathbf{J}) - \frac{1}{2} \cdot (\mathbf{I} - \mathbf{J})$.

6.4 $\overrightarrow{OP} = 4 \cdot \mathbf{I} - 6 \cdot (\mathbf{I} + \mathbf{J}) + 5 \cdot (\mathbf{I} + \mathbf{J} + \mathbf{K})$.

6.6 $r = k$, $s = 11k$, $t = -27k$, and $n = 15k$, for any real number k.

Section 7

7.1 $(-1, 5, 8)$; $(5, -7, 6)$; $(-10, 5, -35)$; $(-18, 36, 6)$; $(13, -20, 11)$; $(-18, 27, -17)$; $(-5, 7, -6)$; $(-20, 55, 45)$.

7.2 $(-7, -\sqrt{2}/2, 15 + 5\sqrt{2})$; $(-1 + \sqrt{2}, -\frac{13}{2}, -2 + \sqrt{3})$; $(-6, -126, 18 + 18\sqrt{3})$; $(\frac{1}{2} + 7\sqrt{2}, 0, -\frac{73}{2} + \sqrt{3}/2)$; $(-2 + \sqrt{2}/3, -\frac{79}{30}, \frac{13}{3} + 2\sqrt{3}/5)$.

7.3 (a) $(-1, 1, 3)$; $(8, 11, 14)$; $(5, 7, 9)$; $(12, 21, 30)$; $(-40, -36, -32)$.

7.6 $(-6, \frac{1}{2}, 5)$; $(-1 + 2\sqrt{5}, \sqrt{2}, 7)$; $(-1 + \sqrt{5}, 2\sqrt{2}, 1)$; $(5 + \sqrt{5}, -\frac{1}{2} + 2\sqrt{2}, 6)$; $(-\sqrt{5}, 3\sqrt{2}, -6)$; $(1 - \sqrt{5}, -2\sqrt{2}, -1)$.

7.7 (a) -34; (b) -4; (c) 10; (d) $10\sqrt{30}$; (e) $-\frac{47}{2}$.

7.8 (a) $\dfrac{-34}{\sqrt{38}\sqrt{41}}$; (b) $\dfrac{-2\sqrt{2}}{3}$; (c) $\dfrac{300}{\sqrt{361}\sqrt{722}}$; (d) 1; (e) $\dfrac{-47}{78}$.

7.9 $\dfrac{-12\sqrt{5}}{35}, \dfrac{-\sqrt{5}}{35}, \dfrac{-2\sqrt{5}}{7}$; $\dfrac{2\sqrt{5}-1}{|RS|}, \dfrac{-\sqrt{2}}{|RS|}, \dfrac{7}{|RS|}$, where

$|RS| = 2\sqrt{18 - \sqrt{5}}$; $\dfrac{1-\sqrt{5}}{|SP|}, \dfrac{-2\sqrt{2}}{|SP|}, \dfrac{-1}{|SP|}$, where

$|SP| = \sqrt{15 - 2\sqrt{5}}$; $\dfrac{10 + 2\sqrt{5}}{2|QS|}, \dfrac{4\sqrt{2}-1}{2|QS|}, \dfrac{12}{2|QS|}$, where

$2|QS| = \sqrt{40\sqrt{5} - 8\sqrt{2} + 297}$; $\dfrac{\sqrt{5}}{\sqrt{59}}, \dfrac{-3\sqrt{2}}{\sqrt{59}}, \dfrac{6}{\sqrt{59}}$;

$\dfrac{\sqrt{5}-1}{|PS|}, \dfrac{2\sqrt{2}}{|PS|}, \dfrac{1}{|PS|}$, where $|PS| = \sqrt{15 - 2\sqrt{5}}$.

7.14 $\overrightarrow{OP} = [\overrightarrow{OP} - (t/r^2) \cdot \overrightarrow{OQ}] + (t/r^2) \cdot \overrightarrow{OQ}$.

7.16 Using Corollary 7.3, $|\overrightarrow{OP} + \overrightarrow{OQ}|^2 = (\overrightarrow{OP} + \overrightarrow{OQ}) \circ (\overrightarrow{OP} + \overrightarrow{OQ}) = \overrightarrow{OP} \circ \overrightarrow{OP} + \overrightarrow{OP} \circ \overrightarrow{OQ} + \overrightarrow{OQ} \circ \overrightarrow{OP} + \overrightarrow{OQ} \circ \overrightarrow{OQ} = |OP|^2 + |OQ|^2 + 2(\overrightarrow{OP} \circ \overrightarrow{OQ})$. Solve for $\overrightarrow{OP} \circ \overrightarrow{OQ}$.

7.18 (a) $\overrightarrow{OP} \times \overrightarrow{OQ} = 8 \cdot \mathbf{I} - 17 \cdot \mathbf{J} - 7 \cdot \mathbf{K}$; (b) $\overrightarrow{OP} \times \overrightarrow{OQ} = \mathbf{I} - \mathbf{J}$;
(c) $\overrightarrow{OP} \times \overrightarrow{OQ} = {}^{98}/_{15} \cdot \mathbf{I} - {}^{117}/_{10} \cdot \mathbf{J} + {}^{19}/_{6} \cdot \mathbf{K}$;
(d) $\overrightarrow{OP} \times \overrightarrow{OQ} = 0 \cdot \mathbf{I} + 0 \cdot \mathbf{J} + 0 \cdot \mathbf{K}$;
(e) $\overrightarrow{OP} \times \overrightarrow{OQ} = {}^{154}/_{5} \cdot \mathbf{I} - 22 \cdot \mathbf{J}$.

7.19 $(\overrightarrow{OP} \times \overrightarrow{OQ}) \circ \overrightarrow{OR} = \overrightarrow{OP} \circ (\overrightarrow{OQ} \times \overrightarrow{OR}) = -35 - 21\sqrt{2} - \sqrt{3}/2$.

Section 8

8.1 (a) $(-{}^7/_4, {}^{19}/_6, {}^{33}/_8)$; (b) $(-{}^2/_5, {}^{22}/_{15}, {}^9/_5)$
(c) $((-16 + 9\sqrt{2})/4, (36 - 17\sqrt{2})/6, (64 - 31\sqrt{2})/8)$.

8.2 $({}^1/_3, {}^1/_3, {}^1/_3)$.

8.5 (a) $x = -1 - 3k, y = 5 + 6k, z = 6 + 3k$; (b) $x = 3 + 8k$, $y = 4 + 2k, z = 2 - 8k$; (c) $x = 6 + 6k, y = 1 + k, z = 7 + 7k$.

8.8 By equation (8-3), $\overrightarrow{OP} = \overrightarrow{OA} + k_1 \cdot \overrightarrow{OV} + k_2 \cdot \overrightarrow{OW}$, where $\overrightarrow{OV}$ and $\overrightarrow{OW}$ are vectors with the same length and direction as the directed line segments $\overrightarrow{AB}$ and $\overrightarrow{AC}$, respectively. If the coordinates of A, B, and C are (x_A, y_A, z_A), (x_B, y_B, z_B), and (x_C, y_C, z_C), respectively, then the coordinates of $\overrightarrow{OA}$, $\overrightarrow{OV}$, and $\overrightarrow{OW}$ are (x_A, y_A, z_A), $(x_B - x_A, y_B - y_A, z_B - z_A)$, and $(x_C - x_A, y_C - y_A, z_C - z_A)$, respectively. Let P have coordinates (x, y, z). The parametric equations of π are obtained by equating the coordinates of the vectors in the two members of equation (8-3).

8.9 (a) $x = 3 - k_1 - 9k_2, y = 1 - 2k_1 + 3k_2, z = 7 - 3k_1 - 9k_2$;
(b) $x = {}^1/_2 + {}^1/_6 k_1 - {}^1/_2 k_2, y = 3 - 3k_1 - 3k_2$,
$z = {}^1/_4 + {}^1/_{12} k_1 + {}^3/_4 k_2$; (c) $x = 2 - 8k_1 + (\sqrt{11} - 2)k_2$,
$y = 3 + (\sqrt{7} - 3)k_1 + k_2, z = 5 - 4k_1 - 5k_2$.

8.11 (a) $ax + by + cz - a^2 - b^2 - c^2 = 0$;
(b) $x^2 + y^2 + z^2 - ax - by - cz = 0$.

8.12 $\cos\theta = 2\sqrt{2}/3$.

8.13 $x = 11k, y = 15k, z = -36k$.

8.14 (a) 15; (b) 23.

Section 9

9.3 (a) $(-{}^9/_2, (3 - 6\sqrt{3})/2, {}^1/_6, (-9 + 4\sqrt{2})/2)$;
(b) $(-3\sqrt{3} + \sqrt{6}, 2\sqrt{6}, (9\sqrt{3} - 4\sqrt{2} - 18\sqrt{6})/18, \sqrt{2})$;
(c) $(-4, -1 - \sqrt{3}, {}^{31}/_{18}, \sqrt{2})$.

9.4 (a), (c), (f), and (g) are in $\mathfrak{F}_\infty$.

9.5 (a) $u(x) = {}^1/_2(e^x - e^{-x})$; (b) $u(x) = x^3 + 7x - 2$;
(c) $u(x) = 2\sqrt{2}e^x + {}^2/_3 x^3 + {}^2/_3 x + {}^2/_3$;
(d) $u(x) = e^x + e^{-x} + 2x^3 + 3$.

9.6 (a) $u(0) = 0,\ u(-1) = \dfrac{1 - e^2}{2e},\ u({}^1/_2) = \dfrac{e - 1}{2\sqrt{e}}$;
(b) $u(0) = -2,\ u(-1) = -10,\ u({}^1/_2) = {}^{13}/_8$;
(c) $u(0) = 2\sqrt{2} + {}^2/_3,\ u(-1) = \dfrac{6\sqrt{2} - 2e}{3e},\ u({}^1/_2) = 2\sqrt{2e} + {}^{13}/_{12}$;

(d) $u(0) = 5$, $u(-1) = \dfrac{e^2 + e + 1}{e}$, $u(\tfrac{1}{2}) = \dfrac{4e + \sqrt{e} + 16}{4\sqrt{e}}$.

9.7 (a) $(-2, 2, 2, -4, -8)$; (b) $(0, 1, -\tfrac{4}{3}, -\tfrac{17}{3}, 0)$;
(c) $(4, 2, 0, -2, -4)$

9.9 (a) and (c) are subspaces.

9.10 (b), (d), (f), and (g) are subspaces.

9.16 Any two vectors in S span $\mathcal{S}$.

9.18 $\{1, x, x^2, \ldots, x^k\}$ is one spanning set.

9.20 $\{(2, 0, 0, 0, 1), (0, 0, 1, 1, 0), (0, 1, 0, 0, 0)\}$ is one set that spans $\mathcal{S}$.

9.25 (a) Let $\mathcal{S}_1$ be the subspace of $\mathcal{V}_4$ spanned by $\{(1, 0, 0, 0,)\}$, $\mathcal{S}_2$ the subspace spanned by $\{(0, 1, 0, 0)\}$, $\mathcal{S}_3$ the subspace spanned by $\{(0, 0, 1, 0)\}$, and $\mathcal{S}_4$ the subspace spanned by $\{(0, 0, 0, 1)\}$. Then $\mathcal{V}_4 = \mathcal{S}_1 \oplus \mathcal{S}_2 \oplus \mathcal{S}_3 \oplus \mathcal{S}_4$.
(b) Let $\mathcal{T}_1$ be the subspace of $\mathcal{V}_4$ spanned by $\{(1, 0, 0, 0), (0, 1, 0, 0)\}$, and let $\mathcal{T}_2$ be the subspace spanned by $\{(0, 0, 1, 0), (0, 0, 0, 1)\}$. Then $\mathcal{V}_4 = \mathcal{T}_1 \oplus \mathcal{T}_2$. There are other solutions for (a) and (b).

Section 10

10.1 (b), (c), (e), and (f) are linearly dependent.

10.2 (a) and (d) are linearly dependent.

10.3 (a), (d), and (e) are linearly independent.

10.6 Use Theorem 10.3.

10.8 (d) is a basis of $\mathcal{E}_3$.

10.9 (a) is a basis of $\mathcal{V}_3$.

10.12 Find an infinite linearly independent set of vectors (functions) in $\mathcal{F}_\infty$.

10.14 (a) Let $\{\mathbf{U}_1, \mathbf{U}_2, \ldots, \mathbf{U}_k\}$ be a basis of $\mathcal{S} \cap \mathcal{T}$. By Exercise 10.13, there are vectors $\mathbf{U}_{k+1}, \ldots, \mathbf{U}_s$ and $\mathbf{V}_{k+1}, \ldots, \mathbf{V}_t$ such that $\{\mathbf{U}_1, \ldots, \mathbf{U}_k, \mathbf{U}_{k+1}, \ldots, \mathbf{U}_s\}$ is a basis of $\mathcal{S}$ and $\{\mathbf{U}_1, \ldots, \mathbf{U}_k, \mathbf{V}_{k+1}, \ldots, \mathbf{V}_t\}$ is a basis of $\mathcal{T}$. Show that $\{\mathbf{U}_1, \ldots, \mathbf{U}_k, \mathbf{U}_{k+1}, \ldots, \mathbf{U}_s, \mathbf{V}_{k+1}, \ldots, \mathbf{V}_t\}$ is a basis of $\mathcal{S} + \mathcal{T}$.

Section 11

11.3 If $\mathbf{U} = x_1 \cdot \mathbf{I} + y_1 \cdot \mathbf{J} + z_1 \cdot \mathbf{K}$ and $\mathbf{V} = x_2 \cdot \mathbf{I} + y_2 \cdot \mathbf{J} + z_2 \cdot \mathbf{K}$, then by Theorem 7.2, $\mathbf{U} \circ \mathbf{V} = x_1x_2 + y_1y_2 + z_1z_2$.

11.6 (a), (b), and (d) are orthonormal.

11.8 If $\mathcal{S} = \{\mathbf{0}\}$, then $\mathcal{T} = \mathcal{V}$, and it is immediate that $\mathcal{V} = \mathcal{S} \oplus \mathcal{T}$. Therefore, suppose that $\mathcal{S} \neq \{\mathbf{0}\}$, and let $\{\mathbf{U}_1, \mathbf{U}_2, \ldots, \mathbf{U}_k\}$ be an orthonormal basis of $\mathcal{S}$. Show that there are vectors $\mathbf{U}_{k+1}, \ldots, \mathbf{U}_n$ such that $\{\mathbf{U}_1, \mathbf{U}_2, \ldots, \mathbf{U}_k, \mathbf{U}_{k+1}, \ldots, \mathbf{U}_n\}$ is an orthonormal basis of $\mathcal{V}$. If $\mathbf{V} \in \mathcal{V}$, then $\mathbf{V} = \sum_{i=1}^k a_i \cdot \mathbf{U}_i + \sum_{i=k+1}^n a_i \cdot \mathbf{U}_i$, where $\sum_{i=1}^k a_i \cdot \mathbf{U}_i \in \mathcal{S}$. Show that $\sum_{i=k+1}^n a_i \cdot \mathbf{U}_i \in \mathcal{T}$. Hence $\mathcal{V} = \mathcal{S} + \mathcal{T}$. Then show that $\mathcal{S} \cap \mathcal{T} = \{\mathbf{0}\}$.

11.9 (a) Dispose of the case $\mathbf{U} = \mathbf{0}$ or $\mathbf{V} = \mathbf{0}$. Otherwise, $\mathcal{G}(x \cdot \mathbf{U} + \mathbf{V}, x \cdot \mathbf{U} + \mathbf{V}) = x^2\mathcal{G}(\mathbf{U}, \mathbf{U}) + 2x\mathcal{G}(\mathbf{U}, \mathbf{V}) + \mathcal{G}(\mathbf{V}, \mathbf{V}) \geq 0$. Now use the fact

that $Ax^2 + Bx + C \geq 0 (A > 0)$ for all x if and only if $B^2 - 4AC \leq 0$.
(b) Show that $|\mathbf{U} + \mathbf{V}| \leq |\mathbf{U}| + |\mathbf{V}|$ if and only if $\mathcal{I}(\mathbf{U}, \mathbf{V}) \leq \sqrt{\mathcal{I}(\mathbf{U}, \mathbf{U})\mathcal{I}(\mathbf{V}, \mathbf{V})}$ and use part (a).

11.10 (a) $\{1/\sqrt{39}) \cdot (3, -5, 2, 1)\}$; (b) $\{(1/\sqrt{53}) \cdot (1, 6, 0, 4), (1/\sqrt{12557}) \cdot (102, -77, 318, 90)\}$; (c) $\{(1/\sqrt{3}) \cdot (1, 0, 1, 1), (1/\sqrt{6}) \cdot (0, 2, 1, -1), (1/\sqrt{22}) \cdot (10/3, 4/3, -3, -1/3)\}$; (d) $\{(1/2, 1/2, 1/2, 1/2), (\sqrt{3}/6, \sqrt{3}/6, \sqrt{3}/6, -\sqrt{3}/2), (\sqrt{6}/6, \sqrt{6}/6, -\sqrt{6}/3, 0)\}$.

Section 12

12.1 (a) $x_1 = 0, x_2 = 0, x_3 = 0$; (b) inconsistent; (c) $x_1 = (10 - 2r)/11, x_2 = (-25 + 5r)/11, x_3 = r$, where r is any real number; (d) inconsistent.

12.4 Multiply $\mathbf{U}_3^{(5)}$ by -466, obtaining $\mathbf{U}_1^{(4)}, \mathbf{U}_2^{(4)}, \mathbf{U}_3^{(4)}$. Multiply $\mathbf{U}_2^{(4)}$ by 30 and add to $\mathbf{U}_3^{(4)}$, obtaining $\mathbf{U}_1^{(3)}, \mathbf{U}_2^{(3)}, \mathbf{U}_3^{(3)}$. Multiply $\mathbf{U}_2^{(3)}$ by $1/3$, obtaining $\mathbf{U}_1^{(2)}, \mathbf{U}_2^{(2)}, \mathbf{U}_3^{(2)}$. Multiply $\mathbf{U}_1^{(2)}$ by 4 and add to $\mathbf{U}_3^{(2)}$, obtaining $\mathbf{U}_1^{(1)}, \mathbf{U}_2^{(1)}, \mathbf{U}_3^{(1)}$. Multiply $\mathbf{U}_1^{(1)}$ by $1/2$, obtaining $\mathbf{U}_1, \mathbf{U}_2, \mathbf{U}_3$.

Section 13

13.2 (a) $k = 2, m_1 = 1, m_2 = 4$; (b) $k = 3, m_1 = 1, m_2 = 2, m_3 = 3$; (c) $k = 3, m_1 = 1, m_2 = 2, m_3 = 3$; (d) $k = 3, m_1 = 1, m_2 = 2, m_3 = 3$; (e) $k = 4, m_1 = 1, m_2 = 2, m_3 = 3, m_4 = 4$; (f) $k = 4, m_1 = 1, m_2 = 2, m_3 = 3, m_4 = 4$.

13.3 $x_1 = 94/679, x_2 = 685/679, x_3 = -30/97, x_4 = 79/97$.

Section 14

14.1 (a) inconsistent; (b) $x_1 = 64/55, x_2 = -69/55, x_3 = -28/55, x_4 = 8/5$; (c) inconsistent; (d) $x_1 = 3/2, x_2 = 25/8, x_3 = -1/8$.

14.2 (a).

14.3 (a) $x_1 = -3r, x_2 = -r, x_3 = r$, where r is any real number. The solution space is spanned by $(-3, -1, 1)$.
(b) Trivial solution. Solution space is $\{(0, 0, 0)\}$.
(c) $x_1 = -r, x_2 = r, x_3 = r$, where r is any real number.
(d) Trivial solution. Solution space is $\{(0, 0, 0)\}$.

14.5 (b) and (c).

14.7 A basis of $\mathfrak{I}$ is $\{(21, 14, 0, 0, 12), (3, 2, 12, 0, 0), (3, 4, 0, 6, 0)\}$. (There are others.)

Section 15

15.2 (a) $(10, 5, 0, -5)$; $(-2, 3, -3/2, 1)$; $(4, 0, -2, \sqrt{6})$;
(b) $4 \cdot \mathbf{I} + 4 \cdot \mathbf{J} - 4 \cdot \mathbf{K}$; $14 \cdot \mathbf{I} - 21 \cdot \mathbf{J} + 35 \cdot \mathbf{K}$; $\mathbf{0}$;
(c) $x^2 + 2x - 3$; $14x^3 + 4x^2 + 2$; $4x^{33} + 2x^{11} - x$.

15.3 For any $\mathbf{V} \in \mathcal{V}$, let $\mathbf{U} = (1/k) \cdot \mathbf{V}$. Then $\mathbf{V} = D_k(\mathbf{U})$. $D_k(\mathbf{U}) = k \cdot \mathbf{U} = \mathbf{0}$ implies $\mathbf{U} = \mathbf{0}$, if $k \neq 0$.

15.5 (a) Use Theorem 7.5;
(b) $(b + 3c) \cdot \mathbf{I} + (-a + 2c) \cdot \mathbf{J} + (-3a - 2b) \cdot \mathbf{K}$;

$(-b - c) \cdot \mathbf{I} + (a + c) \cdot \mathbf{J} + (a - b) \cdot \mathbf{K}$;
$(-5b - 3c) \cdot \mathbf{I} + 5a \cdot \mathbf{J} + 3a \cdot \mathbf{K}$; $-2c \cdot \mathbf{I} - c \cdot \mathbf{J} + (2a + b) \cdot \mathbf{K}$.

15.8 For $\varphi = 45°, 90°, 150°$, and $270°$, $T_\varphi(-\mathbf{I} + \mathbf{J}) = -\sqrt{2} \cdot \mathbf{I}$, $-\mathbf{I} - \mathbf{J}$, $\dfrac{\sqrt{3} - 1}{2} \cdot \mathbf{I} + \dfrac{-1 - \sqrt{3}}{2} \cdot \mathbf{J}$, and $\mathbf{I} + \mathbf{J}$, respectively.

For $\varphi = 45°, 90°, 150°$, and $270°$,
$T_\varphi(2 \cdot \mathbf{I} + 4 \cdot \mathbf{J}) = -\sqrt{2} \cdot \mathbf{I} + 3\sqrt{2} \cdot \mathbf{J}$, $-4 \cdot \mathbf{I} + 2 \cdot \mathbf{J}$,
$(-\sqrt{3} - 2) \cdot \mathbf{I} + (1 - 2\sqrt{3}) \cdot \mathbf{J}$, and $4 \cdot \mathbf{I} - 2 \cdot \mathbf{J}$, respectively.
For $\varphi = 45°, 90°, 150°$, and $270°$, $T_\varphi(3 \cdot \mathbf{I} - 5 \cdot \mathbf{J})$
$= 4\sqrt{2} \cdot \mathbf{I} - \sqrt{2} \cdot \mathbf{J}$, $5 \cdot \mathbf{I} + 3 \cdot \mathbf{J}$, $\dfrac{5 - 3\sqrt{3}}{2} \cdot \mathbf{I} + \dfrac{3 + 5\sqrt{3}}{2} \cdot \mathbf{J}$,
and $-5 \cdot \mathbf{I} - 3 \cdot \mathbf{J}$, respectively.

15.11 Since the system of equations

$$x' \cos \varphi - y' \sin \varphi = x$$
$$x' \sin \varphi + y' \cos \varphi = y$$

has a unique solution (x', y'), it follows that if $\mathbf{V} = x \cdot \mathbf{I} + y \cdot \mathbf{J} \in \mathcal{E}_2$, then there is a unique vector $\mathbf{U} = x' \cdot \mathbf{I} + y' \cdot \mathbf{J} \in \mathcal{E}_2$ such that $T_\varphi(\mathbf{U}) = \mathbf{V}$.

15.14 $\frac{4}{3} \cdot \mathbf{I} + \frac{4}{3} \cdot \mathbf{J} + \frac{4}{3} \cdot \mathbf{K}$; $\mathbf{I} + \mathbf{J} + \mathbf{K}$; $\mathbf{I} + \mathbf{J} + \mathbf{K}$; $3 \cdot \mathbf{I} + 3 \cdot \mathbf{J} + 3 \cdot \mathbf{K}$; $\mathbf{0}$.

15.18 $5 \cdot \mathbf{I} - 3 \cdot \mathbf{J}$; $\mathbf{I} + \mathbf{J}$; $\mathbf{0}$; $7 \cdot \mathbf{I} + 2 \cdot \mathbf{J}$; $-\mathbf{I} + 3 \cdot \mathbf{J}$.

15.20 $-6 \cdot \mathbf{I} - 3 \cdot \mathbf{J} + 15 \cdot \mathbf{K}$; $(\frac{58}{165}) \cdot \mathbf{I} + (\frac{779}{165}) \cdot \mathbf{J} + (\frac{421}{165}) \cdot \mathbf{K}$; $(\frac{31}{55}) \cdot \mathbf{I} + (\frac{131}{110}) \cdot \mathbf{J} + (-\frac{3}{22}) \cdot \mathbf{K}$; $(\frac{56}{165}) \cdot \mathbf{I} + (\frac{281}{330}) \cdot \mathbf{J} + (\frac{7}{66}) \cdot \mathbf{K}$; $\mathbf{0}$.

15.22 Let $m = a^2 + b^2 + c^2$ and $n = d^2 + e^2 + f^2$. Then

$$x' = \left(\frac{na^2 + md^2}{mn}\right)x + \left(\frac{nab + mde}{mn}\right)y + \left(\frac{nac + mdf}{mn}\right)z$$
$$y' = \left(\frac{nab + mde}{mn}\right)x + \left(\frac{nb^2 + me^2}{mn}\right)y + \left(\frac{nbc + mef}{mn}\right)z$$
$$z' = \left(\frac{nac + mdf}{mn}\right)x + \left(\frac{nbc + mef}{mn}\right)y + \left(\frac{nc^2 + mf^2}{mn}\right)z.$$

Section 16

16.4 $L^{-1}(\{\mathbf{0}\}) = \{\mathbf{0}\}$ if and only if the system of equations $ax + by = 0$, $cx + dy = 0$ has only the trivial solution. This is the case if and only if $ad - bc \neq 0$.

16.6 Use the fact that equations (15-5) have a unique solution (x, y) for every given (x', y').

16.7 $P_{\mathbf{V}_0}(\mathcal{E}_3) = \mathcal{S}$, and $P_{\mathbf{V}_0}^{-1}(\{\mathbf{0}\}) = \mathcal{E}_\pi$, where π is the plane through O perpendicular to $\mathbf{V}_0$.

16.9 Use Theorem 16.4.

16.12 Prove that isomorphic vector spaces have the same dimension.

16.13 (a) If $\{\mathbf{U}_1, \mathbf{U}_2, \ldots, \mathbf{U}_k\}$ is a basis of $\mathcal{V}$, then $\{L(\mathbf{U}_1), L(\mathbf{U}_2), \ldots, L(\mathbf{U}_k)\}$ spans $L(\mathcal{V})$.
(b) If $L^{-1}(\{\mathbf{0}\}) = \{\mathbf{0}\}$, use Theorem 16.4. Otherwise, let $\{\mathbf{U}_1, \mathbf{U}_2, \ldots, \mathbf{U}_k\}$ be a basis of $L^{-1}(\{\mathbf{0}\})$ such that $\{\mathbf{U}_1, \mathbf{U}_2, \ldots, \mathbf{U}_k, \mathbf{U}_{k+1}, \ldots, \mathbf{U}_n\}$ is a basis of $\mathcal{V}$. Show that $\{L(\mathbf{U}_{k+1}), \ldots, L(\mathbf{U}_n)\}$ is a basis of $L(\mathcal{V})$.

Section 17

17.3 $\frac{5}{2} \cdot \mathbf{I} + \frac{3}{2} \cdot \mathbf{J}$; $-\frac{5}{2} \cdot \mathbf{I} + \frac{1}{2} \cdot \mathbf{J} + 3 \cdot \mathbf{K}$; $12 \cdot \mathbf{I}$; $\mathbf{0}$; $-6 \cdot \mathbf{I} - 2 \cdot \mathbf{J}$;
$\dfrac{-3x - 3y}{2} \cdot \mathbf{I} + \dfrac{3x + 3y}{2} \cdot \mathbf{J}$; $3 \cdot \mathbf{I}$; $\mathbf{0}$.

17.4 $3\sqrt{2}$; $4\sqrt{2} \cdot \mathbf{I} - 2\sqrt{2} \cdot \mathbf{J}$; $-3\mathbf{I} + (\sqrt{2} - 3) \cdot \mathbf{J}$; $\mathbf{0}$.

17.8 Since $L(\mathcal{V}) \subseteq \mathcal{W}$, it follows that $ML(\mathcal{V}) = M(L(\mathcal{V})) \subseteq M(\mathcal{W})$. If L is a linear transformation of $\mathcal{V}$ onto $\mathcal{W}$, then $L(\mathcal{V}) = \mathcal{W}$, and $ML(\mathcal{V}) = M(\mathcal{W})$. If $\mathbf{V} \in L^{-1}(\{\mathbf{0}\})$, then $L(\mathbf{V}) = \mathbf{0}$, and $ML(\mathbf{V}) = M(L(\mathbf{V})) = M(\mathbf{0}) = \mathbf{0}$. Hence $L^{-1}(\{\mathbf{0}\}) \subseteq (ML)^{-1}(\{\mathbf{0}\})$. If M is nonsingular, then $ML(\mathbf{V}) = M(L(\mathbf{V})) = \mathbf{0}$ implies $L(\mathbf{V}) = \mathbf{0}$. Hence, $(ML)^{-1}(\{\mathbf{0}\}) \subseteq L^{-1}(\{\mathbf{0}\})$.

17.9 $M(\mathcal{V}_5)$ contains (1, 1, 1, 1) and $ML(\mathcal{V}_3)$ does not contain this vector. $L^{-1}(\{\mathbf{0}\}) = \{\mathbf{0}\} = (ML)^{-1}(\{\mathbf{0}\})$. ML is a nonsingular linear transformation of $\mathcal{V}_3$ into $\mathcal{V}_4$. M is a singular transformation of $\mathcal{V}_5$ into $\mathcal{V}_4$. ML is not an isomorphism of $\mathcal{V}_3$ onto $\mathcal{V}_4$.

Section 18

18.1 (a) Let $\mathbf{V}_0 = a \cdot \mathbf{I} + b \cdot \mathbf{J} + c \cdot \mathbf{K}$. The matrix is

$$\begin{bmatrix} \dfrac{a}{a^2 + b^2 + c^2} & \dfrac{b}{a^2 + b^2 + c^2} & \dfrac{c}{a^2 + b^2 + c^2} \end{bmatrix}.$$

(b) $[2 \quad -3 \quad 1]$; (c) $\begin{bmatrix} 1 & 2 & 0 & 0 & 0 \\ 0 & 0 & 1 & 0 & -1 \\ 0 & 3 & 1 & 0 & 0 \\ 0 & 0 & 0 & 0 & 1 \end{bmatrix}$;

(d) $\begin{bmatrix} 1 & -1 & 0 & 0 \\ 0 & 0 & 1 & 1 \\ 1 & 1 & 1 & 0 \\ 0 & 0 & 0 & 1 \\ 1 & 0 & -1 & 1 \end{bmatrix}$.

18.2 (a) The n by n matrix $\begin{bmatrix} k & 0 & \cdots & 0 \\ 0 & k & & \cdot \\ \cdot & & \cdot & \cdot \\ \cdot & & \cdot & \cdot \\ \cdot & & \cdot & 0 \\ 0 & \cdots & 0 & k \end{bmatrix}$;

(b) $\begin{bmatrix} 0 & 1 & 1 \\ -1 & 0 & -1 \\ -1 & 1 & 0 \end{bmatrix}$; (c) $\begin{bmatrix} \frac{\sqrt{3}}{2} & -\frac{1}{2} \\ \frac{1}{2} & \frac{\sqrt{3}}{2} \end{bmatrix}$;

(d) $\begin{bmatrix} 1 & 2 & 0 & 0 & 0 \\ 0 & 0 & 0 & 0 & 0 \\ 0 & 0 & 0 & 1 & -1 \\ 0 & 0 & 0 & 0 & 0 \\ 0 & 0 & 1 & 0 & 0 \end{bmatrix}$; (e) $\begin{bmatrix} 1 & 0 & 0 & 0 & 0 \\ 1 & 1 & 0 & 0 & 0 \\ 1 & 1 & 1 & 0 & 0 \\ 0 & 0 & 0 & 5 & 0 \\ 0 & 0 & 0 & 0 & -1 \end{bmatrix}$.

18.3 $\begin{bmatrix} 0 & 1 & 0 & 0 & 0 \\ 0 & 0 & 2 & 0 & 0 \\ 0 & 0 & 0 & 3 & 0 \\ 0 & 0 & 0 & 0 & 4 \\ 0 & 0 & 0 & 0 & 0 \end{bmatrix}$; $\begin{bmatrix} -1 & 1 & 0 & -3 & -4 \\ -1/2 & 1/2 & 3/2 & -5/2 & -2 \\ 0 & 0 & 0 & 4 & 4 \\ -1/2 & 1/2 & -3/2 & -5/2 & -2 \\ 1/2 & -1/2 & 3/2 & 5/2 & 2 \end{bmatrix}$.

18.4 $\begin{bmatrix} 1 & 2 & -1 & 0 & 1 \\ 0 & 3 & 1 & 0 & 1 \\ 1 & 5 & 2 & 0 & -1 \\ 0 & 0 & 0 & 0 & 1 \\ 1 & -1 & -1 & 0 & 1 \end{bmatrix}$.

18.5 M: $\begin{bmatrix} 1 & 0 & 0 & 0 & 0 \\ 0 & 1 & 0 & 0 & 0 \\ 0 & 0 & 1 & 0 & 0 \\ 0 & 0 & 0 & 1 & 0 \end{bmatrix}$ ML: $\begin{bmatrix} 1 & -1 & 0 \\ 0 & 0 & 0 \\ 1 & 0 & -1 \\ 0 & 1 & 0 \end{bmatrix}$.

18.6 L: $\begin{bmatrix} 1 & 2 & 0 & 0 & 0 \\ 0 & 0 & 0 & 0 & 0 \\ 0 & 0 & 0 & 1 & -1 \\ 0 & 0 & 0 & 0 & 0 \\ 0 & 0 & 1 & 0 & 0 \end{bmatrix}$ M: $\begin{bmatrix} 1 & 0 & 0 & 0 & 0 \\ 1 & 1 & 0 & 0 & 0 \\ 1 & 1 & 1 & 0 & 0 \\ 0 & 0 & 0 & 5 & 0 \\ 0 & 0 & 0 & 0 & -1 \end{bmatrix}$

$L + M$: $\begin{bmatrix} 2 & 2 & 0 & 0 & 0 \\ 1 & 1 & 0 & 0 & 0 \\ 1 & 1 & 1 & 1 & -1 \\ 0 & 0 & 0 & 5 & 0 \\ 0 & 0 & 1 & 0 & -1 \end{bmatrix}$ $10 \cdot L$: $\begin{bmatrix} 10 & 20 & 0 & 0 & 0 \\ 0 & 0 & 0 & 0 & 0 \\ 0 & 0 & 0 & 10 & -10 \\ 0 & 0 & 0 & 0 & 0 \\ 0 & 0 & 10 & 0 & 0 \end{bmatrix}$

ML: $\begin{bmatrix} 1 & 2 & 0 & 0 & 0 \\ 1 & 2 & 0 & 0 & 0 \\ 1 & 2 & 0 & 1 & -1 \\ 0 & 0 & 0 & 0 & 0 \\ 0 & 0 & -1 & 0 & 0 \end{bmatrix}$ LM: $\begin{bmatrix} 3 & 2 & 0 & 0 & 0 \\ 0 & 0 & 0 & 0 & 0 \\ 0 & 0 & 0 & 5 & 1 \\ 0 & 0 & 0 & 0 & 0 \\ 1 & 1 & 1 & 0 & 0 \end{bmatrix}$.

18.9 $-\tfrac{1}{2}\cdot\mathbf{I} - \tfrac{1}{2}\cdot\mathbf{J} + 4\cdot\mathbf{K}$; $\mathbf{0}$; $\tfrac{1}{2}\cdot\mathbf{I} + \tfrac{1}{2}\cdot\mathbf{J} + 5\cdot\mathbf{K}$; $\tfrac{1}{2}\cdot\mathbf{I} + \tfrac{1}{2}\cdot\mathbf{J}$; $5\cdot\mathbf{I} + 5\cdot\mathbf{J} + 10\cdot\mathbf{K}$.

18.11 $(11, 6, 17, -2)$; $(3, 2, 3, 0)$; $(-14, -10, -11, -1)$; $(-1, 0, -4, 1)$; $(-\sqrt{2} + 3\sqrt{3}, 2\sqrt{3}, -4\sqrt{2} + 3\sqrt{3}, \sqrt{2})$.

Section 19

19.3 $\begin{bmatrix} 5/2 & -16/15 & 3 & 51/10 \\ 38/5 & -7 & 33/10 & 13 \end{bmatrix}$; $\begin{bmatrix} 1/2 & -29/15 & -1 & 51/10 \\ 38/5 & -10 & 13/10 & -21 \end{bmatrix}$;

$\begin{bmatrix} 5/2 & 7/5 & -3 & 49/10 \\ 32/5 & -7/2 & 37/10 & 6 \end{bmatrix}$; $\begin{bmatrix} 5/2 & -16/15 & 3 & 51/10 \\ 38/5 & -7 & 33/10 & 13 \end{bmatrix}$;

$$\begin{bmatrix} 15 + \sqrt{2} & \dfrac{42 + \sqrt{2}}{3} & -25 + 2\sqrt{2} & 24 \\ 29 & \dfrac{-10 + 3\sqrt{2}}{2} & 22 + \sqrt{2} & 55 + 17\sqrt{2} \end{bmatrix};$$

$$\begin{bmatrix} 30 & -28 & 20 & 102 \\ 152 & -170 & 46 & -80 \end{bmatrix}.$$

19.6 (a) $\begin{bmatrix} 0 & -1 & 2/5 & 6/5 & -1/5 \\ -2/5 & 0 & 0 & 3/5 & 0 \\ 0 & 2 & 3/5 & 2/5 & -1/5 \end{bmatrix}$;

(b) $\begin{bmatrix} -63 & 119 & -28 & 56 & 7 \\ 7 & 28 & 0 & -7 & 49 \\ -105 & -14 & -126 & -7 & 0 \end{bmatrix}$.

19.8 $[20 \quad 68 \quad -18 \quad 130]$; $\begin{bmatrix} 10 & 32 & -7 & 45 \\ -2 & -9 & 4 & -35 \\ 2 & 6 & -1 & 5 \\ 0 & 6 & -6 & 60 \end{bmatrix}$;

$\begin{bmatrix} 57 & -2 \\ 16 & 3 \end{bmatrix}$; $[2 \quad 7]$; $\begin{bmatrix} 339 & -11 \\ 98 & 23 \end{bmatrix}$.

19.17 For the diagonal matrix $A = [a_{i,j}]$, write $a_{i,j} = a_{i,j}\delta_{i,j}$, where $\delta_{i,j} = 0$ if $i \neq j$ and $\delta_{i,j} = 1$ if $i = j$.

19.20 Suppose $AB = I_n$. Let L and M be linear transformations of $\mathcal{V}_n$ into $\mathcal{V}_n$ such that A is the matrix of L and B is the matrix of M with respect to some basis of $\mathcal{V}_n$. Then $LM = I_{\mathcal{V}_n}$. Show that M is nonsingular, hence an isomorphism of $\mathcal{V}_n$ onto $\mathcal{V}_n$. Then M^* exists, and $M^*M = MM^* = I_{\mathcal{V}_n}$. Show that $L = M^*$. Hence $ML = I_{\mathcal{V}_n}$, and consequently $BA = I_n$.

19.22 (a) $\begin{bmatrix} -\frac{1}{7} & \frac{2}{7} \\ 0 & 1 \end{bmatrix}$; (b) $\begin{bmatrix} -\frac{15}{7} & \frac{30}{7} \\ \frac{24}{7} & -\frac{20}{7} \end{bmatrix}$; (c) $\begin{bmatrix} 1 & 0 & -2 \\ \frac{3}{4} & \frac{1}{4} & -\frac{3}{2} \\ 0 & 0 & 1 \end{bmatrix}$;

(d) $\begin{bmatrix} \dfrac{a}{(a^2+b^2)} & -\dfrac{b}{(a^2+b^2)} \\ \dfrac{b}{(a^2+b^2)} & \dfrac{a}{(a^2+b^2)} \end{bmatrix}$.

Section 20

20.2 (a) Row rank = 3; (b) Row rank = 2; (c) Row rank = 3; (d) Row rank = 4; (e) Row rank = 3.

20.3 (a) Nonhomogeneous system: $x_1 = (10 - 5r - 4s)/7$, $x_2 = (-2 + r - 16s)/14$, $x_3 = s$, $x_4 = (-2 - r)/2$, $x_5 = r$, where r and s are any real numbers. Homogeneous system: $x_1 = (-10r - 5s - 4t)/7$, $x_2 = (2r + s - 16t)/14$, $x_3 = t$, $x_4 = (2r - s)/2$, $x_5 = s$, $x_6 = r$, where r, s, and t are any real numbers.

(b) Nonhomogeneous system: $x_1 = -\frac{12}{5}$, $x_2 = \frac{3}{10}$. Homogeneous system: $x_1 = 12r/5$, $x_2 = -3r/10$, $x_3 = r$, where r is any real number.

(c) Nonhomogeneous system: inconsistent. Homogeneous system: $x_1 = x_2 = x_3 = 0$

(d) Nonhomogeneous system: inconsistent. Homogeneous system: $x_1 = x_2 = x_3 = x_4 = 0$.

(e) Nonhomogeneous system: inconsistent. Homogeneous system: $x_1 = r$, $x_2 = x_3 = x_4 = 0$, where r is any real number.

20.7 (a), (b), and (d).

20.9 (a) $x_1 = \frac{1}{16}$, $x_2 = -\frac{5}{16}$ (b) $x_1 = -1$, $x_2 = -1$, $x_3 = 3$.

20.11 $$\left\{\left(\frac{-10r - 5s - 4t + 5}{7}, \frac{2r + s - 16t - 15}{14}, t, \frac{2r - s - 3}{2}, s, r\right) \,\middle|\, r, s, \text{ and } t \text{ any real numbers}\right\}.$$

20.12 $L^{-1}(\{\mathbf{0}\})$ is the subspace of $\mathcal{V}_3$ spanned by the vector $(\frac{12}{5}, -\frac{3}{10}, 1)$.

20.13 $\{2 \cdot \mathbf{I} - 3 \cdot \mathbf{J} + r \cdot \mathbf{K} \mid r \text{ is any real number}\}$

Section 21

21.1 (a) 2; (b) 3; (c) 2.

21.2 $\begin{bmatrix} -1 & 3 & -1 \\ 4 & -5 & 4 \\ 2 & -1 & 1 \end{bmatrix}$.

21.3 $-2 \cdot \mathbf{I} + 3 \cdot \mathbf{J} + 5 \cdot \mathbf{K} = -2 \cdot (\mathbf{I} + \mathbf{J} + \mathbf{K}) + 5 \cdot (\mathbf{J} + \mathbf{K}) + 2 \cdot \mathbf{K}$
$= \frac{1}{4} \cdot (2 \cdot \mathbf{I} - 3 \cdot \mathbf{J} + 5 \cdot \mathbf{K}) + \frac{5}{4} \cdot (\mathbf{I} - \mathbf{J} + 4 \cdot \mathbf{K})$

$$+ \tfrac{5}{4}\cdot(-3\cdot\mathbf{I} + 4\cdot\mathbf{J} - \mathbf{K}) = \frac{-14-\sqrt{3}}{14}\cdot(\tfrac{1}{2}\cdot\mathbf{I} - \sqrt{3}/2\cdot\mathbf{J})$$

$$+ \frac{1-14\sqrt{3}}{14}\cdot(\sqrt{3}/2\cdot\mathbf{I} + \tfrac{1}{2}\cdot\mathbf{J}) + \tfrac{5}{7}\cdot(4\cdot\mathbf{J} + 7\cdot\mathbf{K}).$$

Section 22

22.1 $\begin{bmatrix} 16 & 11 & 1 \\ 9/2 & 7/2 & 3/2 \end{bmatrix}$.

22.2 $\{(0, -1/3, -1), (1, 0, -1), (0, 0, 1)\}$ and $\{(1, 0, 0, 0), (5, 1, 0, 0), (4, 1/3, 1, 0), (-1, -1/3, 0, 1)\}$.

22.3 $\begin{bmatrix} 1 & 0 \\ 0 & 1 \\ 0 & 0 \end{bmatrix}$.

22.4 $(2, 5)$, $(-1, \pi)$, $(1/2, \sqrt{2})$.

22.5 (a), (c).

22.7 $[0] = \{km \mid k \in Z\}$, $[1] = \{km + 1 \mid k \in Z\}, \ldots,$
$[m-1] = \{km + m - 1 \mid k \in Z\}$.

22.10 $PAQ = B$ if and only if $Q'A'P' = B'$.

22.11 $\begin{bmatrix} 7/4 & 1/4 & 3/4 \\ 29/4 & 27/4 & 17/4 \\ -19/2 & -13/2 & -11/2 \end{bmatrix}$

$L(\mathbf{E}_1^{(3)})$: $(3, 1, 0)$ and $(-1/4, 5/4, -1/2)$;
$L(\mathbf{E}_2^{(3)})$: $(1, -1, 0)$ and $(1/4, 3/4, -3/2)$;
$L(\mathbf{E}_3^{(3)})$: $(3, 0, 1)$ and $(1/2, 3/2, -2)$.

22.14

$$AP = P\begin{bmatrix} k_1 & 0 & \cdots & 0 \\ 0 & k_2 & & \vdots \\ \vdots & & \ddots & 0 \\ 0 & \cdots & 0 & k_n \end{bmatrix}.$$

Section 23

23.1 $\begin{bmatrix} 0 & 0 & 1 & 0 \\ 0 & 1 & 0 & 0 \\ 1 & 0 & 0 & 0 \\ 0 & 0 & 0 & 1 \end{bmatrix}$, $\begin{bmatrix} 1 & 0 & 0 & 0 & 0 \\ 0 & 1 & 0 & 0 & 0 \\ 0 & 0 & 1 & 0 & 0 \\ 0 & 1/2 & 0 & 1 & 0 \\ 0 & 0 & 0 & 0 & 1 \end{bmatrix}$, $\begin{bmatrix} 1 & 0 & 0 \\ 0 & \sqrt{2} & 0 \\ 0 & 0 & 1 \end{bmatrix}$,

$$\begin{bmatrix} 0 & 1 & 0 \\ 1 & 0 & 0 \\ 0 & 0 & 1 \end{bmatrix}, \quad \begin{bmatrix} 1 & 0 & 0 & 0 \\ 0 & 1 & 0 & 0 \\ 0 & 0 & 1 & 0 \\ 10 & 0 & 0 & 1 \end{bmatrix}. \text{ The inverses are } \begin{bmatrix} 0 & 0 & 1 & 0 \\ 0 & 1 & 0 & 0 \\ 1 & 0 & 0 & 0 \\ 0 & 0 & 0 & 1 \end{bmatrix},$$

$$\begin{bmatrix} 1 & 0 & 0 & 0 & 0 \\ 0 & 1 & 0 & 0 & 0 \\ 0 & 0 & 1 & 0 & 0 \\ 0 & -\tfrac{1}{2} & 0 & 1 & 0 \\ 0 & 0 & 0 & 0 & 1 \end{bmatrix}, \quad \begin{bmatrix} 1 & 0 & 0 \\ 0 & \dfrac{1}{\sqrt{2}} & 0 \\ 0 & 0 & 1 \end{bmatrix}, \quad \begin{bmatrix} 0 & 1 & 0 \\ 1 & 0 & 0 \\ 0 & 0 & 1 \end{bmatrix},$$

$$\begin{bmatrix} 1 & 0 & 0 & 0 \\ 0 & 1 & 0 & 0 \\ 0 & 0 & 1 & 0 \\ -10 & 0 & 0 & 1 \end{bmatrix}.$$

23.2 $$\begin{bmatrix} 7 & 5 & 1 \\ 2 & -1 & 0 \\ -3 & 4 & 2 \\ 1 & 3 & 1 \end{bmatrix}, \quad \begin{bmatrix} 2 & -\tfrac{1}{3} & 0 \\ 7 & \tfrac{5}{3} & 1 \\ -3 & \tfrac{4}{3} & 2 \\ 1 & 1 & 1 \end{bmatrix}, \quad \begin{bmatrix} \tfrac{1}{2} & \tfrac{2}{3} & -1 & 4 & \tfrac{3}{2} \\ -\tfrac{5}{3} & 0 & \tfrac{1}{3} & \tfrac{1}{6} & 2 \\ \tfrac{5}{3} & 2 & -3 & 0 & 1 \end{bmatrix},$$

$$\begin{bmatrix} -\tfrac{11}{2} & \tfrac{2}{3} & -1 & 4 & \tfrac{3}{2} \\ 1 & 0 & 1 & \tfrac{1}{2} & 6 \\ -\tfrac{49}{3} & 2 & -3 & 0 & 1 \end{bmatrix}, \quad \begin{bmatrix} 4 & -3 & 2 \\ 5 & 7 & 1 \\ -1 & 2 & 0 \\ 3 & 1 & 1 \end{bmatrix},$$

$$\begin{bmatrix} -5 & 0 & -1 & \tfrac{1}{2} & 6 \\ \tfrac{1}{2} & \tfrac{2}{3} & 1 & 4 & \tfrac{3}{2} \\ \tfrac{5}{3} & 2 & 3 & 0 & 1 \end{bmatrix}.$$

23.5 $$\begin{bmatrix} 0 & 0 & 0 & \tfrac{1}{4} \\ 0 & 0 & \tfrac{1}{3} & 0 \\ 0 & \tfrac{1}{2} & 0 & 0 \\ 1 & 0 & 0 & 0 \end{bmatrix}, \quad \begin{bmatrix} \tfrac{10}{19} & -\tfrac{1}{19} & \tfrac{14}{19} \\ -\tfrac{2}{19} & \tfrac{4}{19} & \tfrac{1}{19} \\ -\tfrac{1}{19} & \tfrac{2}{19} & -\tfrac{9}{19} \end{bmatrix}, \quad \begin{bmatrix} 1 & 0 & 0 & 0 \\ -\tfrac{1}{2} & 3 & 0 & 0 \\ 2 & 0 & 1 & 0 \\ -\tfrac{1}{3} & 0 & 0 & 2 \end{bmatrix},$$

$$\begin{bmatrix} 0 & 1 & -1 \\ 1 & -1 & 0 \\ 0 & 0 & 1 \end{bmatrix}.$$

Section 24

24.4 (a) 2; (b) 3; (c) 1; (d) 1; (e) 3; (f) 5; (g) 3.

24.7 The mapping L defined by $L[(a_1, a_2, \ldots, a_k, 0, 0, \ldots, 0)] = (a_1, a_2, \ldots, a_k)$ is an isomorphism.

Section 25

25.1 (a) $\begin{bmatrix} 1 & 0 & 0 & 0 & 0 \\ 0 & 1 & 0 & 0 & 0 \\ 0 & 0 & 1 & 0 & 0 \end{bmatrix}$; (b) $\begin{bmatrix} 1 & 0 & 0 & 0 & 0 \\ 0 & 1 & 0 & 0 & 0 \\ 0 & 0 & 1 & 0 & 0 \\ 0 & 0 & 0 & 1 & 0 \\ 0 & 0 & 0 & 0 & 1 \end{bmatrix}$;

(c) $\begin{bmatrix} 1 & 0 \\ 0 & 1 \\ 0 & 0 \\ 0 & 0 \end{bmatrix}$; (d) $\begin{bmatrix} 1 & 0 & 0 & 0 \\ 0 & 0 & 0 & 0 \\ 0 & 0 & 0 & 0 \end{bmatrix}$.

25.2 $s + 1$, where s is the minimum of m and n.

Section 26

26.1 14; 6; 13.

26.2 Both terms have minus signs.

26.5 (a) $-x^6 - x^5 + x^4 + 6x^3 - 4x^2 - x - 12$; (b) 116; (c) 1; (d) $1/210$; (e) $\sqrt{210}$.

26.7 t precedes $t - s$ smaller numbers and s is preceded by $t - s - 1$ larger numbers other than t.

Section 27

27.1 (a) -1932; (b) 0; (c) $3(x + y)(y + z)(z + w) - (x + y)^2 - (y + z)^2 - (z + w)^2$; (d) -112; (e) 0; (f) $a^2bc^2(1 - b)(c - b)$; (g) $(b - a)(c - a)(c - b)(d - a)(d - b)(d - c)$.

27.2 (a) 27.1 and 27.2; (b) 27.2 and 27.4; (c) 27.4; (d) 27.4 and 27.3.

27.6 144; 7500; 144; $1/192$; 1728.

Section 28

28.1 -364, -81, -47; -364, 81, -47.

28.2 85; 89; 94; 41; -82.

Section 29

29.2 There are 40 three-rowed minors. Twenty-eight are zero and there are two with each of the following values: 1, -1, 2, -2, 3, -3.

29.5 (a) $\begin{bmatrix} 3 & -1 & 2 \\ 0 & 1 & -5 \\ 6 & 7 & 4 \end{bmatrix}$; (b) $\begin{bmatrix} 0 & 0 & 1 & 0 \\ 0 & 1 & -2 & 0 \\ 0 & 0 & 0 & 1 \\ 1 & 0 & 0 & 0 \end{bmatrix}$; (c) $\begin{bmatrix} 1 & 1 & 1 & 1 \\ 2 & 2 & 2 & 2 \\ 3 & 3 & 3 & 3 \\ 4 & 4 & 4 & 4 \end{bmatrix}$.

29.7 Use Theorem 29.2 and consider two cases, $|A| \neq 0$, and $|A| = 0$. Show that $|A| = 0$ implies $|A^{\text{adj}}| = 0$.

29.8 (a) $x_1 = 4/15$, $x_2 = 1/10$, $x_3 = 1/2$, $x_4 = 2/15$; (b) $x_1 = 3$, $x_2 = 1$, $x_3 = -3$; (c) $x_1 = x_2 = x_3 = 0$.

29.9 (a) $\begin{bmatrix} 13/45 & 6/45 & 1/45 \\ -10/45 & 0 & 5/45 \\ -2/45 & -9/45 & 1/45 \end{bmatrix}$; (b) $\begin{bmatrix} 0 & 0 & 0 & 1 \\ 2 & 1 & 0 & 0 \\ 1 & 0 & 0 & 0 \\ 0 & 0 & 1 & 0 \end{bmatrix}$.

Section 30

30.1 2, 3; $(r, 0, 0, 0)$, $(-3s, s, 0, 0)$.

30.2 $([2 - \sqrt{2}]s, s, 0)$.

30.4 1, 4, -2; $r \cdot \mathbf{I} + r \cdot \mathbf{J} + r \cdot \mathbf{K}$, $s \cdot \mathbf{I} + 4s \cdot \mathbf{J} + 16s \cdot \mathbf{K}$, $t \cdot \mathbf{I} - 2t \cdot \mathbf{J} + 4t \cdot \mathbf{K}$. A basis of characteristic vectors is $\{\mathbf{I} + \mathbf{J} + \mathbf{K}, \mathbf{I} + 4 \cdot \mathbf{J} + 16 \cdot \mathbf{K}, \mathbf{I} - 2 \cdot \mathbf{J} + 4 \cdot \mathbf{K}\}$. The matrix of L with respect to this basis of characteristic vectors is

$$\begin{bmatrix} 1 & 0 & 0 \\ 0 & 4 & 0 \\ 0 & 0 & -2 \end{bmatrix}.$$

30.10 (a) $(33 \pm \sqrt{633})/12$; (b) $1, -1, i, -i$; (c) $1, 1 + 2i, 1 - 2i$.

Section 31

31.2

$$Q^{-1} = \begin{bmatrix} 0 & 0 & -2/3 & 4/3 \\ 0 & 0 & 2/3 & -1/3 \\ -1/3 & 2/3 & 0 & 1/3 \\ 1/3 & 1/3 & -2 & 8/3 \end{bmatrix}.$$

31.3 (a), (c).

31.4 (a) $\begin{bmatrix} 1 & -1 & 0 & 0 \\ 1 & 2 & 0 & 0 \\ -6 & 0 & 1 & -2 \\ 8 & 1 & -2 & 1 \end{bmatrix}$; (c) $\begin{bmatrix} 1 & 0 & -1 & 0 \\ 0 & 1 & 0 & -1 \\ 0 & 1 & 0 & 1 \\ 1 & 0 & 1 & 0 \end{bmatrix}$.

(There are other solutions.)

Section 32

32.2 $L(x \cdot \mathbf{I} + y \cdot \mathbf{J} + z \cdot \mathbf{K}) \circ L(x \cdot \mathbf{I} + y \cdot \mathbf{J} + z \cdot \mathbf{K})$
$= (x')^2 + (y')^2 + (z')^2$
$= (x \cos \varphi - y \sin \varphi)^2 + (x \sin \varphi + y \cos \varphi)^2 + z^2$
$= x^2 + y^2 + z^2 = (x \cdot \mathbf{I} + y \cdot \mathbf{J} + z \cdot \mathbf{K}) \circ (x \cdot \mathbf{I} + y \cdot \mathbf{J} + z \cdot \mathbf{K})$.

32.8 (a) $\begin{bmatrix} \frac{-3}{\sqrt{13}} & \frac{2}{\sqrt{13}\sqrt{14}} & \frac{2}{\sqrt{14}} \\ \frac{2}{\sqrt{13}} & \frac{3}{\sqrt{13}\sqrt{14}} & \frac{3}{\sqrt{14}} \\ 0 & \frac{-\sqrt{13}}{\sqrt{14}} & \frac{1}{\sqrt{14}} \end{bmatrix}$; (b) $\begin{bmatrix} \frac{-2}{\sqrt{13}} & \frac{3}{\sqrt{13}} \\ \frac{3}{\sqrt{13}} & \frac{2}{\sqrt{13}} \end{bmatrix}$;

(c) $\begin{bmatrix} \frac{5}{\sqrt{86}} & \frac{3}{\sqrt{43}} & \frac{1}{\sqrt{2}} \\ \frac{6}{\sqrt{86}} & \frac{-5}{\sqrt{43}} & 0 \\ \frac{5}{\sqrt{86}} & \frac{3}{\sqrt{43}} & \frac{-1}{\sqrt{2}} \end{bmatrix}$.

Section 33

33.2 If $\mathcal{B}$ is symmetric, then

$$[r_1 \quad r_2 \quad \cdots \quad r_n]A\begin{bmatrix} s_1 \\ s_2 \\ \vdots \\ s_n \end{bmatrix} = [s_1 \quad s_2 \quad \cdots \quad s_n]A\begin{bmatrix} r_1 \\ r_2 \\ \vdots \\ r_n \end{bmatrix}$$

for all real numbers r_i and s_j. By making suitable choices for $[r_1 \quad r_2 \quad \cdots \quad r_n]$ and $[s_1 \quad s_2 \quad \cdots \quad s_n]$, show that this implies that A is symmetric.

33.3 $\mathcal{B}(\mathbf{E}_4^{(4)}, \mathbf{E}_3^{(4)}) = -1$, $\mathcal{B}(\mathbf{E}_3^{(4)}, \mathbf{E}_4^{(4)}) = 0$;

$$\begin{bmatrix} 0 & 1 & 0 & 0 \\ 0 & 0 & 0 & 0 \\ 0 & 0 & 0 & 0 \\ 0 & 0 & -1 & 0 \end{bmatrix}.$$

33.5 All three matrices are matrices of the quadratic form $-3x_1^2 + 8x_1x_2 + 4x_2x_3 + 4x_3^2$. Matrix (c) is the symmetric matrix of the form.

33.9 $\begin{bmatrix} 1 & \frac{\sqrt{2}}{2} \\ \frac{\sqrt{2}}{2} & -5 \end{bmatrix}$; $\begin{bmatrix} 3 & 5/2 & -7/2 \\ 5/2 & -4 & 1/2 \\ -7/2 & 1/2 & -1 \end{bmatrix}$; $\begin{bmatrix} 0 & 1/2 & 1/2 & 1/2 \\ 1/2 & 0 & 1/2 & 1/2 \\ 1/2 & 1/2 & 0 & 1/2 \\ 1/2 & 1/2 & 1/2 & 0 \end{bmatrix}$.

Section 34

34.1 Express the characteristic vectors **U** and **V** as n by 1 matrices U and V in terms of their coordinates. Then $\mathbf{U} \circ \mathbf{V} = U'V = V'U$. Let **U**

belong to r and **V** belong to s. Show that $(AU)'V = (AV)'U$. Then show that $r \cdot U'V = s \cdot V'U$.

34.3 Let Q be a matrix such that $Q^{-1}AQ$ is in diagonal form. Use the fact that $Q^{-1}(A - r \cdot I_n)Q$ has the same rank as $A - r \cdot I_n$.

34.4 (a) $\begin{bmatrix} 1 & 0 & 0 \\ 0 & \frac{-5+\sqrt{29}}{r} & \frac{2}{r} \\ 0 & \frac{2}{r} & \frac{5-\sqrt{29}}{r} \end{bmatrix}$, where $r = \sqrt{58 - 10\sqrt{29}}$;

(b) $\begin{bmatrix} \frac{1}{\sqrt{2}} & \frac{-1}{\sqrt{2}} & 0 & 0 \\ 0 & 0 & \frac{1}{\sqrt{2}} & \frac{1}{\sqrt{2}} \\ 0 & 0 & \frac{1}{\sqrt{2}} & \frac{-1}{\sqrt{2}} \\ \frac{1}{\sqrt{2}} & \frac{1}{\sqrt{2}} & 0 & 0 \end{bmatrix}$; (c) $\begin{bmatrix} \frac{\sqrt{3}}{3} & 0 & \frac{\sqrt{6}}{3} \\ 6 & \frac{-\sqrt{3}}{2} & \frac{-\sqrt{3}}{6} \\ \frac{\sqrt{2}}{2} & \frac{1}{2} & \frac{-1}{2} \end{bmatrix}$;

(d) $\begin{bmatrix} \frac{2}{\sqrt{5}} & \frac{-1}{\sqrt{5}} \\ \frac{1}{\sqrt{5}} & \frac{2}{\sqrt{5}} \end{bmatrix}$; (e) $\begin{bmatrix} \frac{-1}{\sqrt{5}} & \frac{2}{\sqrt{5}} & 0 & 0 \\ \frac{2}{\sqrt{5}} & \frac{1}{\sqrt{5}} & 0 & 0 \\ 0 & 0 & \frac{1}{\sqrt{5}} & \frac{-2}{\sqrt{5}} \\ 0 & 0 & \frac{2}{\sqrt{5}} & \frac{1}{\sqrt{5}} \end{bmatrix}$;

(f) $\begin{bmatrix} \frac{1}{\sqrt{10}} & \frac{3}{\sqrt{10}} \\ \frac{-3}{\sqrt{10}} & \frac{1}{\sqrt{10}} \end{bmatrix}$.

(There are other solutions.)

34.5 (a) The symmetric matrix of Q is

$$\begin{bmatrix} 0 & 1/2 & 1/2 & 1/2 \\ 1/2 & 0 & 1/2 & 1/2 \\ 1/2 & 1/2 & 0 & 1/2 \\ 1/2 & 1/2 & 1/2 & 0 \end{bmatrix}.$$

The vectors $\mathbf{W}_1 = (1/\sqrt{2}, 0, 0, -1/\sqrt{2})$,
$\mathbf{W}_2 = (0, 1/\sqrt{2}, 0, -1\sqrt{2})$, $\mathbf{W}_3 = (0, 0, 1/\sqrt{2}, -1/\sqrt{2})$, and
$\mathbf{W}_4 = (\frac{1}{2}, \frac{1}{2}, \frac{1}{2}, \frac{1}{2})$ are an orthonormal basis of $\mathcal{V}_4$. For
$\mathbf{U} = a_1 \cdot \mathbf{W}_1 + a_2 \cdot \mathbf{W}_2 + a_3 \cdot \mathbf{W}_3 + a_4 \cdot \mathbf{W}_4$,
$\mathcal{Q}(U) = -\frac{1}{2}a_1^2 - \frac{1}{2}a_2^2 - \frac{1}{2}a_3^2 + \frac{3}{2}a_4^2$.
(b) The symmetric matrix of $\mathcal{Q}$ is

$$\begin{bmatrix} 1 & 1 & 1 & 1 \\ 1 & 1 & 1 & 1 \\ 1 & 1 & 1 & 1 \\ 1 & 1 & 1 & 1 \end{bmatrix}.$$

Using the same orthonormal basis of $\mathcal{V}_4$ as in part (a), if
$\mathbf{U} = a_1 \cdot \mathbf{W}_1 + a_2 \cdot \mathbf{W}_2 + a_3 \cdot \mathbf{W}_3 + a_4 \cdot \mathbf{W}_4$, then $\mathcal{Q}(\mathbf{U}) = 4a_4^2$.

34.7 (a) $11\bar{x}^2 + 6\bar{y}^2 = 100$; (b) $\frac{3}{2}\bar{x}^2 - \frac{23}{2}\bar{y}^2 = 4$;
(c) $\bar{x}^2 + (1 + 3\sqrt{2})\bar{y}^2 + (1 - 3\sqrt{2})\bar{z}^2 = 25$;
(d) $-2\bar{x}^2 + (2 + \sqrt{6})\bar{y}^2 + (2 - \sqrt{6})\bar{z}^2 = 1$.

Section 35

35.5 (a) $$\begin{bmatrix} 0 & 0 & 1 \\ 0 & \frac{1}{6} & -1 \\ \frac{12}{7\sqrt{15}} & \frac{1}{6\sqrt{15}} & \frac{1}{7\sqrt{15}} \end{bmatrix},$$ rank = 3, index = 1.

(b) $$\begin{bmatrix} 0 & \frac{\sqrt{3}}{3} & 0 & 0 \\ \frac{\sqrt{3}}{r} & \frac{\sqrt{3}}{3r} & \frac{6\sqrt{3}}{r} & 0 \\ 0 & 0 & 1 & 0 \\ \frac{-29\sqrt{3}}{rs} & \frac{-208\sqrt{3}}{rs} & \frac{-55\sqrt{3}}{rs} & \frac{119\sqrt{3}}{rs} \end{bmatrix},$$

where $r = \sqrt{119}$, $s = 3\sqrt{662}$. Rank = 4, index = 2.

(c) $$\begin{bmatrix} 1 & 0 & 0 & 0 \\ -1 & 1 & 0 & 0 \\ -1 & 0 & 1 & 0 \\ -1 & 0 & 0 & 1 \end{bmatrix},$$ rank = 1, index = 1.

(There are other solutions for the matrices.)

35.6 (f) is positive definite.

35.9 (a) $\begin{bmatrix} 0 & 0 & 0 & 1 \\ \frac{1}{2} & 0 & 2 & -1 \\ \frac{1}{2} & 0 & 0 & -1 \\ \frac{-1}{\sqrt{6}} & \frac{2}{\sqrt{6}} & \frac{-5}{\sqrt{6}} & \frac{2}{\sqrt{6}} \end{bmatrix}$, $y_1^2 + y_2^2 - y_3^2 - y_4^2$;

(b) $\begin{bmatrix} 1 & 0 & 0 \\ -1 & 1 & 0 \\ -1 & 0 & 1 \end{bmatrix}$, y_1^2, positive semidefinite;

(c) $\begin{bmatrix} \frac{1}{\sqrt{2}} & 0 & 0 \\ -2 & 1 & 0 \\ \sqrt{3} & 0 & \frac{1}{\sqrt{3}} \end{bmatrix}$, $y_1^2 - y_2^2 - y_3^2$;

(d) $\begin{bmatrix} 1 & 0 & 0 \\ \frac{\sqrt{6}}{6} & \frac{\sqrt{6}}{6} & 0 \\ \frac{3\sqrt{3}}{5} & \frac{\sqrt{3}}{5} & \frac{\sqrt{3}}{15} \end{bmatrix}$, $y_1^2 + y_2^2 - y_3^2$;

(e) $\begin{bmatrix} 1 & 0 & 0 \\ 6 & 1 & 0 \\ -3 & -1 & 1 \end{bmatrix}$, $y_1^2 + y_2^2$, positive semidefinite;

(f) $\begin{bmatrix} 1 & 1 & 0 & 0 \\ -1 & 1 & 0 & 0 \\ -1 & -1 & 1 & 0 \\ \frac{-\sqrt{3}}{3} & \frac{-\sqrt{3}}{3} & \frac{-\sqrt{3}}{3} & \frac{2\sqrt{3}}{3} \end{bmatrix}$, $y_1^2 - y_2^2 - y_3^2 - y_4^2$.

index

A 2
B 3
C 4
D 5
E 6
F 7
G 8
H 9
I 0
J 1